High Levels of Natural Radiation 1996
Radiation Dose and Health Effects

High Levels of Natural Radiation 1996 Radiation Dose and Health Effects

Proceedings of the 4th International Conference on
High Levels of Natural Radiation held in Beijing, China on
October 21 to 25, 1996

Editors:

WEI, Luxin

Professor of Laboratory of Industrial Hygiene
Ministry of Health
Beijing, China

SUGAHARA, Tsutomu

Chairman of Health Research Foundation
Emeritus Professor of Kyoto University
Kyoto, Japan

TAO, Zufan

Professor of Laboratory of Industrial Hygiene
Ministry of Health
Beijing, China

 1997

Elsevier

Amsterdam – Lausanne – New York – Oxford – Shannon – Tokyo

© 1997 Elsevier Science B.V. All rights reserved.

No part of this publication may be reproduced, stored in a retrieval system, or transmitted in any form or by any means, electronic, mechanical, photocopying, recording or otherwise without the prior written permission of the publisher, Elsevier Science B.V., Permissions Department, P.O. Box 521, 1000 AM Amsterdam, The Netherlands.

No responsibility is assumed by the Publisher for any injury and/or damage to persons or property as a matter of products liability, negligence or otherwise, or from use or operation of any methods, products, instructions or ideas contained in the material herein. Because of rapid advances in the medical sciences, the Publisher recommends that independent verification of diagnoses and drug dosages should be made.

Special regulations for readers in the USA – This publication has been registered with the Copyright Clearance Center Inc. (CCC), 222 Rosewood Drive, Danvers, MA 01923, USA. Information can be obtained from the CCC about conditions under which photocopies of parts of this publication may be made in the USA. All other copyright questions, including photocopying outside of the USA, should be referred to the copyright owner, Elsevier Science B.V., unless otherwise specified.

International Congress Series No. 1136
ISBN 0-444-82630-0

This book is printed on acid-free paper.

Published by:
Elsevier Science B.V.
P.O. Box 211
1000 AE Amsterdam
The Netherlands

Library of Congress Cataloging in Publication Data:

Printed in the Netherlands

Preface

There are regions in the world where terrestrial background gamma radiation levels appreciably exceed the normal range. Such regions exist in Brazil, India, Iran, China, Italy, France, Madagascar, Nigeria and in some other countries. In addition to the higher background gamma radiation levels, there are places in the world where the concentrations of radon (Rn-222, Rn-220) and their decay products in the air are remarkably high, such as the Austrian spa of Badgastein, located near Salzburg, the Helsinki area of Finland, some area of France, Germany, Sweden, Switzerland and the United States of America. It was reported there are also places in the world having higher concentrations of natural radionuclides in drinking water. These areas of high natural radioactivity have attracted scientists of multi-discipline to study their intensity of radioactivity and their relation with the health of human beings. In recent years, these data have been used for exploring the stochastic effects of low dose exposure. These investigations have been conducted since the decade of 1950s. For exchanging information and experience of investigations, the first international symposium on areas of high natural radioactivity was held in Poços de caldas, Brazil in 1975. As a continuation of the first symposium, the second one was held in Bombay, India in 1981, and the third one was in Ramsar of the Islamic Republic of Iran in 1990. The present Proceedings are for the fourth International Conference on High Levels of Natural Radiation (4 th ICHLNR) which was held in Beijing of China on October 21-25, 1996.

In addition to the topics presented and discussed in previous ICHLNRs (world natural radiation, world high level natural radiation areas, environmental transfer pathways, technologically enhanced natural radiation environment, radon in the environment, cytogenetic studies, epidemiological studies and radiation measurement methods, in this conference there were some new topics for presentation and discussion: radioactive contamination by nuclear tests, Chernobyl accident and some related biological studies; methodology and quality control for low dose measurements, methodology of epidemiology in low dose radiation exposure; stimulatory effects of low dose radiation, adaptive response and genetic effects of low dose irradiation and radiation paradigm regarding risk from exposure to low dose radiation. It should be noted that many papers concerning radiobiology presented in this conference reflected the increasing interest in the dose-effects relationship of long term low dose exposure and the need of the combination of epidemiology and biology. It should be also noted that in this conference there are papers of international cooperative research work, which is very helpful to promote the progress of this kind of investigation, and widen the field of vision of the scientists who participate in the cooperation. Many participants of this conference also emphasized the importance of indoor radon and its relation to the health effects on human beings.

Through the Conference the necessity of international cooperation and harmonization of research technique and protocol was recognized by all participants. Significance of this kind of studies for the health of world population was emphasized by Dr. Bennett representing the United Nations Scientific Committee on the Effects of Atomic Radiation. Thus the International Committee as indicated in Appendix was established. We believe that the study on high level of natural radiation and its health effects would develop very rapidly from now on.

WEI, Luxin
SUGAHARA, Tsutomu
TAO , Zufan

Acknowledgments

The editors would like to express their appreciation to all the contributors for their cooperation, especially for submitting their manuscripts very quickly, so that the editors are able to publish the Proceedings as scheduled, and to Drs. K.B. Li , C.-S.Zhu, M. Sohrabi, L.J.Appleby, S. Akiba, P. C. Kesavan and C. Satoh as reporters in summarizing the scientific sessions and as reviewers to review the relevant papers with excellent work. The editors also wish to express their sincere gratitude to Prof. J. X. Yao for his conscientious and effective work in reviewing a lot of manuscripts and in assisting editorial work. We are grateful to Ms. S. Yasui, Drs. M. Watanabe, Q. F. Sun, D. Q. Chen, R. Zhao, Q. J. Liu, J. Li, and Mr. Y. S. Liu for assisting the editors' work throughout the whole editorial process. Special thanks to Dr. Watanabe and his colleagues in Nagasaki University who have done a beautiful work for making all camera-ready materials for printing.

The editors also wish to express their thankfulness to the International Cooperation Department of Ministry of Health, PRC and Health Research Foundation, Japan for their financial support to the Conference. The supports of IAEA, WHO, UNEP, UNSCEAR and CMA to the Conference are acknowledged as well.

Editors:
WEI, Luxin
SUGAHARA, Tsutomu
TAO, Zufan

This is also the commemorative publication of the Health Research Foundation, Japan for its 55th Anniversary.

Conference officers

Honorary President:
 CHEN Minzhang, Minister of Health, PRC

President:
 WEI Luxin, Prof. and Chief of High Background Radiation Research Group, PRC

Vice President:
 ZHAO Tongbin   WEI Kedao   WU Dechang

Secretary General: *Vice Secretary General:*
 TAO Zufan   CAI Dongqiang
                                 LI Wenyuan

International Scientific Committee:
Chairman: *Vice Chairman:*
 Tsutomu SUGAHARA, Japan   LI Kaibao, China

Program Committee:
Chairman: *Vice Chairman:*
 CHEN Deqing   ZHAO Ru

Local Committee:
Chairman: *Vice Chairman:*
 ZHAO Tongbin   XING Wenting

Secretariat:
 TAO Zufan   CAI Dongqiang   LI Wenyuan   ZHANG Shaoyi
 LI Jia   WAN Xiang   SUN Quanfu   ZHOU Vivian
 ZHANG Jihui   LIU Yusheng

Organized by
 International Health Exchange Center of Ministry of Health, PRC
 Laboratory of Industrial Hygiene of Ministry of Health, PRC

In cooperation with
 The International Atomic Energy Agency
 The World Health Organization
 The United Nations Environment Program
 The United Nations Scientific Committee on the Effects of Atomic Radiation
 Health Research Foundation, Japan
 Society of Radiological Medicine and Protection, Chinese Medical Association

Contents

Preface	v
Acknowledgments	vii

I Dose estimation

High background radiation area – an important source of exploring the health effects of low dose ionizing radiation
    L.-X. Wei   1
Exposures to natural radiation worldwide
    B.G. Bennett   15
Cosmic ray contribution in the measurement of environmental gamma ray dose
    K. Nagaoka, K. Honda and K. Miyano   25
Survey of natural radiation background and assessment of its population dose in China
    K.-D. Wei, G.-Z. Cui, S.-A. Zhao and J.-H. Gao   31
Investigation of outdoor γ-radiation level in Suzhou urban and rural areas
    Y. Tu and D.-Z. Jiang   39
Environment ionizing radiation level and dose to population in Qinghai Province
    Z.-Y. Cao, Z.-Q. Liang, B.-C. Li, J.-Q. Ni and P. Zhang   43
Methodology of external radiation dosimetry and measurement assurance applied in epidemiology in high background radiation area in Yangjiang, China
    K.-B. Li and S.-A. Zhao   49

II Radioactive aerosols

World high level natural radiation and/or radon-prone areas with special regard to dwellings
    M. Sohrabi   57
Origin of a new high level natural radiation area in the hot spring region on Mahalat, central Iran
    M. Sohrabi, M.M. Beitollahi, Y. Lasemi and E. Amin Sobhani   69
Level of ^{222}Rn and its daughters in the environment of Xinglong district in Hainan Province
    Q.-X. Zeng, X.-S. Cheng, Y. Li, Z.-H. Huang, Z.-H. Deng and A.-H. Liu   75

Radon concentration and radiation dose to the population exposed in the high elevation areas of Sichuan
    Y.-G. Liu and Y.-H. Mao   81

^{222}Rn concentration in different types of cave dwellings in Gansu Province
    P.-Y. Gao, W. Sun, D.-Y. Tian, J.-L. Huo, Y.-F. Tao, S.-Z. Zhang, Y. Wang, B.-R. Xu and S.-X. Guo   85

Concentration of radon and its daughters in environment and internal doses to population in Qinghai Province
    J.-Q. Ni, Z.-Y. Cao, Q.-L. Ma and Z.-Q. Liang   89

The size distribution of atmospheric radioactive aerosols and its measurement
    Y.-H. Jin, M. Shimo, W.-H. Zhuo, L. Xu and G.-Q. Fang   97

A new sub-nationwide survey of outdoor and indoor ^{222}Rn concentrations in China
    Y.-H. Jin, T. Iida, Z.-Y. Wang and S. Abe   103

III Natural radionuclides

Indian research on high levels of natural radiation: pertinent observations for further studies
    P.C. Kesavan   111

Brazilian research in areas of high natural radioactivity
    A.R. Oliveira, C.E.V. de Almeid, H.E. da Silva, J.H. Javaroni, A.M. Castanho, M.J. Coelho and R.N. Alves   119

Natural radioactivity of soil samples in some high level natural radiation areas of Iran
    M. Sohrabi, M. Bolourchi, M.M. Beitollahi and J. Amidi   129

Investigations about the radioactive disequilibrium among the main radionuclides of the thorium decay chain in foodstuffs
    D.C. Lauria, Y. Nouilhetas, L.F. Conti, M.L.P. Godoy, L. Julião and S. Hacon   133

Transfer of ^{226}Ra, ^{228}Ra, ^{210}Pb, ^{210}Po in the aquatic organism and the foodstuff chain
    X.-T. Yang, D.-T. Weng, W.-Y. Chen, X.-Y. Chen, J.-X. Chen and S.-M. Zhao   141

Intake of food and internal doses by ingestion in two high radiation areas in China
    H.-D. Zhu   149

Biosphere effects of natural radionuclides released from coal-fired power plants
    Y.-H. Mao and Y.G. Liu   155

IV Fallout from nuclear test

Overview of scope – RADTEST project
    R. Kirchmann, Sir F. Warner and L. Appleby   161

Assessment of risks from combined exposures to radiation and other agents at environmental levels
   T. Jung and W. Burkart   167

Exposure dose from natural radiation and other sources to the population in China
   C.-S. Zhu, Y. Liu and L.-A. Zhang   179

Transfer of nuclear fission products to crops and soils and countermeasures
   S.-M. Xu, W.-H. Zhao, L.-X. Hou and Z.-R. Shang   185

Dosimetry study of residents near Semipalatinsk nuclear test site
   J. Takada, M. Hoshi, S. Endo, M. Yamamoto, T. Nagatomo, B.I. Gusev, R.I. Rozenson, K.N. Apsalikov and N.J. Tchaijunusova   191

Radioactive contamination in North Xinjiang resulting from former USSR atmospheric nuclear tests in 1962
   J.-L. Hou, S.-Y. He, S.-R. Li, Z.-Z. Pi, F.-W. Liu, T.-G. Xu, X.-Z. Wang, X.-F. Liu and G.-M. Liu   197

Observations in Kyoto, Japan of radioactive fallout from Chinese atmospheric nuclear explosion tests
   N. Fujinami   207

V Epidemiological study

Methodology of epidemiology in low dose radiation exposure
   H. Kato   215

Recent advances in dosimetry investigation in the high background radiation area in Yangjiang, China
   Y.-L. Yuan, H. Morishima, H. Shen, T. Koga, Q.-F. Sun, K. Tatsumi, Y.-R. Zha, S. Nakai, L.-X. Wei and T. Sugahara   223

Study of the indirect method of personal dose assessment for the inhabitants in HBRA of China
   H. Morishima, T. Koga, K. Tatsumi, S. Nakai, T. Sugahara, Y.-L. Yuan, Q.-F. Sun and L.-X. Wei   235

Databases and statistical methods of cohort studies (1979–90) in Yangjiang
   Q.-F. Sun, S. Akiba, J.-M. Zou, Y.-S. Liu, Z.-F. Tao and H. Kato   241

Study on cancer mortality among the residents in high background radiation area of Yangjiang, China
   Z.-F. Tao, H. Kato, Y.-R. Zha, S. Akiba, Q.-F. Sun, W.-H. He, Z.-X. Lin, J.-M. Zou, S.-Z. Zhang, Y.-S. Liu, T. Sugahara and L.-X. Wei   249

Infant leukemia mortality among the residents in high-background-radiation areas in Guangdong, China
   S. Akiba, Q.-F. Sun, Z.-F. Tao, Y.-R. Zha, Z.-X. Lin, W.-H. He and H. Kato   255

Confounding factors in radiation epidemiology and their comparability between the high background radiation areas and control areas in Guangdong, China
   Y.-R. Zha, J.-M. Zou, Z.-X. Lin, W.-H. He, J.-M.Lin, Y.-H. Yang and H. Kato   263

Epidemiologic study of cancer in the high background radiation area in Kerala
   M. Krishnan Nair, N. Sreedevi Amma, P. Gangadharan, V. Padmanabhan,
   P. Jayalekshmi, S. Jayadevan and K.S. Mani   271

Radiation associated lung cancer in northern Iran
   R.M. Raie and Z. Taherzadeh Boroujeny   277

A survey of senile dementia in the high background radiation areas in Yangjiang, Guangdong Province, China
   J. Li, H. Sugasaki, Y.-H. Yang, M. Mine, Y.-R. Zha, M. Kishikawa,
   Q.-F. Sun, Y. Nakane, J.-M. Zou, M. Tomonaga, Z.-F. Tao and L.-X. Wei   283

Advanced cytogenetical techniques necessary for the study of low dose exposures
   I. Hayata   293

Preliminary report on quantitative study of chromosome aberrations following life time exposure by high background radiation in China
   T. Jiang, C.-Y. Wang, D.-Q. Chen, Y.-R. Yuan, L.-X. Wei, I. Hayata,
   H. Morishima, S. Nakai and T. Sugahara   301

Effect of low dose rates on the production of chromosome aberration under lifetime exposure to high background radiation
   S. Nakai, T. Jiang, D.-Q. Chen, I. Hayata, Y.-L. Yuan, N. Morishima,
   S. Fujita, T. Sugahara and L.-X. Wei   307

Cytogenetic findings in inhabitants of different ages in high background radiation areas of Yangjiang, China
   D.-Q. Chen, S.-Y. Yao, C.-Y. Zhang, L.-L. Dai, T. Jiang, C.-Y. Wang, S. Li,
   Y.-L. Yuan, L.-X. Wei, S. Nakai and T. Sugahara   317

Cytogenetic studies in lymphocytes of healthy inhabitants of Mazandaran a high level natural radiation area in Iran
   R.M. Raie and M. Aledavoud   325

VI Radiation hormesis

The radiation paradigm regarding health risk from exposures to low dose radiation
   T. Sugahara   331

Cellular and molecular basis of the stimulatory effect of low dose radiation on immunity
   S.-Z. Liu   341

A survey of long-term and low dose radiation effect on SIL-2R in residents
   Z.-Y. Zeng, Y.-Y. Cheng, G. Zeng, W.-T. Lu, Z. Li and S. Zhang   355

Immune competence and immune response to virus in high background radiation area, Yangjiang, China
   J.-M. Zou, J. Yao, N.-G. Chen and Y.-R. Zha   359

Stimulatory effect of *in vivo* exposure to enriched uranium oxyfluoride on DNA excision repair in mouse lymphocytes
   Z.-S. Yang, S.-P. Zhu and S.-Q. Yang   365

VII Adaptive response and genetic effects

Adaptive response in human beings living in the area contaminated by
Chernobyl accident
   I. Pelevina, G. Afanasjev, A. Aleschenko, V. Gotlib, O. Kudrjachova,
   L. Semenova and A. Serebryanyi   373
Investigation on the adaptive response to ionizing radiation induced by high
background exposure to spa workers
   S.-Y. Yao, T. Jiang, J. Yamashita and Y. Takizawa   379
Techniques for detecting the genetic effects of the atomic bombings
applicable to the study of the genetic effects of natural radiation
   C. Satoh, M. Kodaira, J. Asakawa, N. Takahashi, R. Kuick, S.M. Hanash
   and J.V. Neel   385
Genetic instability induced by low-dose radiation
   M. Watanabe, S. Kayata, S. Kodama, K. Suzuki and T. Sugahara   391
Experimental study of the genetic effects of high levels of natural radiation
in South-France
   M. Delpoux, A. Léonard, H. Dulieu and M. Dalebroux   397

VIII Summary and general discussion   407

IX Concluding remarks   419

X Start of International Committee on High Levels of Natural Radiation and Radon Areas (ICHLNRRAs)   421

XI Appendix

Ap1 Opening remarks   423
Ap2 Closing remarks   426
Ap3 Speeches at closing ceremony   429

Index of authors   433
Subject index   437

I Dose estimation

High background radiation area -- an important source of exploring the health effects of low dose ionizing radiation

Wei, Luxin

Laboratory of Industrial Hygiene, Ministry of Health, China, 2 Xinkang Street, Deshengmenwai, Beijing 100088 China

1. INTRODUCTION

Health effects of low dose ionizing radiation is a topic of long term debate and a problem of public concern. To clarify these effects, both epidemiological work and radiobiological exploration are needed. One of the great advantages of radiation epidemiology in the high background radiation areas is the possibility of obtaining results from direct observation on human beings, needless to extrapolate the dose-effects relationship from high to low dose levels, and from animal to man. This is especially true when the families of the subjects have lived in the investigated areas for many generations. In several high background radiation areas in the world, such as in Brazil. India, Iran and China the radiation levels that the local inhabitants exposed to are similar to or above those exposed by the workers of nuclear industry,[1] and by most inhabitants in the radionuclide contaminated areas after serious nuclear accident.[2]

Certainly, there are difficulties in doing health research with so low dose range, especially if the research is for the dose-stochastic effects relationships: (1) large cohort size as well as long period of observation and follow up to fulfill the need of statistical power; (2) estimate of the individual dose accumulated for long time (even for several decades); (3) to distinguish the minor signals from noise (confounding factors), and (4) to interpret various results of observation which may be controversial.

As other research work, radiation-epidemiology in the high background radiation areas has its advantages and limitation. To understand these points is helpful for reasonable design of research plan and adoption of suitable methodology, so that we can fully utilize the important source of high background radiation areas for exploring the health effects of low dose ionizing radiation. In this presentation I will take an instance of many research work in this field to discuss the possibility of utilization of this important source.

2. AN INSTANCE OF MANY:

Epidemiological investigation in the high background radiation areas (HBRA) of Yangjiang, China (in brief)

2.1 The High Background Radiation Research Group (HBRRG, China)started its epidemiology in HBRA since 1972, the work conducted before 1991 had been reported in the last ICHLNR and elsewhere[3-5]. Since 1991, the research group has cooperated with the Health Research Foundation (HRF) of Japan. In this presentation, the recent advances of this work will be cited. However, as this epidemiology is a work of continuation, the previous investigation will be also related.

2.2 Since1991, additional measurements of environmental gamma radiation were conducted for indoor, out door, field environment; occupancy factors and the measurements for individual cumulative doses were also replenished. The dose groups were re-classified according to the external gamma radiation levels.[6] The newly classified dose groups by the external radiation levels, number of the cohort members and the person-years of observation (1979-1990) were listed in Table 1. It should be noted that if we consider the HBRA as a whole (including three groups of "high", "medial" and "low") thus the ratio of external radiation dose rates (absorbed dose rate in air) between HBRA and CA was 3.5 for indoor, and 3.7 for outdoor.

Table 1

Classification of Dose Groups (Based on the measurements of external gamma-radiation, indoor and outdoor, in 526 hamlets of the investigated areas)*

Dose Group	Average of external radiation level (range) mGy a^{-1}	No. of subjects**(M+F)	Person-years of observation(1979-1990)**
1. High	2.46(2.24-3.08)	19564	212540.27
2. Medial	2.10(1.98-2.22)	23359	254551.97
3. Low	1.83(1.25-1.98)	21147	229088.90
4. Control	0.68(0.50-0.95)	24876	252836.86
Total	(0.50-3.08)	88946	949018.00

* Based on Ref. [6]

** Calculated only for fixed cohort

2.3 Cancer mortality study

For the purpose of obtaining data from stable populations, a fixed cohort was established in 1991. The people those registered in January first of 1987 are the cohort members. The data available for cancer mortality analysis of the fixed cohort were from the year of 1979. The Research Group has collected the data up to 1995. However, at present, only 1979-1990 data were confirmed and analyzed.

As Table1 shows, 88,946 persons (64070 in HBRA, 24876 in CA) were observed and followed up, accumulating 949,018 person-years in the period of 1979 to 1990.

Table 2 shows the relative risks (RRs, adjusted with age groups and genders) for some cancers in HBRA (1979-1990) (RR=1.00 for CA). The RRs were around 1.00, mostly lower than one, only the RR for nasopharyngeal cancer was somewhat higher. However, all the RRs were not significant statistically.

Table 3 demonstrates the adjusted RRs (90% CI) of site-specific cancers for the four dose groups (1979-1990 data), still it shows no statistical difference for all the site-specific cancers. It also shows that there is no increase of any cancer mortality with the increase of annual dose of external radiation.[7]

Table 2

RRs (90% CI) adjusted with age groups and sex for some cancers in HBRA (1979-1990) (RR=1.00 for CA)

Site of cancer	RR(90% CI)
All cancers	0.9959(0.8641-1.148)
All except leuk.	0.9954(0.8613-1.150)
liver	0.7658(0.5819-1.008)
lung	0.7298(0.4517-1.179)
Stomach	0.8972(0.5787-1.391)
Bone-marrow(leuk.)	1.010 (0.4867-2.097)
Nasopharynx	1.206(0.8590-1.693)

Table 3

Adjusted RRs (90% CI) of site-specific cancers for dose groups (1979-1990 data) of a fixed cohort

Site of cancer	CA RR	RR for HBRA			
		Low Group	Medial group	High group	Subtotal
All cancers	1.00	1.121(0.9398-1.337)	0.9551(0.8002-1.140)	0.9177(0.7613-1.106)	0.9959(0.8641-1.148)
All Cancers except leuk.	1.00	1.150(0.9615-1.374)	0.9336(0.7785-1.120)	0.9126(0.7540-1.104)	0.9954(0.8613-1.150)
Leukaemia	1.00	0.4884(0.1566-1.523)	1.447(0.6415-3.262)	1.041(0.4159-2.605)	1.010(0.4867-2.097)
Nasopharynx	1.00	1.314(0.8674-1.990)	1.173(0.7769-1.770)	1.137(0.7395-1.748)	1.206(0.8590-1.693)
Lung	1.00	0.6683(0.3432-1.301)	0.6396(0.3363-1.217)	0.8972(0.4889-1.646)	0.7298(0.4517-1.179)
Liver	1.00	0.9351(0.6631-1.319)	0.7655(0.5401-1.085)	0.5979(0.4024-0.8884)	0.7658(0.5819-1.008)
Stomach	1.00	0.7873(0.4338-1.429)	1.195(0.7189-1.986)	0.6617(0.3515-1.245)	0.8972(0.5787-1.391)

2.4 Incidence of hereditary diseases and congenital deformity

The results listed in Table 4 and 5 have been reported in the third conference of ICHLNR in Ramsar, Iran in details. The prevalence rate of 31 kinds of hereditary diseases and congenital defects in children ($\leqslant$ 12 years old) in HBRA was compared with those in CA, which showed the prevalence rates between HBRA and CA were almost identical except the rate of Down Syndrome (higher in HBRA statistically), however, the prevalence rate of Down Syndrome was in the normal range of spontaneous rate if compare it with other places in the same Province (Guangdong Province) and some other places in China [5] (Tables 4 and 5).

Table 4

Prevalence Rate of 31 Hereditary Diseases and Congenital Defects in Children of HBRA and CA ($\leqslant$12 years old)

Area	Total of 31 diseases			Down Syndrome		
	No. of examinees	No. of Cases	Rate per thousand	No. of examinees	No. of Cases	Rate per thousand
HBRA	13425	304	22.64*	25258	22	0.87**
CA	13087	295	22.54*	21837	4	0.18**

*$P > 0.05$ **$P < 0.05$

Table 5

Frequencies of Down syndrome in some places in China

Place and year of investigation	Age of examinees	No. of examinees	No. of Down's syndrome cases	Frequency (10^{-3})
Wuhan, 1980	0-12	177057	48	0.27
Hebei Prov. 1981-82	0-14	275642	106	0.39
Hunan Prov. 1982	0-7	4880	3	0.62
Guiyang, 1981	0-11	7894	16	2.02
Guangdong Prov.				
Zhanjiang, 1982	0-14	20084	12	0.60
Foshan, 1982-83	0-14	6420	4	0.62
HBRA 1975,79,85	0-12	25258	22	0.87
CA 1975, 79, 85	0-12	21837	4	0.18

2.5 Frequency of chromosomal aberrations

For detecting the frequency of chromosomal aberrations in the HBRA and CA, in recent five years the HBRRG has adopted two approaches for sample preparation and microscopic analysis: (1) conventional method, to study both stable and unstable aberrations for subjects of various age groups and dose groups. All the 37 donors had their own dose records based on direct measurements with TLD. (2) improved method (Hayata's method)[8]: 28 subjects were selected from 10 households (families) (6 in HBRA, 4 in CA) of three generations (grand parents, parents, children). Separated lymphocytes cultures and a microscope with an auto-stage fixed in it.[9] Only the unstable aberrations (dicentrics and centric rings) were observed.

The results obtained from the convention method demonstrated that both the chromosomal aberrations of dicentrics with fragments and those of translocations were higher significantly in HBRA. [10](Table 6). The results also show that there was a tendency that the frequency of these aberrations increased with the increase of accumulated dose when compared with corresponding CA. The results obtained from the improved method show that the frequency of chromosomal aberrations of dicentrics and centric rings was higher in HBRA compared to CA for all age groups but except the teenagers (Table 7).[9]Since the frequency of dicentric and centric rings does not increase significantly with the increase of age in CA, it seems that the increment of these aberrations were mainly caused by the dose elevation of background radiation which accumulates with age when the subjects were long term residents (settled in the HBRA).

Table 6

Chromosome aberrations of inhabitants with different age from HBRA and CA.

Age groups (yrs)	No. of donors	Cum. dose average (mGy)	Cells scored	No. of aberrations (/1000 cells)					
				Dic + CR +f	Dic + CR -f	AF	Tr	Total	Hyper-diploidy
HBRA									
10	10	24.9	11910	0.42	0.08	0.84	1.43	2.77	1.13
40	10	94.1	10200	0.69	0.49	1.27	2.45	4.90	1.04
55	8	131.5	10000	1.00	0.20	1.50	2.80	5.50	1.76
70	10	161.4	9700	1.03	0.10	1.96	2.89	5.98	2.96
Total	38	—	41810	0.77**	0.22	1.36	2.34*	4.69***	1.52
CA									
10	10	6.4	11700	0.09	0.17	1.03	1.37	2.65	0.69
40	10	23.9	11090	0.45	0.18	0.90	0.90	2.43	0.92
55	7	34.4	7330	0.55	0.27	0.95	1.91	3.68	1.18
70	10	44.5	9200	0.11	0.11	1.30	2.72	4.24	2.29
Total	37	—	39320	0.28**	0.18	1.04	1.65*	3.15***	1.11

Dic = dicentric, CR= centric ring, +f or -f = with or without fragment, AF = excess acentric fragment, acentric ring and minute, Tr = translocation. * u = 2.196, $P < 0.05$, ** u = 3.000, $P < 0.01$, *** u = 3.498, $P < 0.001$

Table 7

Frequencies of chromosome aberrations (dicentric and centric ring) detected in different age groups of inhabitants from HBRA and Control area*

Area	No. of Cases	Mean age (years) ± SD	Cells Scored	Dicentric+centric ring in 1000 cells	P-value
HBRA	3	79.00 ± 1.4	8505	4.23	
	3	67.33 ± 3.1	9155	3.60	<0.001
	3	34.66 ±1.7	7812	2.56	<0.05
	6	10.8 3 ± 1.2	12418	1.05	
CA	5	61.80 ± 5.6	9829	1.22	
	4	31.25 ± 4.0	12029	1.25	
	4	8.75 ± 0.8	10222	1.08	

* Donors were selected from 10 families, 6 from HBRA, 4 from CA.

2.6 Confounding factors

Three kinds of confounding factors were surveyed (with laboratory examination and analysis) in this investigation: (1) common factors: race, nationality, population structure (especially for age distribution), mortalities of all kinds of diseases and their sequence, especially the sequence of malignancies. (2) local factors: local mutagens, trace elements in soil, human hair and placenta (including selenium and manganese), local epidemic infections (e.g. hepatitis B type). (3) main factors: diet, especially intake of fruits and vegetables, cigarette smoking.

Ames et al. reported (1995) [11] a review of epidemiological studies showing protection by fruits and vegetables against cancer: consumption of adequate fruits and vegetables is associated with a lowered risk of degenerative diseases such as cancer, cardiovascular disease, cataracts, and brain and immune dysfunction. Nearly 200 studies in the epidemiological literature have been reviewed and relate, with great consistency, the lack of adequate consumption of fruits and vegetables to cancer incidence. Thus the dietary components and quantities of daily intake influence the incidence or mortality of cancer, and then influence the results of epidemiology.

Table 8 shows the dietary components and quantities of daily intake in HBRA and CA surveyed by the HBRRG of China [12] which demonstrated that they were comparable in the four dose groups.

Table 9 shows the percentage of cigarette smokers in different dose groups in the investigated areas of HBRA and CA, which indicates that most smokers were male (above 16 years of age), around 55% to 60% of male population, very small number of smokers in females (2% of female population). The percentages of smokers for both sexes of the four dose groups were comparable.

Table 8

Dietary Components and Quantities of Daily Intake in HBRA and CA

Component*	Dietary Intake (g/person. day)**			
	Group 1	Group 2	Group 3	Group 4
Rice, wheat flour	434.0	450.0	447.0	456.0
Sweet potato and other grains	93.0	124.0	165.0	---
Bean and bean products	37.8	35.7	51.4	9.4
Vegetables	423.0	427.0	447.0	480.0
Fruits	39.7	37.1	48.7	49.4
Sugar	7.5	11.5	8.5	6.1
Milk and milk products	Trace	Trace	Trace	Trace
Eggs	5.1	13.3	8.4	7.3
Meats	84.3	89.9	105.5	109.7
Aquatic products	71.7	69.6	84.4	86.7
Edible Vegetable oils	6.8	7.0	7.5	13.2
Soy sauce	4.7	3.4	2.9	5.7
Salt	9.2	9.5	8.8	9.3

*The classification was based on 《Diet, Life-style and Mortality in China》 Editors: Chen Junshi et al.. People's Medical Publishing House, Beijing, 1991.
** Classification into four groups according to the radiation levels, to which the cohort members exposed. Group 1, 2 and 3 were in HBRA; Group 4 was in CA. Samples were taken from 120 households (30 from each group).

Table 9

Percentage of Cigarette Smokers in different dose groups

	No. of persons inquired			No. of smokers			percentage		
Dose group	M	F	Subtotal	M	F	Subtotal	M	F	Subtotal
High	52	49	101	31	0	31	59.6	0	30.7
Medial	51	50	101	28	0	28	54.9	0	27.7
Low	50	50	100	29	1	30	58.0	2.0	30.0
Control	45	38	83	26	0	26	57.8	0	31.3

3. HOW TO INTERPRET THE RESULTS WHICH SEEM TO BE CONTRADICTORY?

Some organizations and authors, based on the "no-threshold, linear" hypothesis estimated the risk for low dose exposure and stated "no matter how small the dose, there will be increment of cancer induction." But the results obtained in the HBRA of China have not demonstrated any increment of cancer mortality. Another question is: if the chromosomal aberration (especially the stable type) is related to mutation, why it appears that both beneficial effects (no increase of canter mortality found) and "detrimental" effects (increase of chromosomal aberrations) exist in a same population? to understand and interpret the results above mentioned, two approaches of analysis were made: (1) microdosimetric considerations (2) hypothesis on "benefit-detriment competition".

(1) Microdosimetric considerations:

The dose and dose-rate range of low LET exposure in HBRA is characterized by small average numbers of tracks per cell with long intervals of time between them. Effects are, therefore, likely to be dominated by individual track events, acting alone.[13] Based on the theory of dual radiation action, and taking into account the repair effect of sublesion, it is assumed that the stochastic effect (if any) would be too small to be detected in an epidemiological study, in case of HBRA.[14] However, an upper limit of risk estimate can be obtained.

(2) Hypothesis on benefit-detriment competition

The High Background Radiation Research Group stated in 1985:[5]

"However, the fact is that the increase of frequency of chromosome aberrations were observed in the high background radiation areas (HBRA), and the possibility of effects induced by ionizing radiation can not be excluded, although the doses were low. Nevertheless, on the contrary, cell-mediated immunity examination revealed that there is a tendency of strengthen of immune functions among the HBRA inhabitants. If we say the former is disadvantageous, thus the latter is beneficial. so, there is a competition between these two kinds of effects. As above mentioned, up to now we have not found the increase of mutation-based diseases. Possibly, it implies that the beneficial effects are superior to the detrimental effects in case of low level radiation of HBRA."

Recent literature described the fate of damaged DNA and the process of multistep colorectal carcinogenesis[15], In case of radiation, the DNA damage may be caused by direct action or indirect action. Besides the apoptosis, damaged DNA may be balanced by elaborate defense and repair process, although it is not complete. Thus, there is probability of converting DNA lesions to mutation, if the increase of rate of cell division occurs. The process of carcinogenesis needs multisteps, and it needs several mutations.[15] In case of low dose exposure, the stimulatory effects strengthen (defense and immune functions strengthen) for inhibiting and preventing the formation of mutations. Many reports of epidemiology and radiobiology and the results of investigation in HBRA of Yangjiang denoted that in case of low dose exposure the defense and repair system might be superior to the disadvantageous effects. Table 10 and 11 show the results of immulogical function obtained from the examinations of inhabitants in HBRA and CA, which demonstrated the strengthen of these functions. [16, 17]Table 12 shows the results obtained from laboratory experiment. which shows low dose ionizing radiation may reduce the incidence of mammary tumor. [18]

Table 10

Morphological Transformation of Peripheral Blood Lymphocytes in Response to PHA in Inhabitants of High Background Radiation Area in Yangjiang

Age (years)	Morphological Transformation Rate (%)		P-value
	HBR Area	Control area	
16-55	75.0±0.90(148)*	70.6±0.99(148)*	<0.01
16-25	77.2±1.12(82)*	71.3±1.28(84)*	<0.01
45-55	72.2±1.40(66)*	69.8±1.57(64)*	>0.05

From: Liu Shu-Zheng 1986. Stimulatory effects of low-dose ionizing radiation on some immunology parameters. Proc. "International Symposium on Biological Effects of Low Level Radiation" Nanjing, China, PP.143-165.

* Number in parenthesis is the number of examinees

Table 11

Comparison between frequency of PBL IL-2SC* in HBRA and CA inhabitants

Age (years)	HBRA in Yangjiang		CA in Enping		P value
	N	Frequency(%)	N	Frequency(%)	
< 20	9	20.11±0.66	9	17.02±0.69	>0.05
20-50	9	19.89±0.58	9	17.17±0.79	<0.05
> 50	7	20.71±0.32	9	15.78±0.52	<0.01
Total	25	20.20±0.32	27	16.63±0.53	<0.01

From: Yao J. et al. 1993. Ref.[17]

* peripheral blood lymphocyte interleukin 2-secreting cells

Table 12

The incidence of spontaneously occurring mamary tumor of mice subjected to LDR* and CRD**

Group	Incidence in percentage
Untreated	73%
LDR	43%
CRD	27%
LDR-CRD	16%

Based on T. Makinodan 1992, in "Low dose irradiation and biological defense mechanisms", Sugahara, T., Sagan, L.A. and Aoyama, T. (eds). Elseveir Science Publishers, Amsterdam, pp. 233-237

* LDR--low dose ionizing radiation, in this case, LDR means: 8-month old female C3H\He mice, chronically exposed to γ-radiation (4rad\exposure, 3 exposures\week, for four weeks) mice were observed for 35 weeks.

** CRD--chronically restricted diet (calorically 70% of ad-labium diet)

SUMMARY

1. One of the great advantages of radiation-epidemiology in the high background radiation areas is the possibility of obtaining the results from direct observation on human beings, needless to do extrapolation of dose-effects relationship from high to low dose levels. This is especially true when the families of the subjects have lived in the investigated areas for many generations. In several high background radiation areas in the world, the radiation levels that the local inhabitants exposed to are similar to, or above those exposed by the workers of nuclear industry, and by most people in the radionuclide contaminated areas during and after serious nuclear accidents.

2. However, there are great difficulties in obtaining accurate, available data to determine the dose-stochastic effects relationship in such a low dose range: (1) large cohort size and long period of observation and follow-up, (2) estimate of the individual dose accumulated for long time, (3) to distinguish the minor signals from noise (confounding factors), and (4) to interpret various results of observation which may be contradictory.

3. The long term epidemiological study (1972-1990) in the high background radiation areas (HBRA) in Yangjiang, China and the nearby control areas (CA) demonstrated that no increase of cancer mortality, or incidence of hereditary diseases and congenital deformities was found in the HBRA after careful study on

comparability between HBRA and CA and between the dose groups within the cohort (1970-1986 data, and the 1979-1990 data which were used for fixed cohort analysis on cancer mortality). However, a minor increase of frequency of chromosomal aberrations was found in HBRA, the results were reproducible.

4. To understand and interpret the results above mentioned, two approaches of analysis were made: (1) microdosimetric considerations; (2) hypothesis on benefit-detriment competition.

5. Based on the theory of dual radiation action, and taking into account the repair effect of sublesion, it is assumed that the stochastic effect (if any) would be too small to be detected. However, an upper limit of risk estimate can be obtained in epidemiology.

6. Literature on mechanisms of carcinogeneses revealed that DNA lesions have a certain probability of giving rise to mutations when the cell divides, and mutations in several critical genes can lead to tumors. However, in the meanwhile there exist defense and repair system for inhibiting and preventing the formation of mutations. Perhaps, this system even effective during the Promotion stage of carcinogenesis. Thus these two kinds of system constitute a series of competition. In case of radiation exposure, the results will be dependent on the dose and dose rate. Many reports and the results of investigation in HBRA of Yangjiang denote that in case of low dose exposure the defense and repair system may be superior to the disadvantageous effects. More evidences (both epidemiological and biological) are needed to prove this hypothesis. Thus, international cooperation is very important to corroborate whether the findings above mentioned are reproducible, and to know whether the findings can be interpreted by radiobiology.

7. Investigation on high background radiation areas has already provided, and still will provide important information for evaluating the health effects on human beings exposed to low dose ionizing radiation.

REFERENCES

1. United Nations. Sources and effects of ionizing radiation. United Nations Scientific Committee on the Effects of Atomic Radiation, 1993 Report to the General Assembly, Annex D: Occupational radiation exposures. United Nations, New York, 1993.
2. United Nations, Sources and effects of ionizing radiation. United Nations Scientific Committee on the Effects of Atomic Radiation, 1993 Report to the General Assembly, New York, 1993.
3. Wei, Luxin; Zha, Yongru; Tao, Zufan et al., Epedemiological investigation in high background radiation areas in Yangjiang, China. In "High Levels of Natural Radiation, Proceedings of an International Conference Ramsar, 3-7 November 1990." IAEA, Vienna 1993.
4. Luxin Wei. Health Effects on Populations Exposed to Low-Level Radiation in China. In "Radiation and Public Perception: Benefits and Risks", chapter 16. Editors: J. P. Young and R.S. Yalow. American chemical Society, Washington, DC 1995.

5. Wei, Luxin et al., High Background Radiation Research in Yangjiang China. Atomic Energy Press (China), Beijing 1996.
6. Y. L. Yuan et al.. Recent advances of dosimetry investigation in the high background radiation area in Yangjiang, China. In this Proceedings. Editors: Wei, Luxin; Sugahara, Tsutomu and Tao, Zufan. Elsevier Science Publishers B.V. Amsterdam 1997.
7. Z.F. Tao, H.Kato, Y. R. Zha et al.. Study on cancer mortality among the residents in high background radiation area of Yangjiang, China. In this Proceedings. Editors: Wei, Luxin; Sugahara, Tsatomu and Tao, Zufan. Elsevier Science Publishers B.V. Amsterdam 1997.
8. I. Hayata. Advanced cytogenetical techniques necessary for the study of low dose exposures. In this Proceedings. Editors: Wei, Luxin; Sugahara, Tsutomu and Tao, Zufan. Elsevier Science Publishers B.V. Amsterdam 1997.
9. T.Jiang, C. Y. Wang, D.Q. Chen et al.. Preliminary report on quantitative study of unstable chromosome aberration following life time exposure by high background radiation in Yangjiang , China. In this Proceedings. Editors: Wei, Luxin; Sugahara, Tsutomu and Tao, Zufan. Elsevier Science Publishers B. V. Amsterdam 1997.
10. D. Q. Chen, S. Y. Yao, C. Y. Zhang et al.. Cytogenetic findings from inhabitants with different age in high background radiation areas of Yangjiang, China. In this Proceedings. Editors: Wei, Luxin; Sugahara, Tsutomu and Tao, Zufan. Elsevier Science Publishers B. V. Amsterdam 1997.
11. Bruce N. Ames, Lois Swirsky Gold, and Walter C. Willett. The causes and prevention of cancer. Proc. Natl. Sci. USA. (date 4/28/95).
12. Y. R. Zha and J. M. Zou. Confounding factors in radiation epidemiology and their comparability between the high background radiation areas and control areas in Guangdong, China. In this Proceedings, Editors: Wei, Luxin; Sugahara, Tsutomu and Tao, Zufan. Elsevier Science Publishers B. V. Amsterdam 1997.
13. United Nations. Sources and effects of ionizing radiation. United Nations Scientific Committee on the Effects of Atomic Radiation, 1993 Report to the General Assembly, Annex F: Influence of dose and dose rate on stochastic effects of radiation. United Nations, New York, 1993.
14. Li, Kaibao and Zhao, Zhaoluo. Application of microdosimetric concept and theory to radiation epidemiological study. In 《 High Background Radiation Research in Yangjiang China 》, Chapter 6,. Chief Editor: Wei, Luxin. Atomic Energy Press, Beijing 1996.
15. United Nations. Sources and effects of ionizing radiation. United Nations Scientific Committee on the Effects of Atomic Radiation, 1993 Report to the General Assembly, Annex E: Mechanisms of radiation oncogenesis. United Nations, New York, 1993.
16. S.Z. Liu. Stimulatory effect of low radiation on immunity. In this Proceedings. Editors: Wei, Luxin; Sugahara, Tsutomu and Tao, Zufan. Elsevier Science Publishers B.V. Amsterdam 1997.
17. J.M. Zou, Y.R. Zha and J. Yao. Studies on immune competence and immune response to virus in high background radiation area, Yangjiang, China. In this

Proceedings. Editors: Wei, Luxin; Sugahara, Tsutomu and Tao, Zufan. Elsevier Science Publishers B.V. Amsterdam 1997.
18. Makinodan, T. cellular and subcellular alterations in immune cells induced by chronic, intermittent exposure in vivo to very low doses of ionizing radiation and its ameliorating effects on progression of autoimmune disease and mammary tumor growth. In: Sugahara, T., Sagan, L.A. and Aoyama, T. (eds), Low Dose Irradiation and Biological Defense Mechanisms. Elsevier Science Publishers, Amsterdam. Pp.233-237. 1992.

Exposures to natural radiation worldwide

B.G. Bennett

United Nations Scientific Committee on the Effects of Atomic Radiation (UNSCEAR), Wagramerstrasse 5, 1400 Vienna, Austria

1. INTRODUCTION

A background level of ionizing radiation from natural sources has always existed on earth, and all living organisms are inescapably exposed. The radiation originates from cosmic rays, which enter the earth's atmosphere from outer space, and from radionuclides in the environment on earth. The main terrestrial radionuclides are ^{40}K, ^{238}U and decay products and ^{232}Th and decay products, which are present in soil, water, food and in the body. Although most persons are more concerned about man made sources of radiation arising from releases of radionuclides to the environment from nuclear installations or from accidents, the dominant exposures are, in fact, almost always from the natural sources.

A great deal of attention has been directed to the natural radiation background, and measurements have been made in most all countries. The United Nations Scientific Committee onto Effects of Atomic Radiation (UNSCEAR) has compiled these results and evaluated the representative exposures to the world's population from this source of radiation. This paper summarizes the latest assessment by UNSCEAR of exposures worldwide from natural radiation sources.

There is considerable variation in exposures to natural radiation sources. This is due to altitude for cosmic ray exposures and to varying content of uranium and thorium in soils, radium and polonium in foods and water and radon in indoor air. Several areas of extreme variations have been identified as high natural radiation background areas. These areas have been carefully studied to determine the extent and magnitude of the exposures and to evaluate the health of populations living in these areas. This conference will be a further contribution to knowledge of this interesting and important subject.

2. EXPOSURES IN AREAS OF NORMAL BACKGROUND

The total exposure to natural radiation in areas of normal background averages 2.4 ms a^{-1}. One third of this is from external exposure and two thirds from internal exposure. The largest and most variable component is from radon and its decay products in indoor air. The contributions to the average exposures in normal areas are listed in Table 1.

Table 1
Representative exposure to natural background radiation

Component of exposure	Annual effective dose (mSv)	
	Areas of normal background	Areas of elevated background
External exposure		
Cosmic rays	0.38	2.0
Terrestrial radionuclides	0.46 ^{a}	4.3
Internal exposure		
Cosmogenic radionuclides	0.01	0.01
Terrestrial radionuclides	0.23 ^{b}	0.6
Radon	1.2	10
Thoron	0.07	0.1
TOTAL	2.4	

a Apportional as follows: ^{40}K 0.12 mSv, ^{232}Th series radionuclides 0.21 mSv, ^{238}U series radionuclides 0.13 mSv.
b Includes 0.17 mSv from ^{40}K and 0.06 mSv (Table 2) from uranium and thorium series radionuclides.

2.1 EXTERNAL EXPOSURE

2.1.1 Cosmic rays

The most significant feature of exposure to cosmic rays is the variation with altitude. Cosmic rays are attenuated by the earth's atmosphere. Therefore, the highest exposures are at aircraft altitudes or at high elevations above sea level. The absorbed dose rate in air at sea level at mid and high latitudes is 32 nGy h^{-1}. This approximately doubles with every 1,500 meters of altitude. The dose rate is slightly less at lower latitudes in the equatorial region. This is the exposure rate from the ionizing component of cosmic rays, due mainly to muons. The radiation weighing factor taken as one, giving the same value for the average effective dose rate.

A non ionizing component of cosmic ray exposures is contributed by neutrons. Greater uncertainty is present in estimating the average effective dose rate from this component. The value at sea level was estimated to be 3.6 nSv h^{-1} in the UNSCEAR 1993 Report [1].

The exposure rates from cosmic rays are slightly less indoors due to the shielding effect of buildings. This will depend on materials used in construction; there is greater shielding in a large apartment building with concrete ceilings than in a small house with a wooden roof. UNSCEAR has assumed an average reduction in the ionizing component of cosmic rays of 20% indoors, corresponding to a shielding factor of 0.8. Because of uncertainties, a shielding factor has not been applied to the non-ionizing component of cosmic rays.

From the above values of exposure rates and shielding factor and assuming 80% of time is spent indoors and 20% outdoors, the annual average effective dose from cosmic rays has been estimated. The result is 0.24 mSv from the ionizing component and 0.03 mSv from neutrons at sea level. Most of the world's population lives at or near sea level. When account is taken of the distribution of the world's population with altitude and the variation in the exposure rate, an average annual effective dose of 0.38 mSv is determined for exposure to cosmic rays.

2.1.2 Terrestrial radiation

The natural, long lived radionuclides ^{40}K, ^{238}U and ^{232}Th with the decay products of the latter two, are present in all soils and building materials, contributing to external exposure rates both outdoors and indoors. Measurements have been made in most areas of the world; the data compiled by UNSCEAR accounts for three fifths of the world's population. Country average outdoor absorbed dose rates in air ranged from 24 to 160 nGy h^{-1}, with most values between 40 and 80 nGy h^{-1}. The population weighted mean value was 57 nGy h^{-1} [1].

Surveys have also been made in many countries of the external radiation levels indoors, although these have not been quite so extensive as outdoor surveys. The levels depend primarily on the composition of the building materials used in the construction of houses. Country average indoor absorbed dose rates in air ranged from 20-190 nGy h^{-1}, with most values within a narrower range of 60-100 nGy h^{-1}. The population weighted mean value was 80 nGy h^{-1}[1]. In most areas, the building materials enhance exposure rates indoors by 40% - 50% above outdoor levels. The indoor to outdoor ratios in countries varied from 0.8 to 2. For wooden and lightweight houses, a ratio near one would be expected. For houses of brick, stone or concrete, the concentrations in the building materials and the surrounding source geometry would give the indoor-outdoor absorbed dose rate ratio above 1 and up to 2.

The value of 0.7 Sv Gy^{-1} is used to convert the absorbed dose in air to effective dose. This accounts for the differences in energy absorption characteristics of air and tissue and the shielding of organs by overlying tissues. With indoor and outdoor occupancy of 0.8 and 0.2, respectively, and the mean dose rates quoted above, the average annual effective dose from external exposure to terrestrial gamma radiation is 0.46 mSv.

2.2 INTERNAL EXPOSURE

2.2.1 Cosmogenic radionuclides

Several cosmogenic radionuclides are produced in the upper atmosphere by the interaction of cosmic rays with the nuclei of atoms present in air (e.g. nitrogen, oxygen and argon). These contribute quite low and relatively uniform radiation exposures over the earth's surface. The most significant cosmogenic radionuclides and the annual effective doses caused by internal exposure are ^{3}H (0.01 mSv), ^{7}Be (0.03 mSv), ^{14}C (12 mSv) and ^{22}Na (0.2 mSv).

2.2.2 Terrestrial radionuclides

Potassium is an essential element that is under close homeostatic control in the body. The radioactive isotope of potassium, ^{40}K, is present in the ratio 1.18 10^{-4} to stable potassium. The average concentration of ^{40}K in the body is about 55 Bq kg^{-1}. The average annual effective dose from internal exposure to ^{40}K is estimated to be 0.17 mSv.

The doses from radionuclides in the uranium and thorium series can vary widely, reflecting the intakes to the body with diet and air. Previous estimates of doses estimated from concentrations of radionuclides in the body have been supplemented in the UNSCEAR 1993 Report with estimates of doses derived from concentrations of the radionuclides in food, water and air, which when combined with values of the dose per unit intake, can give more widely based dose estimates. These results are summarized in Table 2. The intake amounts are only representative; considerable variations are present in different locations. The dose estimates are also only approximate. Some values of the dose per unit intake have subsequently been revised [3], and these will bring consequential changes in the dose estimates.

Radionuclides in the ^{238}U series (other than radon) that contribute significant exposure include ^{226}Ra, ^{210}Pb and ^{210}Po. Ingestion is the primary route for transfer of ^{226}Ra to the human body. Annual intakes in normal areas are estimated to be 19 Bq by ingestion and 3.5 mBq by inhalation. Intake from drinking water derived from surface water sources is generally small, however when the drinking water source is ground water, the intake may be quite significant. The average annual effective dose is estimated to be 3.8 mSv from the ingestion intake and 0.01 mSv from the inhalation intake. Intakes of ^{210}Pb and ^{210}Po are variable, and both ingestion and inhalation contribute to the effective dose. An important source of ^{210}Po in diet is seafood.

Table 2
Internal exposure from terrestrial radionuclides of the uranium and thorium series

Radionuclide	Annual intake (Bq)		Annual effective dose (μSv)	
	Ingestion	Inhalation	Ingestion	Inhalation
^{238}U	4.9	0.0069	0.12	0.21
^{234}U	4.9	0.0069	0.15	0.21
^{230}Th	2.5	0.0035	0.18	0.18
^{226}Ra	19	0.0035	3.8	0.01
^{210}Pb	32	3.5	32	7.0
^{210}Po	55	0.35	11	0.35
^{232}Th	1.3	0.0069	0.52	1.4
^{228}Ra	13	0.0069	3.9	0.01
^{228}Th	1.3	0.0069	0.09	0.69
^{235}U	0.2	0.0004	0.01	0.01
TOTAL			52	10

Both inhalation and ingestion of ^{232}Th are relatively low, and only ^{228}Ra in this series is somewhat more available to plants and animals. Dietary intake is, thus, more important to body contents than decay of ^{232}Th in the body. The shorter-lived decay products of ^{228}Ra, namely ^{228}Th (half-life 1.9 years) and ^{224}Ra (half-life 3.66 days), both alpha emitters are the most important contributors to the dose.

Radionuclides in the ^{235}U series are less abundant in the environment, and the doses from ingestion or inhalation of these radionuclides are negligible.

2.3 Radon

The concentrations of radon, ^{222}Rn, and its decay products in air are extremely variable. The main source of radon gas in the atmosphere is the emanation from soil. The content of radium in the soil and the soil properties are the primary determining factors, but moisture content and meteorological conditions are also of influence. Diurnal variations in radon concentrations in air at ground level occur with a minimum at noon and a maximum two times higher in the night. Seasonal variations occur as well, with a minimum in winter and a maximum three times higher in summer.

Higher concentrations of in air are encountered at continental sites and lower concentrations on islands and at coastal sites. A population-weighted average value of ^{222}Rn in outdoor air is estimated to be 10 Bq m^{-3} [1]. With an assumed equilibrium factor of 0.8, the average equilibrium equivalent concentration of radon is 8 Bq m^{-3}, with a typical range, excluding extreme values, of 2 to 12 Bq m^{-3}.

Radon enters indoor air from the soil or rock under or surrounding the buildings, from building materials, water supplies, natural gas and outdoor air. The ventilation rate is an important factor in determining the build-up of concentrations, which often exceed those outdoors. The results of large-scale surveys of indoor radon concentrations have been reported by UNSCEAR[1]. The results from countries generally fit a log-normal distribution with a geometric standard deviation of around 3. The representative average is about 40 Bq m^{-3}. With the equilibrium factor taken to be 0.4 indoors, the equilibrium equivalent concentration of radon indoors is estimated to be 16 Bq m^{-3} [1].

The dosimetry of radon following inhalation is still under discussion. In the UNSCEAR 1993 Report [1], the effective dose rate factor was taken to be 9 nSv h^{-1} per Bq m^{-3} equilibrium equivalent concentration. Accounting for average indoor and outdoor radon concentrations with indoor occupancy of 0.8, the annual effective dose is determined to be 130 mSv from outdoor exposure, 1000 mSv from indoor exposure.

2.4 Thoron

The radon isotope in the ^{232}Th decay series is ^{220}Rn. It is often referred to by its historical name, thoron, to distinguish it from the more important radon isotope, ^{222}Rn. The short half-life of ^{220}Rn (55.6 s) prevents its build up in the environment and limits the dose from it and its decay products. Concentrations in outdoor air at ground level are around 10 Bq m^{-3}. Indoor concentrations are less subject to the ventilation rate but are more determined by the exhalation rate from building materials. Indoor concentrations of ^{220}Rn are less variable than ^{222}Rn, based on the limited monitoring experience available. The representative concentration of ^{220}Rn indoors is taken to be 3 Bq m^{-3}. The average equilibrium equivalent concentrations are 0.1 Bq m^{-3} outdoor sand 0.3 Bq m^{-3} indoors. The annual effective dose is 69 mSv [1].

3. EXPOSURES IN AREAS OF HIGH BACKGROUND

Areas of high levels of natural radiation have been identified in several areas of the world. These areas have been mentioned by UNSCEAR as representing the extremes in variations in the various components of exposure. There has been, however, no systematic evaluations of the exposures in these areas, primarily because of the limited populations.

In most cases, the extremes in any one location have been in only one component of exposure. Some of the variations encountered are included in Table 1. There is no total given for the areas of high background, since only one or a few components are to be combined with the otherwise normal exposures from the other components. For example, the external exposures in monazite sand coastal regions are high, but the cosmic ray exposures, being at sea level, are at the low end of the scale.

3.1 EXTERNAL EXPOSURE

3.1.1 High altitude areas

Because of the clear increase in cosmic ray exposure with altitude, the inhabitants of areas at high elevations are more exposed than at sea level (0.27 mSv a^{-1}) or the world average (0.38 mSv a^{-1}). Some large cities are at high elevations, e.g. Mexico City (17.3 million population at 2,240 m), Quito, Ecuador (11 million population at 2,840 m) and La Paz, Bolivia (1 million population at 3,900 m). The annual exposures at these locations are 3, 4.2 and 7.5 times higher than that at sea level. There are no indications that exposures from the other components of natural radiation exposure are abnormal at these locations.

3.1.2 Areas of uranium and thorium mineralization

The most well known areas of high natural radiation background are Kerala, India and Espirito Santo and Minas Gerais, Brazil. The 55 km long coastal strip in Kerala province in India and the 2.5 km strip in Tamil Nadu province to the south have high external radiation levels because of the high thorium content of monazite sands in these areas. Dose rates in air of 200-4,000 nGy h^{-1} have been reported, which are 3.5 to 70 times the representative world average(57 nGy h^{-1}). There has been some surface refilling in inhabited areas and changes in construction of houses that have led to substantial reductions in exposures both outdoors and indoors [7].

The coastal region of Espirito Santo in Brazil is also a location of monazite sand deposits. Dose rates in air 130-1,200 nGy h^{-1} were reported in the city of Guarapari and 220-4200 nGy h^{-1} in the village of Meaipe [8]. The inland plateau of the Minas Gerais region is an area with thorium and uranium mineralization in soil and absorbed dose rates in air of 100-4,000 nGy h^{-1} [8]. The less inhabited area of Pocos de Caldos is well known for the highest external exposure rate in this region.

There are many other areas with elevated external exposures to natural radiation, including areas with granitic and uraniferous rocks (central France, Italy, Sweden), thorium deposits(Kenya), sedimenting regions (Nile Delta, Ganges Delta), mineral spring waters (Ramsar, Iran)and phosphate deposits (Florida, United States).

3.2 INTERNAL EXPOSURE

3.2.1 Special foodchains

There is a degree of enhanced internal exposure to natural radiation in all areas of high background due to the transfer of the terrestrial radionuclides in soil to plants and animals. More complete evaluations of both external and internal exposures have been completed for high background areas in India and Brazil [5, 6].

Special recognition has been given to the food chain in the Arctic and sub-Arctic regions involving lichens-caribou/reindeer-man. Long-term retention and accumulation of particles containing ^{210}Pb and ^{210}Po in lichens occurs in these areas, leading to high concentrations in the meat of animals. Intakes by humans may be of the order of 140 Bq of ^{210}Pb and 1,400 Bq of ^{210}Po [2], many times higher than normal. The population receiving higher exposures is uncertain, but may involve some tens of thousands of persons.

Levels of ^{210}Po are also high in some marine foods. These foods are not always substantial parts of diets, but the intakes in the Marshall islands seem to offset the otherwise lower than normal natural background levels [9].

3.2.2 Radon indoors

The widespread occurrence of high levels of radon that build up in indoor spaces is probably the most common cause of elevated exposures to natural radiation. The concentrations of radon indoors averages 40 Bq m^{-3} [1], as determined by surveys in many countries, but extremely wide variations are encountered. The general soil conditions can define radon-prone areas, but many factors of the house construction and particularly the ventilation rate determine the actual levels. Measurements are necessary to determine the specific levels in any particular house.

The radon levels are particularly high in some northern countries, Sweden for example, where buildings are tightly sealed and ventilation may be low. It has been thought that in warmer, tropical countries, higher ventilation rates would result in indoor levels being similar to the lower outdoor levels. Many houses in these areas, however, are more commonly closed and air-conditioned, so higher levels of radon can occur here as well. Underground dwellings and caves, where people may live, visit or work, also frequently have elevated levels of radon.

Radon concentrations of several thousands of Bq m^{-3} have been measured in houses in many countries. The ICRP has recommended that remedial measures be taken when the levels exceed 200-600 Bq m^{-3}, corresponding to contributions to the annual effective dose of 3-10 mSv [4].

4. SUMMARY

All persons, no matter where they live in the world, are exposed to background radiation from natural sources. The average exposures have been assessed by UNSCEAR from the many measurements that have been made throughout the world. The mean values for the various components of external and internal exposure are given in Table 1. These define the normal areas of background exposure, where the representative annual effective dose is 2.4 mSv.

Some indications of the variations in these exposures are also listed in Table 1. Variations about the mean levels of factors of 2 or 3, which begin to define high background areas, are common, and factor of 5 to 10 are not unusual. The high background areas may be defined by the altitude, the soil or rock conditions, particular food chains and the building materials and construction features.

Natural background radiation exposure determines the baseline upon which exposures from many man-made sources occur. Continued surveys of natural background radiation levels will provide more representative average values and better define the high background areas to which many populations are exposed.

REFERENCES

1. United Nations. Sources and Effects of Ionizing Radiation. United Nations Scientific Committee on the Effects of Atomic Radiation, 1993 Report to the General Assembly, with scientific annexes. United Nations sales publication E.94.IX.2. United Nations, New York, 1993.
2. United Nations. Sources, Effects and Risks of Ionizing Radiation. United Nations Scientific Committee on the Effects of Atomic Radiation, 1988 Report to the General Assembly, with annexes. United Nations sales publication E.88.IX.7. United Nations, New York, 1988.
3. International Commission on Radiological Protection. Age-dependent doses to members oft public from intake of radionuclides: Part 5. Compilation of ingestion and inhalation dose coefficients. ICRP Publication 72. Annals of the ICRP 26(1). Pergamon Press, Oxford, 1996.
4. International Commission on Radiological Protection. Protection against radon-222 at home and at work. ICRP Publication 65. Annals of the ICRP 23(2). Pergamon Press, Oxford, 1993.
5. Amral, E., E. Rochado, H. Paretzke et al. The radiological impact of agricultural activities in an area of high natural radioactivity. Radiat. Prot. Dosim. 45 (1992) 289-292.
6. Paul, A., P. Pillai, K. Pillai et al. Pathways of internal exposures in a high background area. p. 135-145 in: Proceedings of International Conference on High Levels of Natural Radiation, Ramsar, Iran, November 1990.
7. Paul, A.C., T. Velayudhan, P.M.B. Pillai et al. Radiation exposure at the high background area at Manavalakurichi - the changing trends. J. Environ. Radioact. 22 (1994) 243-250.
8. Pfeiffer, W.C., E. Penna-Franca, C. Costa Ribeiro et al. Measurements of environmental radiation exposure dose rates at selected sites in Brazil. An. Acad. Bras. Cienc. 53 (1981) 683-691.
9. Noshkin, V., W. Robinson and K. Wong. Concentration of ^{210}Po and ^{210}Pb in the diet of the Marshall Islands. Sci. Total Environ. 155 (1994) 87-104.

Cosmic ray Contribution in the Measurement of Environmental Gamma ray Dose

K.Nagaoka, K.Honda and K.Miyano

Japan Chemical Analysis Center, 295-3 Sanno-cho, Inage-ku, Chiba-shi 263, Japan.

1. INTRODUCTION

Nowadays, several kinds of dosimeters are being used for environmental γ ray monitoring around nuclear facilities in Japan. However, the results obtained from those dosimeters are not always in agreement. It is caused by several different characteristics of the dosimeters, e.g. response to cosmic rays, energy response for γ rays, angular response, fading effect and others. Those effects in environmental dose measurements vary with the dosimeters. In order to estimate environmental radiation dose, we have to know the characteristics of each dosimeter.

In particular different responses of the dosimeters to cosmic rays influence the results. Therefore the response of each dosimeter to cosmic rays has been studied. It was well known that dose of cosmic rays vary with altitude, latitude and shielding effect. Then measurements were carried out at various altitudes and underground facilities at about the same latitude.

2. EXPERIMENTAL

2.1 Dosimeters used in this work

(1) NaI scintillation spectrometer (NaI-sp) : 3 inch ϕ spherical NaI(Tl) detector with quarts optical window
(2) Air-equivalent ionization chamber (IC) : 21 liter cylindrical chamber of air-equivalent plastic wall filled with air
(3) Pressurized ionization chamber (PIC) : 14 liter spherical chamber of 3.2mm thick stainless steel wall filled with Ar gas at 4 atm
(4) Thermoluminescence dosimeter (TLD) : $CaSO_4$(Tm) with Pb and Sn filter for energy response compensation
(5) Radiophotoluminescence glass dosimeter (RPLD) : silver activated phosphate glass with Sn and Al filter for energy response compensation
(6) NaI scintillation surveymeters
 (TCS-166) : 1 inch $\phi \times$ 1 inch cylindrical NaI(Tl) detector, energy response compensated by the Discriminator Bias Modulation Method (DBM)
 (TCS-121) : 1 inch $\phi \times$ 1 inch cylindrical NaI(Tl) detector, energy response compensated with 1mm and 50mm thickness Pb covers

2.2 Methods

Simultaneous measurements with NaI-sp and other dosimeters were performed on Mt.Fuji varying altitude by 200m, and in a shield box at underground in Nokogiriyama facilities of the Institute for Cosmic Ray Research, University of Tokyo. The shield box was made of about 400 sheets of Pb boards piled up. The box wall was about 40 cm in thickness. The innerface of the Pb box was coverd with Cu plates in order to shield Pb- X rays. Measurements were carried out in 1993 and 1994.

Cosmic ray contribution and self dose were determined by subtracting γ ray dose from measured values of IC, PIC, TLD, RPLD, TCS166 and TCS121. γ ray dose was determined with NaI-sp at same place. Cosmic ray contributions of NaI-sp were estimated[1] from counting rates over 3MeV and subtracted. Self dose of NaI-sp was estimated by measurement in the shield box and subtracted.

2.3 Measurement conditions

(1) NaI-sp

Pulse height spectra were accumlated for 30 min on Mt.Fuji, and for 4 hr at Nokogiriyama facilities with a 512 channel analyzer. The pulse height spectra were converted to γ ray dose by means of the G(E) function method [2)3)4)]. The energy region for the conversion was from 50keV to 3MeV.

(2) IC [5)]

The IC was calibrated with ^{137}Cs γ rays by Electrotechnical Laboratory (ETL) Agency of Industrial Science, Ministry of International Trade and Industry that was primary standard dosimetry laboratory (PSDL). This IC had been sealed to eliminate alpha contamination due to radon and thoron in air before this work. But the sealing was not complete, therefore air bag was connected through a silicon tube with IC. In field measurement, air density was corrected with atmospheric temperature and atmospheric pressure, and the sensitivity were checked at all measured points by a ^{226}Ra γ ray source. The measurements were carried out 6 times for 200s on Mt.Fuji, and 10 times for 2000s in the shield box.

(3) PIC [6)]

The PIC was also calibrated by ETL. The measurements were carried out about 5 times for 5min on Mt.Fuji, and 13 times for 1 hr in the shield box. The energy response characteristic should be taken into consideration in environmental measurement. Energy response is shown in Figure 1. The cosmic ray contribution and self dose of the PIC are calculated as follows:

Cosmic ray contribution and self dose = measured value - (γ ray dose $\times$ F)

$F = (\Sigma (D_E \times R_E) / \Sigma D_E)$

Where D_E is dose rate due to γ rays of energy E, and R_E is response of PIC to γ rays of energy E. γ ray energy distributions were obtained from pulse height spectra with the NaI-sp by means of the stripping method.

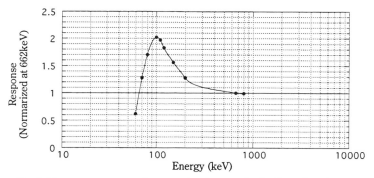

Fig.1 Energy response of pressurized argon ionization chamber (stainless steel wall thickness: 3.2mm)

(4) TLD and RPLD

6 pieces of TLDs (3 pieces of TLD had been irradiated to correct fading effect) and 1 or 5 pieces of RPLD were installed on Mt.Fuji for about 2 months and in the shield box for about 7 months. TLDs were read with TLD reader, and calibrated with ^{226}Ra γ rays at Japan Chemical Analysis Center (JCAC). Fading effects were corrected by measurements of the irradiated TLDs. RPLD measurements were performed by Toshiba Glass Co., Ltd.

(5) NaI(Tl)scintillation surveymeters

Two types of NaI(Tl)scintillation survey meters (TCS-166 and TCS-121) were used. To compensate the response of NaI(Tl)scintillator on photon energy, TCS-166 uses the method of DBM circuit, and TCS-121 uses Pb covers. Calculations of TCS-121 were as follows:

$$D = [(A-B)/20) + (B - C)] \times G$$

where A is an average of five times reading without Pb cover, B is with Pb cover (1mm), C is with Pb cover (50mm), G is calibration factor for B using a ^{137}Cs γ ray source.

3. RESULTS AND DISCUSSION

3.1 γ ray dose rate

The measured values are shown with absorbed dose in air (Gy or Gy h^{-1}).

γ ray dose rates ranged from 6 to 20 nGy h^{-1}, and the average was 13 nGy h^{-1} on Mt. Fuji. These were considerably low as compared with several country averages which range from 24 to 160 nGy h^{-1} [7]. Figure 2 shows examples of pulse height spectrum with NaI(Tl) scintillation detector. In the spectrum at 2400 m above sea level, a peak of the ^{137}Cs was observed.

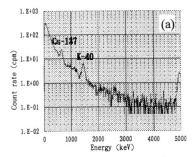
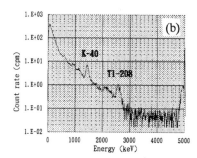

Fig.2 Pulse height spectrum with NaI(Tl) scintillation detector at 2,400m above sea level on Mt.Fuji (a), at 30m above sea level (JCAC) (b)

3.2 The responses to cosmic rays

Cosmic ray contribution and self dose for all dosimeters were correlated with counting rate over 3MeV with the NaI-sp. The results of measurements are shown in Fig.3. Slopes of the regression lines show the responses to cosmic rays, intercepts of the regression lines correspond to self doses. IC and PIC have nearly the same response to cosmic rays. TLD and RPLD have nearly the same response, but the responses are slightly lower than IC. NaI scintillation survey meters had very low response. Relative sensitivities of PIC, TLD, RPLD, and NaI(Tl) scintillation surveymeters (TCS166 and TCS121) in comparison with IC were 0.97, 0.76, 0.73, 0.18 and -0.03, respectively.

Cosmic ray contributions were obtained by subtracting the intercepts value from previous data. Relations between cosmic ray contribution and the counting rate over 3MeV with NaI-sp, altitude and atmospheric pressure are shown in Fig.4, 5 and 6. Conversion factors from the counting rate to the contribution (nGy h^{-1} / cpm) for IC, PIC, TLD, RPLD, and NaI(Tl) scintillation survey meters (TCS166 and TCS121) were 0.33, 0.32, 0.25, 0.24, 0.06 and -0.01, respectively.

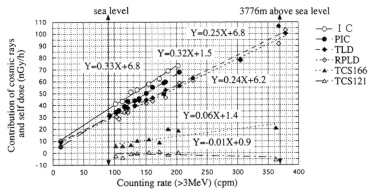

Fig.3 Relations between contributions of cosmic rays for various detectors and counting rate(>3MeV) with 3" ϕ spherical NaI(Tl)scintillation detector. These contributions include self dose.

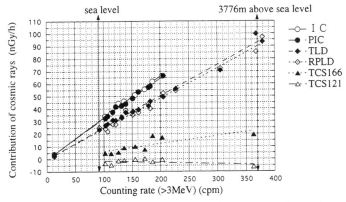

Fig.4 Relations between contributions of cosmic rays for various detectors and counting rate(>3MeV) with 3"φ spherical NaI(Tl)scintillation detector.

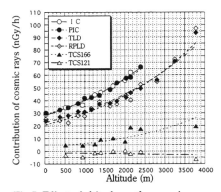

Fig.5 Effect of altitude on the cosmic ray contribution

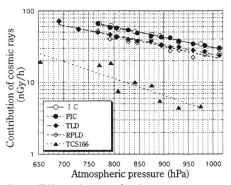

Fig.6 Effect of atmospheric pressure on the cosmic ray contribution

4. CONCLUSION

IC and PIC have nearly the same response to cosmic rays. TLD and RPLD have nearly the same response, but the responses are slightly lower than IC. NaI scintillation survey meters have very low response. Relative sensitivities of PIC, TLD, RPLD, and NaI(Tl) scintillation surveymeters (TCS166 and TCS121) as compared with IC are 0.97, 0.76, 0.73, 0.18 and -0.03, respectively.

The Cosmic ray contribution can be estimated from the altitude, the atmospheric pressure and the counting rate over 3MeV with NaI. As cosmic ray intensity depends upon latitude and shielding effect, furthermore 11 years modulation is informed[7]. The variation should be considered. The variation may be estimated from the counting rate over 3MeV with NaI. It is considered that the counting rate is the most useful to estimate the cosmic ray contribution at ground level.

ACKNOWLEDGMENT

We thank to Dr. T. Nakajima, Dr. M. Noguchi, Dr. Y. Takata and Mr. M. Uesugi, of Japan Chemical Analysis Center for helpful discussion and advice.

REFERENCES

1. T. Nagaoka, Analysis of Response Characteristics of NaI(Tl) Scintillation Detectors to Cosmic Ray, JAERI-M 85-058 (1985) (in Japanese)
2. S. Moriuchi and I. Miyanaga, A Method of Pulse Height Weighting Using the Discrimination Bias Modulation, Health Phys. 12, 1481-1487 (1966)
3. S. Moriuchi, A Dosimetric Instrument Based on the Spectrum Weighting Function Method for Environmental Radiation Measurements, JAERI-M 7066 (1977) (in Japanese)
4. T. Nagaoka, Intercomparison between EML Method and JAERI Method for the Measurement of Environmental Gamma Ray Exposure Rates, Radiat. Prot. Dosim. 18(2),81-88 (1981)
5. H.Kobayashi, M. Tsukuda and A. Sasaki, Studies of Absolute Measurement of Low-Level Natural Environmental Radiation using the Normal-Pressure Ionization chamber, Natural Radiation Environment III (1980)
6. J. A. DeCampo, H. L. Beck and P. D. Raft, High Pressure Argon Ionization Chamber Systems for the Measurement of Environmental Radiation Exposure Rates, USAEC Report HASL-260 (1972)
7. UNSCEAR. Sources and Effects of Ionizing Radiation, Report of the United Nations Scientific Committee on the Effects of Atomic Radiation (1993)

Survey of natural radiation background and assessment of its population dose in China

WEI Kedao, CUI Guangzhi, ZHAO Shi'an and GAO Junhai

Laboratory of Industrial Hygiene, Ministry of Health, 2 Xinkang Street, Deshengmenwai, Beijing 100088, China

ABSTRACT

A nationwide survey of natural environmental radiation was carried out during the period 1984--1988 in China, in which the integrating dosimeters of model ETLD-80 with $CaSO_4$:Dy were used. The results showed that the population-weighted average absorbed dose rates in air from terrestrial γ–radiation outdoors and indoors were 67 and 89 nGy·h-1, respectively. Moreover, the annual effective doses per capita and collective effective doses to population of China from natural environmental external radiation were estimated to be 780 µSv, and 8.6×10^5 man·Sv, based on the model recommended by UNSCEAR 1988 Report.

1. INTRODUCTION

In recent years, accurate and reliable measurements of natural environmental radiation have been carried out by different approaches in many countries and regions of the world. These baseline data are conducive to the determination of changes in concentration of certain radionuclides and to the identification of long-term trends of environmental radioactivity arising from the nuclear fuel cycle, man's use of radioactive materials and man's extensive modification of the earth surface.

This paper presents the outline and results of a nationwide survey of natural radiation background with integrating dosimeters (TLD) during the period 1984--1988, in which 27 institutes [1] of Environmental Radiation Monitoring Network(ERMN)participated under the sponsorship of Laboratory of Industrial Hygiene(LIH), Ministry of Health, China.

The integrating measurement of TLD used for the survey of natural environmental radiation background is convenient and reliable because there is a relatively long-term monitoring period for placing TLDs in measured sites, and

TLDs are little affected by meteorological condition as well as the evaluation of TLDs can be done in the laboratory. In the nationwide survey of natural radiation background in China, 82% of total area all over the country were covered and 84% of its population were involved [1].

2. INSTRUMENTS AND METHODS

2.1. Integrating dosimeters and readers

Integrating dosimeters of model ETLD-80 with $CaSO_4$:Dy were specially developed according to recommendation of ISO and ANSI. Dosimetry performance of ETLD-80 was tested and the results showed that ETLD-80 met all the requirements for measurement of natural radiation background. TLD's readers of FJ-377 made in China were used in most institutes and UD-502A(B) (made in Japan) and Pitman 654 (made in England) were used as well.

2.2. Methods

According to the scope and scale of investigation, the survey was categorized into two parts, i.e., a general survey and a local survey. The general survey means that each province or autonomy region was taken as a statistical sample so as to estimate the dose distribution in China from natural environmental radiation. The local survey means that one city and one countryside in each province were investigated and taken as statistical samples respectively, so that the information about the difference of natural radiation levels between rural and urban areas could be obtained.

Besides the indoor and outdoor site measurements mentioned above, individual-wearing measurements were conducted in 18 provinces to find out the occupancy factors of inhabitants in comparisons with measurements of TLDs indoors and outdoors.

The sample size for the estimation of natural environmental radiation levels was large enough to meet the statistical requirements within 10% sampling error(95% confidential level)[1].

During the investigation period of three months, TLDs were placed in thermometer screen at meteorological stations for outdoor sites and in the desk drawers for indoors at 1.0 ± 0.3m above ground.

Control TLDs were used in the survey to reduce the contribution to dose as low as possible from confounding factors, such as TLD's transportation, storage, fading and self-irradiation. The calculating procedure and uncertainty of estimation for TLD's readings were based on the recommendation by de Planqe [2]. The cosmic-ray fraction of natural environmental radiation was determined by barometers since the relationship of altitude and ionizing component of cosmic-rays was well established. Also the effect of geomagnetic latitude on the cosmic-rays all over the country was taken into account in the calculation.

3. QUALITY ASSURANCE PROGRAM

It is very important to conduct a quality assurance (QA) program with the aim of making the data obtained in the survey reliable and consistent, especially because as many as 27 institutes of Environmental Radiation Monitoring Network participated in this activity.

The measures taken for the QA were as follows:

Developing special model TL dosimeters which were designed for the measurement of natural radiation background in survey.

Making out a unified investigation protocol on the basis of discussions by experts, which was accepted in practice by participating institutes.

Organizing two TLD intercomparisons in 1984 and 1985 in the laboratory by QA group for improvement of TLD's measurement and calibration techniques.

Assigning competent staff to be responsible for measurement and evaluation of TLDs.

Participating in site investigations using TLDs by QA group in two provinces so as to test the agreement of measured data with those by local institutes.

Collecting all the data from the provinces and autonomy regions and checking and processing them by unified procedure.

4. RESULTS

4.1. Results of general survey

A total of 1568 outdoor sites and 2452 indoor sites were investigated in the general survey. The results are shown in Table 1.

Table 1
National average values of absorbed dose rates in air from terrestrial γ-radiation

Location	No. of measurements	Arithmetic mean (nGy $\cdot$ h^{-1})	Population weighted average (nGy $\cdot$ h^{-1})
Outdoors	1568	69	67
Indoors	2452	92	89

It can be seen from Table 1 that the ratio of dose rate indoors to that outdoors from terrestrial γ-radiation was 1.33. In addition, the nationwide average values of dose rates in air from cosmic rays outdoors and indoors were 33 and 29 nGy$\cdot$h^{-1}, respectively, and the ratio of them was 0.88 [1].

To observe the mean values and distributions of dose rates in the areas investigated, the frequency distributions of them from terrestrial γ-radiation are demonstrated in Fig. 1 and Fig. 2 [1].

The map of Fig. 3 shows the distribution of outdoor terrestrial γ-radiation levels, indicating that the level of natural environmental radiation in South China seems to be higher than that in North China. It may be attributed to the geological difference in the two regions.

In order to estimate the contribution of building materials to indoor γ-radiation, a statistical analysis for 2452 dwellings built of brick, concrete, mud and other materials was done, and the results are listed in Table 2.

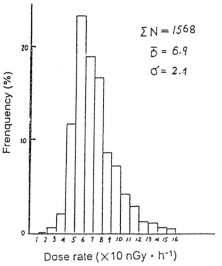
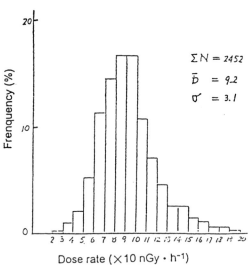

Figure 1. Frequency distribution of outdoor absorbed dose rates in air from terrestrial γ-radiation

Figure 2. Frequency distribution of indoor absorbed dose rates in air from terrestrial γ-radiation

Table 2
Indoor terrestrial γ-radiation levels of different building materials (nGy·h^{-1})

Building materials	Number of dwellings	Range	Mean±S.D.
Brick	1536	20--246	89±28
Concrete	355	20--209	90±28
Mud	115	49--171	84±20
Others	446	30--289	112±36

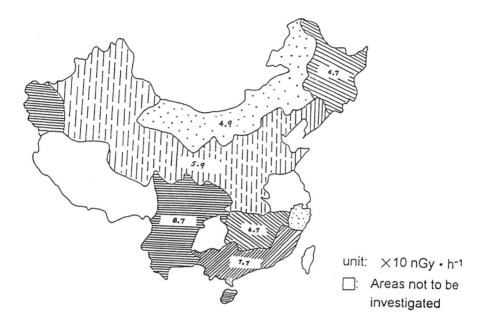

Figure 3 Distribution of outdoor absorbed dose rates in air from terrestrial γ-radiation in China

4.2. Results of local survey

Table 3 and Table 4 show the assignment of TLDs and the results of dose rates from terrestrial γ-radiation in rural and urban areas, respectively.

With regard to individual--wearing, TLDs were worn in the waist of adults and when TLDs were distributed to adults, their occupations and sex were considered.

The total uncertainty for TLD measurement in the survey of natural radiation background was estimated to be about 16% (95% confidential level).

Table 3
Assignment of measured sites in local survey

Areas	Indoors	Outdoors	Individual wearing
Rural	705	569	807
Urban	1105	523	1014

Table 4
Comparison of terrestrial γ-radiation levels between rural and urban areas (nGy·h^{-1})

Areas	Indoors	Outdoors	Individual--wearing
Rural	96	70	86
Urban	86	74	81

5. ASSESSMENT OF POPULATION DOSE

On the basis of the model recommended by UNSCEAR 1988 Report [3], annual effective dose per capita was calculated separately for terrestrial γ–radiation and for cosmic-rays. Formula 1 and formula 2 were used respectively for assessment of effective dose from cosmic-rays and from terrestrial γ-radiation.

$$H_{eff(c)} = 8.76 \times (q_1 D_{c1} + q_2 D_{c2}) \qquad (1)$$
$$H_{eff(\gamma)} = 6.13 \times (q_1 D_{\gamma 1} + q_2 D_{\gamma 2}) \qquad (2)$$

Where
D_{c1}, D_{c2} -- Absorbed dose rates(nGy ·h^{-1}) in air indoors and outdoors from cosmic rays, respectively;
$D_{\gamma 1}$, $D_{\gamma 2}$ -- Absorbed dose rates(nGy ·h^{-1}) in air indoors and outdoors terrestrial γ–radiation, respectively;
q_1, q_2 -- Occupancy factors of inhabitants for indoors (q_1=0.8) and outdoors (q_2=0.2);
8.76 and 6.13 -- Conversion factors to annual effective dose (μSv) from absorbed dose rates of Dc_1 and Dc_2 for cosmic-rays and $D_{\gamma 1}$ and $D_{\gamma 2}$- for terrestrial γ-radiation. It is noted that there are 8760 hours in a year, which are involved in conversion factors.

Table 5
Annual effective doses(AED) per capita and collective effective doses (CED) from natural environmental radiation in China.

Types of radiation	AED(μ Sv)	CED(man · Sv)
Cosmic rays*	261	2.9×10^5
Terrestrial γ - radiation	519	5.7×10^5

The results of assessment of population dose are listed in Table 5.
When the population dose was assessed in the light of individual-wearing of TLDs, correction factor of 0.95 was taken, including backscattering and shielding effect on the dosimeter-wearing body and the time TLD's stayed away from the body as people slept at night.

6. CONCLUSION

The population-weighted average values of absorbed dose rates in air from terrestrial γ-radiation outdoors and indoors are 67 and 89 nGy $\cdot$ h^{-1}, respectively. By comparison with the results obtained by means of different approaches[4], it is found that outdoor terrestrial γ-radiation dose rate measured by TLDs is close to that from soil analysis, and 20% higher than the mean γ-radiation level in the world[4].

A marked characteristic is noted that the level of natural γ-radiation in South China seems to be higher than in North China. It may be attributed to the geological difference in the earth crust.

The results of investigation either in the general survey or in the local survey reveal a fact that the γ-radiation level indoors is higher than that outdoors because of the contributions of building materials. A comparison of outdoor γ-radiation levels in the local survey shows that the level in rural area is slightly lower than that in urban area and is close to the result obtained in the general survey.

The ratios of dose rate in air indoors to that outdoors are 1.33 for terrestrial γ-radiation and 0.88 for cosmic rays, which is of moderate degree in the world. The time-weighted average absorbed dose rates for outdoors ($q=0.2$) and indoors ($q=0.8$) agree with those by individual-wearing measurement.

REFERENCES

1. An investigation group for natural environmental radiation of MH, China. Investigation of natural radiation background and assessment of its population dose in China. Chin. J. Radiol. Med. Prot. 9 (1989) 225. (in Chinese)
2. de Planque G, et al. ,Error analysis of environmental radiation measurements made with Integrating detectors, NBS, SP 456, 1976.
3. UNSCEAR, Source, effect and risk of Ionizing Radiation, Report to the General Assembly with annexes, New York, UN. 1988.
4. D Wei et al. ,An overview of surveillance of environmental radiation in China, JCAC M-9401 Tokyo, Japan, June 15, 1994.

Investigation of outdoor γ-radiation level in Suzhou urban and rural areas

Tu Yu and Jiang Dezhi

Suzhou Medical College, Suzhou 215007, China

1. INTRODUCTION

Suzhou city, lying in latitude 31°19' N and longitude 120° 27' E, with the altitude of 3 meters, total area 119 km², population about 1000 thousands, is located in the southeast of Jiangsu province, to the south of Changjiang river, east of Taihu lake, north of Zhejiang province, and west of Shanghai city. Qinshan nuclear power station is located at 80 km to the southeast of the Suzhou city. Suzhou urban area lies in the leeward of the nuclear power station, where southeast wind prevails in spring, summer and autumn. In order to predict the environment quality, monitor continuously the air γ-radiation level for 24 hours and detect the abnormal, we need to know the natural background of γ-radiation. With the starting of the Qinshan nuclear power station second stage construction, it was very important to monitor the γ-radiation of environment; hence, we measured the environmental radiation in Suzhou during period 1990-1995, the data measured are reported as follows.

2. INSTRUMENT AND METHOD

2.1. Instrument

The data were obtained with the portable BH 3103a-type x, γ dose meter which was manufactured in Beijing nuclear instrument factory and calibrated by the China atomic energy institute, the coefficient of calibration being 1.01. The data were checked every hour by radiation source CS-137 in the midst of measurement, with a unit of 10^{-8} Gy/h.

2.2. Method

Measurement spots included road surface in urban and suburb districts, field and lawn over 10 meters from the buildings were chosen at random, the means of the three readings at every spot were regarded as the actual measurement data.

The dose rate of cosmic rays was measured with the portable HD high pressure ionization chamber. The measurement spots were on the surface of gold cock lake, with 3-metre deep water and over 1 km to the bank, which is 3 km from Suzhou City. The measured value was 2.0×10^{-8} Gy/h. On account of the dose rate of Rn as low as 0.002×10^{-8} Gy/h, we did not take its influence into account.

3. RESULTS AND ANALYSIS

3.1. Results
See Table 1.

Table 1
γ-Radiation dose rates in Suzhou outdoor (10^{-8} Gy/H)

Place	No. of Spots	Measured Values	X±Sd
Granite Road Surface	150	8.8-19.2	11.4±0.30
Cobble Road Surface	150	4.8-11.3	6.3±0.26
Cement Road Surface	167	2.7-12.5	6.1±0.25
Urban Lawn	160	2.8- 6.8	5.0±0.28
Field	225	3.6- 9.0	4.5±0.27
Soil Surface	150	1.2- 7.7	4.3±0.27
Asphalt Road Surface	160	1.2- 7.7	3.8±0.24
Total	1162	1.2-19.2	5.8±0.29

Note: the values of cosmic rays were subtracted from numerical values in the table.

3.2. Analysis

As shown in table 1, the main outdoor places in Suzhou outdoor ranked in decreasing order of γ-radiation dose rate as follows: granite road, cobble road surface, cement road surface, urban lawn, field, soil surface and asphalt road surface. There are significant differences among the different places ($p<0.01$), the highest radiation dose rate being 11.4×10^{-8} Gy/h in the granite road surface and the lowest being 3.8×10^{-8} Gy/h in the asphalt road surface. The former is three times the latter. On the contrary, there are non significant differences among the same places ($p<0.05$).

3.3. Calculation of outdoor γ-radiation dose to inhabitants

The outdoor γ-radiation dose to inhabitants was calculated by the following formula.

$$H_{E(\gamma)} = C \cdot Q \cdot D_A \qquad (1)$$

$H_{E(\gamma)}$: Means annual individual effective dose resulting from the terrestrial radiation (Sv);
$C = 0.7$ stands for the conversion coefficient from the air absorbed dose to the effective dose of population received the earth surface γ-radiation from the radionulides of the same activity;
$Q = 0.2$ denotes the occupancy factor;
D_A means air absorbed dose rate (10^{-8} Gy/h). The results are listed in the Table 2.

Table 2
Population dose equivalent in Suzhou

Space	Air absorbed dose rate(10^{-8} gy/h)	Annual individual annual dose equivalent (10^{-4} sv)	Collective dose equivalent (sv)
Outdoors	5.28	0.71	71
Cosmic Rays	2.00	0.45	45
Total		1.16	116

4. DISCUSSION

In the old urban area of Suzhou, many main roads are made of granite, which contains many radionuclides, so the γ-radiation of the granite road surface is high, and the roads paved with cobble also have a rather high level of γ-radiation. The cement road surface is composed of silicate material rich in radionuclides and has high level radiation. But, the asphalt road surface shows the lowest radiation level, for the asphalt roads are made of organic substance which comes from oil refining and contains less radionuclides. The reason why γ-radiation level on the urban lawn is higher than the field is that there exists the radiation scattering from the buildings around urban lawn. The appearance of Suzhou changes with each passing day, the traditional granite construction materials being gradually taken place by the steel and cement, the height of buildings being increased, the new type reflectorized decoration materials being widely used, and the area of lawn, open ground and fields being decreased; all that is bound to affect the level of γ-radiation.

Environment ionizing radiation level and dose to population in Qinghai province

Cao Zhiyou, Liang Zhuqi, Li Bacui, Ni Jianqing and Zhang Peng

Qinghai Occupational Disease Prevention and Treatment Center, 69 Nanchuan West Road, Xinig 810012, Chin

Qinghai Province is situated in the northeast of Qinghai-Tibet Plateau between latitude 30040'-39020'N and longitude 89°30'-103°E to the north of Sichuan-Tibet, south of Hexi corridor, east of Xinjiang and west of Longdong. There are one city, 38 counties and 8 townships in this province. The whole area is about 723,692 km^2. There are 3,895,700 people in the province. 67% of the population live in the east agricultural region. Population density is 5.38 per km^2.

Qinghai-Tibet Plateau is a landlocked area. Qinhai is lofty, on the whole, with a terrain varying from mountains, highlands, basins to valleys. Except for the eastern part (1,700 meters high), the elevation of most regions lies between 2,500-4,500 meters above sea level, of which the region over 4,000 meters above sea level comprises over 50% of the area in Qinhai. According to its topography, the Qinhai Plateau can be divided into three natural regions: the northeastern mountain region, the northwestern basin and the southern highland. The Kunlun Mountains traverse the middle part of this province, constituting the natural demarcation line between the basin and the highland. Qilian Mountain lies to the north of Kunlun Mountains. The latter extends eastward to Ela Mountain, Anyemaqen Mountains and further eastern region.

1. NATURAL EXTERNAL RADIATION LEVEL

About 320 indoor and outdoor points and roads each were investigated. The natural external radiation absorbed dose rates in air followed a log-normal distribution. The ranges, means and standard deviations of values are shown in Table 1.

For the total natural radiation average dose rates, the ratio of indoor to outdoor values is 1.23, and the ratio of road to outdoor values is 0.98. The average dose rates in this province from high to low are those indoors, outdoors and on roads. The difference among which is statistically significant(P<0.05). The level of

Table1
Natural external absorbed dose rates in air ($\times 10^{-8} Gyh^{-1}$)

Location	No. of points investigated	Range	Mean±standard
Outdoors	345	10.49-26.92	17.49±2.24
Indoors	319	15.09-29.73	21.51±2.58
Roads	324	9.54-27.12	17.23±2.39

natural external radiation is higher than those in majority of areas in China because of the greater contribution of cosmic ionizing radiation. The level of terrestrial gamma radiation is similar to the nationwide country level, being on background level. Man-made radioactive pollution has not been found. There are wide variations in cosmic-ray absorbed dose rate in air. The low value is 5.17×10^{-8} Gyh^{-1} in Minghe Guantin town, the high value is 14.95×10^{-8} Gyh^{-1} in Qilian asbestos open pit, the average value in the whole province is $10.09 \times 10^{-8} Gyh^{-1}$.

2. TERRESTRIAL GAMMA RADIATION LEVEL IN DIFFERENT ENVIRONMENT

Terrestrial gamma radiation levels in 41 investigated sites from high to low are those indoors, outdoors and on roads (Table 2). There is statistically significant difference among them ($P<0.05$).

Table 2
Terrestrial gamma radiation levels in different environment ($\times 10^{-8} Gyh^{-1}$)

Location	No. of points investigated	Range	Mean±standard deviation
Indoors	319	5.90-18.88	11.52±0.90
Outdoors	345	2.65-11.79	7.40±0.96
Roads	324	1.74-12.29	7.17±1.05

Terrestrial gamma radiation levels indoors are higher than those outdoors because the irradiation angle indoors is bigger than that outdoors and the radon concentration indoors is higher. Since 27% of investigated points at bituminous roads, salinized alkaline roads and salt bridges, on which gamma-ray levels are lower, the ratio of radiation levels on roads to those outdoors 0.98.

2.1 Distribution of terrestrial gamma absorbed dose rates in air

According to the geologic map of Qinhai Province (scales 1:4000,000) and observation on the surveyed sites, the distribution of gamma radiation levels tend to be lower in Caidamu Basin and higher in eastern mountains(Figure 1). The range of variation is minor in the former and greater in the latter regions. This result is attribution to the characteristics of the earth's surface. The salinized-alkaline soil and sand dune cover a wide area. U, Th, Ra contents in salt shell, bittern and slag in Caerhan district are lower than those in soil. In the eastern mountain area, however, the igneous rock, metamorphic rock and New Tertiary red rocks contain high contents of U and Th. The minimum gamma radiation dose rate outdoors is $2.65 \times 10^{-8} Gyh^{-1}$ in Qilian asbestos open pit

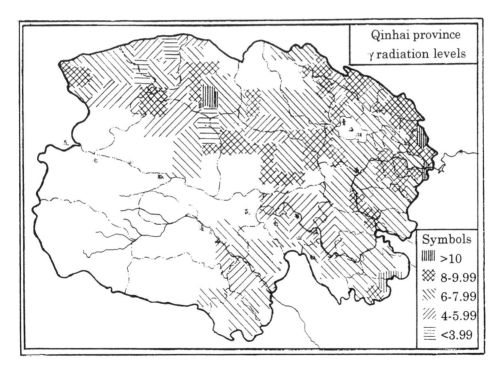

Figure 1. Geologic map of Qinhai Province (scales 1:4000,000) and observation on the surveyed sites, the distribution of gamma radiation levels.

rate is 11.79 x $10^{-8}Gyh^{-1}$ in Xunhua Qitaibo, the ratio being 4.44. The lowest mean is 5.03x $10^{-8}Gyh^{-1}$ in Zhiduo County and the highest mean is 9.53x$10^{-8}Gyh^{-1}$ in Guide and the maximal dose County, the ratio being 1.90. This radiation levels are in the range of natural background.

2.2 Radiation levels in rooms made of different building materials

The radiation levels from high to low are those in houses built of brick and tile, mud and wood, brick and concrete, and tents (Table 3). The houses built of mud and wood are predominant in this province. The majority of herdsmen live in ox hair tents. There is no statistically significant difference in radiation levels between the houses built of brick and tile and those of mud and wood($P>0.05$). The statistical analysis between the houses built of brick and concrete and the tents was not done. The ratios of average radiation levels in different dwellings to that outdoors are 1.48, 1.59, 1.56, and 1.0, averaging 1.43.

Table 3
Radiation levels in different building materials(x $10^{-8}Gyh^{-1}$)

Building materials	No. of points	Range	Mean±Standard deviation
Mud and wood	128	7.95-14.82	11.73±1.49
Brick and tile	86	9.36-15.83	11.92±1.22
Brick and concrete	14	8.56-14.24	11.15±1.45
Ox hair tents	3	6.55-8.71	7.61±0.88

2.3 Gamma radiation levels on different roads

324 points were chosen on the roads made of different materials. Gamma ray levels from high to low are those on mud, sand stone, bituminous, salinized alkaline roads, and salt bridge (Table 4). The ratios of average radiation levels on each of them to that outdoors are 1.07, 0.98, 0.98, 0.90, 0.28, respectively. There is statistically significant difference between the former three values ($P<0.05$). Comparison was not made between the latter two values for the number of points was too small.

Table 4
Gamma-ray levels in different roads (x 10^{-8}Gyh^{-1})

Road type	No. of points	Range	Mean±standard deviation
Sand stone	209	2.91-12.29	7.23±1.02
Bituminous	83	3.07-10.07	7.24±1.02
Mud	25	6.25-10.09	7.92±0.99
Salinized alkaline	5	5.13-7.38	6.63±0.90
Salt bridge	2	1.74-3.43	2.09±0.49

3. NATURAL EXTERNAL RADIATION DOSES TO POPULATION

The annual average effective dose equivalents to population range from 1,170 - 1,901 μSv in 41 investigated sites. The lowest value is in Huangzhong County; the highest value is in Qumalai County, 1.6 times greater than the former.

The per capita annual external effective dose equivalent for the population in whole province is 1,531 μSv (of them the terrestrial gamma-ray dose is 656 μSv). The average per capita annual effective dose equivalent from natural radiation for the population throughout the world is 2,000 μSv (the external dose constitutes 1/3). The natural external radiation dose to the population in Qinghai is at background level. The collective effective dose equivalent is 59.66 x 10^2 man·Sv.

4. CONCLUSION

4.1 The natural external radiation level and the population dose to lie in the range of background radiation. The cosmic radiation dose rate is rather high, the mean value being 10.09 x 10^{-8}Gyh^{-1} in whole province.

4.2 The air absorbed dose rate is in normal range. The dose rates for outdoors, indoors and roads are (10.49-26.92), (15.09-31.87) and (9.54-27.12) x 10^{-8}Gyh^{-1}, respectively. The means values being 17.49 x 10^{-8}, 21.51 x 10^{-8}, 17.23 x 10^{-8}Gyh^{-1}, respectively.

4.3 The absorbed doses in air in different areas are closely related to geological structure.

4.4 Gamma-ray levels in Qinghai Province from high to low are those indoors, outdoors and on roads. The ratios between them are 1.56, 1.00, 0.97.

4.5 The average annual effective dose equivalent per capita from natural radiation is 1,531 µSv. The collective dose equivalent to population is 59.66 x 10^2 man·Sv.

ACKNOWLEDGMENTS

We wish to express our thanks to Dr. Ysuo Watanabe and Yuusaku Kagaya for their help with the calculation of dose estimations. This work was supported by the Department of Public Health, Akita University School of Medicine, Dr. Chen Deqing and Mrs. Kyoko Shibata.

REFERENCES

1. Liu Shuzheng et al. ,Chin. J. Radiol. Med. Prot., 5 (1992) 299.
2. Chen Deqing. Chin. J. Radiol. Med. Prot., 4 (1986) 269.
3. United Nations UNSCEAR 1994 Report, New York, 1994.
4. ICRP. ICRP Publication 60. Pergamon Press. Oxford, 1990.
5. Yao Suyan, Yukio Takizawa, Junsuke Yamashita et al. , J. Radiat. Res. Radiat. Proc, 11 (1993) 242.
6. Bosi A. and Olivieri G., Mutat. Res., I (1989) 13.
7. Bai Yongli and Chen Deqing. ,J. Radiat. Res. Radiat. Proc., 12 (1994) 34 .
8. Fan S. et al. ,Mutat. Res.,243(1990)53 .
9. Ikushima T. ,Mutat. Res., 4 (1989) 241.
10. Wolff S. et al. ,Mutat. Res., 1/2 (1991) 299.
11. Barquinero JF, et al. ,Int. J. Radiat. Biol., 2 (1995) 187.

Methodology of external radiation dosimetry and measurement assurance applied in epidemiology in high background radiation area in Yangjiang, China

Li Kaibao and Zhao Shi'an

Laboratory of Industrial Hygiene, Ministry of Health, Beijing 100088, China

1. INTRODUCTION

Exposures from external sources of radiation have been reviewed by UNSCEAR in its series reports. The changes in the estimates of doses in successive reports have reflected greater knowledge in the field of radiation protection dosimetry, including the dosimetric models, quantities and units as well as measuring techniques. Some important points will be refreshed which are related to this presentation.

In the UNSCEAR 1982 Report [1], one important aspect different from previous reports was that the Committee combined the dose in all organs and tissues in an expression of the effective dose equivalent instead of estimating the absorbed dose to only a limited number of important organs, for example, gonads. The Committee believed the effective dose equivalent to be better representing the whole risk incurred by the exposed populations. In the UNSCEAR 1988 Report [2], the absorbed dose rate in air from the ionizing component of cosmic rays was taken to be numerically equal to the effective dose equivalent rate. Since the UNSCEAR 1993 Report [3] the conversion coefficients from Kair (kerma to air) to the effective dose have been recommended for terrestrial gamma radiation, which are 0.72 SvGy^{-1} for adults, 0.82 SvGy^{-1} for children, 0.93 SvGy^{-1} for infants. It has been assumed that the effective dose is numerically equal to the effective dose equivalent. In addition, the effective dose rate of neutron component of cosmic rays is increased by about 50% at sea level owing to the changes of radiation weighted factor of neutrons in ICRP 1991 Report.

2. INDIVIDUAL-RELATED DOSE ASSESSMENTS

External radiation sources comprise radiation of cosmic rays and terrestrial gamma radiation. To calculate the dose delivered by the external radiation sources to exposed individuals, it is necessary to use dosimetric models linking the measured quantities, such as the exposure, absorbed dose in air and kerma to air to the resulting effective dose in the exposed subjects. The effective dose as defined by ICRP is suitable quantity which is likely correlated well with the total of the stochastic effects.

The total annual effective dose rate (outdoor plus indoor), , from terrestrial gamma radiation can be calculated by the following equation:

$$E = (K_{in} \cdot O_{in} + K_{out} \cdot O_{out}) \cdot C \qquad (1)$$

Where

K_{in}-Kerma rate to air, indoor, Gya^{-1};
O_{in}-the indoor occupancy factor;
K_{out}-Kerma rate to air, outdoor, Gya^{-1};
K_{out}-the outdoor occupancy factor;
C--the conversion coefficients from K to E, $SvGy^{-1}$.

The conversion coefficient, C, takes into account the correction parameters: the mass energy absorption, the depth transmission, the back scattered radiation and the isotropy of radiation field when a person (receptor) is located in the region of interest.

Two methods of direct measurement have been used to determine the values of K. One is to measure it by means of environmental dose rate meter, such as high pressure ionizing chamber (Model RSS-111), and scintillation counter (Model FD-71). When the reading scale of instrument is in old exposure unit (mRh^{-1}) or in the units of the absorbed dose in air, $m\ radh^{-1}$ or Gyh^{-1}, it is convenient to convert them into K in Gyh^{-1} by the following equations:

$$K = X(W/e) \cdot (1/1-g) \qquad (2)$$

Where

X--exposure rate in Ckg^{-1};
W/e--the mean energy expended in air per ion pair formed and per electron charge, with the value of 33.97 JC^{-1};
g--fraction of energy of secondary charged particles that is lost to bremsstrahlung, $g \approx 0$ for terrestrial gamma radiation.

Under the conditions of the charged particle equilibrium,

$$K = D/(1-g) = D \quad (g \approx 0) \qquad (3)$$

Another method used to determine K is by means of passive accumulative TLDs, such as LiF (Mg, P, Cu), $CaSO_4$ (Tm) and $CaSO_4$ (Dy) as well as RPL (radiophotoluminescent) dosimeters. The dosimeters are placed in either indoor or outdoor environment as environmental dosimeters or worn on the body as personal dosimeters for a given period of duration.

In the latter case, if dosimeter badges are worn on the trunk of the body, the personal dosimeter should be calibrated using the 30 cm x 30 cm x 15 cm slab phantom recommended by ISO. When the effective dose is calculated using determined value of by equation (1), corrections have to be made for the readings of the personal dosimeters with regard to body attenuation to incident gamma rays and the sleeping time on bed when the dosimeters may be away from the body. In case that the personal dosimeter are calibrated in free air without phantom, the correction for backscatter of body surface should be added for the readings of dosimeters.

An estimated dose value of 32 Gyh^{-1} at ground level from ionizing components of cosmic rays has been adopted by the Committee since 1977. This value is taken to be numerically equal to effective dose. A mean indoor shielding factor of 0.8 has been assumed by the Committee for this ionizing components. An average effective dose rate of 3.6 $nSvh^{-1}$ at sea level from neutron component has been adopted, neglecting the shielding effect of building structures.

3. QUALITY ASSURANCE AND QUALITY CONTROL

Quality assurance (QA) and quality control (QC) in dose measurements are essential to maintain quality and are of increasing importance in order to meet the requirements of national regulation and international standards and guidelines. The aspects of QA/QC used include performance tests of dosimeters, quality calibrations, internal/external controls in operating procedures.

An important part of a QA program is performance tests of the dosimeters used in dose measurements. The characteristics of environmental dose rate meters have been tested in the SSDL in Laboratory of Industrial Hygiene (LIH) according to the National Verification Regulation of X or Radiation Meters for Environmental Monitoring (JJG 521-88). The performance of TLDs has been verified in the SSDL according to the requirements of National Standards of Thermoluminecence Dosimetry for Personal and Environmental Monitoring (GB10264-88).

Calibrations of the dosimeters have been made against the secondary standard of LS-10 ionizing chamber in the SSDL traceable to the National Primary Standard and Austrian Primary Standard. As one of external quality control measures, standard irradiation test comparisons between LIH and JCAC (Japan Chemical Analysis Center) have been conducted using TLDs, of which the results are shown in Table 1[4-5].

It can be seen that measured values are in good agreement with the standard irradiation values within about 5%.

Table 1
Results of standard irradiation tests

Year	LIH measured values (10^{-6} Ckg^{-1})	JCAC irradiation values (10^{-6} Ckg^{-1})	JCAC/LIH
1995	7.84 (30.4 mR)	8.26 (32 mR)	1.05
	17.7 (68.7 mR)	18.6 (72 mR)	1.05
1996	5.49 (21.3 mR)	5.78 (22.4 mR)	1.05
	9.86 (38.2 mR)	10.2 (39.7 mR)	1.03
	18.3 (71.0 mR)	18.4 (71.5 mR)	1.01
	21.7 (84.1 mR)	23.1 (89.4mR)	1.06
Average			1.04

Internal QC measures adopted by the investigation group include the routine checks and calibration of the dosimetry system, training and qualifying the group members. In the case of TLDs, dosimeters issued for the first time to a individual included detailed instructions on how handle, wear and store the dosimeters. A QA Group has been set up with the responsibility of carrying out quality inspection to ensure the continuing constant in dose measurements. Two measures have been conducted by the QA Group for the purpose:

3.1 Blind sample tests

In order to examine the controls over process, equipment and operations conducted by the investigation group, the blind sample tests were carried out in 1992-1993. Four batches of TLD samples were irradiated in the SSDL with the exposure range of 1.032×10^{-6} -- 2.064×10^{-5} Ckg^{-1} (4-80 mR). These irradiated samples were taken to be as blind testing samples sent to the investigation group for measurements, and then the results were sent to the QA Group for comparison with the known standard irradiation values. The results of blind sample tests are shown in Table 2. Except for batch 3, the results are acceptable.

Table 2
The results of blind sample tests

No. of batches	Dose points	Ave. dev. (%)
1	6	+3.7
2	6	+3.8
3	6	+11.8
4	6	-1.2

3.2 On-site measurement check

QA group issued a certain number of TLDs placed together with the investigation dosimeters in indoor and outdoor environments at the same conditions. QA dosimeters (QAD) and investigation dosimeters (ID) were evaluated by QA group and investigation group respectively. The results are shown in Table 3 [6]. The results are acceptable.

Table 3
The results of on-site check

Duration	No. of check points		ID/QAD	
	HBRA	CA	HBRA	CA
Nov. 22, 1991- Jan. 01, 1992	48	16	0.95	0.86
Dec. 14, 1992- Feb. 18, 1993	30	12	0.93	1.07

4. EVALUATION OF UNCERTAINTY IN DOSE MEASUREMENT

Uncertainties introduced in dose measurement are estimated for TLD method and survey meters respectively. They are listed in Table 4 and Table 5 [7].

Table 4
Uncertainty of TLD Measurements

Error sources	Estimated uncertainty (%)
Calibration factor	6
Energy response	4
Direction response	4
Fading	4
Processing and readout (annealing, batch homogeneity, etc.,)	10
Transit and storage dose measurements	10
Total	16.9

Table 5
Uncertainty of survey meter measurements

Error sources	Estimated uncertainty (%)
Calibration factor	6
Repeatability	7
Linearity	5
Energy response correction	10
Long-term stability	5
Total	15.3

5. CONCLUSIONS

The primary objective of the dose investigation is to provide appropriate dosimetric data for the analysis of biological effects of radiation incurred by the exposed population in the high background radiation area. The methodology of dose estimation used follows the lines of UNSCEAR Reports. Methods and means of introducing QA/QC program have been found useful in maintaining the quality of the dose measurements with adequate accuracy and reliability.

REFERENCES

1. UNSCEAR Report, 7-16, 1982.
2. UNSCEAR Report, Annex A , 50-57, 1988.
3. UNSEAR Report, Annex A (Chinese version) , 33-44, 1993.
4. Ou Xiangming et al. Comparison of TLD results between Laboratory of Industrial Hygiene, China, and Japan Chemical Analysis Center. Chin J Radiol Med Prot, 16 (1996) 245.
5. Private communication, 1996.
6. Zhao Shi'an et al., to be published in Chinese Journal of Radiological Medicine and Protection.
7. Cui Guangzhi et al. Working report ,1989.

II Radioactive aerosols

World high level natural radiation and/or radon-prone areas with special regard to dwellings

M. Sohrabi

National Radiation Protection Department
Atomic Energy Organization of Iran
P. O. Box 14155-4494, Tehran
Islamic Republic of Iran

I. INTRODUCTION

An elevated natural radiation area (ENRA), what is commonly known as a "high level natural radiation area" (HLNRA), may be defined here for working purposes as "an area where at least one of its existing natural radioactivity and/or exposures in the environment (soil, water, air, etc.) or in dwellings has potentials to lead to internal and/or external exposures of the public higher than a predetermined level or a limit with a potentially increased health risk". Austria [1], Brazil [2], China [3], India [4,5] and Iran [6] as well as areas in some other countries have been subjects of long-term dosimetric and epidemiological studies with a view to estimating the risk factors which are at present only based on the available data on Hiroshima and Nagasaki [7].

A series of conferences on HLNRAs have been held over the last two decades: the first in Pocos de Caldas, Brazil in 1975 [2], the second in Bombay, India in 1981 [8], the third in Ramsar, Iran in 1990 [9] and the fourth in Beijing, China in October, 1996 (this conference). Many high-level radon areas (or radon-prone areas) and dwellings have also been discovered during national surveys within the last decade. Owing to the importance of radon indoors and its remedial actions, many topical conferences and workshops at national and international levels have been organized; for example the first and second International Workshops on Radon Monitoring in Radioprotection, Environmental and/or Earth Sciences in 1989 and 1991 at the International Center for Theoretical Physics [10, 11]; the UK National Conference on "Radon 2000" [12] and the International Conference on "Indoor Radon Remedial Actions" [13].

From 2.4 mSv average annual effective equivalent dose from natural sources, about 1.2 mSv is due to radon and its decay products while the rest is due to cosmic rays, terrestrial gamma rays and radionuclides in the body (except radon). The average effective dose due to radon, as stated above, corresponds to an average global population-weighted concentration of about 40 Bq m^{-3} for indoors and 10 Bq m^{-3} for outdoors [14]. By using equilibrium factors of 0.4 for indoors and 0.8 for outdoors, the population-weighted global average equilibrium equivalent concentration (EEC) is therefore estimated to be 16 Bq m^{-3} indoors and 8 Bq m^{-3} outdoors. The equivalent dose 1.2 mSv y^{-1} can be easily calculated by using the conversion factors recommended by UNSCEAR [14]; i.e. the effective equivalent dose (E) of 9 nSv h^{-1}

per 1 Bq m^{-3} for EEC of radon and for both indoors and outdoors, and 1.5 μSv y^{-1} from inhalation of 1 Bq m^{-3} radon dissolved in tissues.

The problem arises from the fact that many people in the world live in some areas and in dwellings with radiation and radon levels even as much as 100 times the global average and receive doses higher than the present ICRP-60 dose limit of 20 mSv y^{-1} or than even the older dose limit of 50 mSv y^{-1} for radiation workers [7]. For example, in one dwelling in Ramsar, Iran, the effective dose E due to external and internal exposures including radon has been measured to be at least 160 mSv y^{-1} which is 8 times higher than the dose limit for radiation workers [6,15]. Therefore, dwellings having such high levels of radiation require implementation of immediate remedial actions. Considering the known health risks from ionizing radiation, the above can easily justify why studies of high level natural radiation and/or radon-prone areas worldwide, as well as mitigation of their radiation levels, are of importance and concern to the regulatory authorities and the public. Although many conferences have been held and extensive researches have been carried out during the past two decades and vast amounts of data have been accumulated on the above subjects, harmonized and standardized definitions of elevated natural radiation areas and its criteria for their classifications according to different levels of exposures or radioactivity have not yet been established. Only some definitions separating a high level natural radiation and/or radon-prone area from a normal natural background area has been proposed so far, based on levels of different parameters such radioactivity level in soil, air, water, etc. [e.g. 2,4]. Therefore, a new general definition for an elevated natural radiation area as well as new criteria for their classifications according to the level of exposures and/or radioactivity are proposed in this paper bearing the following aims and purposes in mind:

1. To have harmonized and standardized definitions and criteria for elevated natural radiation areas including radon.
2. To classify natural radiation areas in particular ENRAs worldwide for epidemiological studies and for risk assessments.
3. To provide a "system of dose limitation" for public living in such areas.
4. To establish a regulatory framework for implementation of remedial actions.
5. To prevent radiophobia in particular among public not living in a real HLNRA.

In this paper, some new definitions and criteria for natural radiation areas are proposed followed by a review of some important radiation and radon-prone areas worldwide with their tentative classifications accordingly, based on the data available.

II. DEFINITIONS AND CRITERIA FOR NATURAL RADIATION AREAS

It is usually a common practice that natural radiation areas having radiation levels a few times higher than that of normal background levels to be called a HLNRA. This is in fact a wrong practice. Such areas have a wide range of radiation levels and they should be precisely classified to prevent any radiophobia especially among public not living in a real HLNRA. Some definitions and criteria differentiating an elevated natural radiation area or a radon-prone area from a normal background radiation area have been proposed in the literature [2,4,16,17,18].

Cullen and Penna-Franca [2] have characterized a HLNRA by one or more of the following conditions:

1. The exposure rate from external terrestrial sources, over extended areas, is greater than 200 mR y^{-1} (2 mGy y^{-1}).
2. The long-lived alpha activity ingested through the local diet and water is greater than 50 pCi d^{-1} (1.85 Bq d^{-1}).
3. The radon concentration of potable water is greater than 5000 pCi l^{-1} (185 kBq m^{-3}).
4. The ^{222}Rn and ^{220}Rn concentrations in the atmosphere are greater than 1 pCi l^{-1} (37 Bq m^{-3}).

Mishra [4], based on the above guidelines and the UNSCEAR results on average global external dose from terrestrial sources and by introduction of a population factor, has proposed the following criteria for a HLNRA.

1. The exposure rate from external terrestrial sources, over extended areas, is greater than 4 mSv y^{-1} (10 times the average global value).
2. The long-lived alpha activity ingested with a local diet and water is greater than 2 Bq d^{-1}.
3. The ^{222}Rn concentration of the potable water is greater than 200 kBq m^{-3}.
4. The ^{220}Rn and ^{222}Rn concentrations of the atmosphere is greater than 40 Bq m^{-3}.
5. The population receiving radiation dose from one or more of the above sources is greater than 1000 so that data obtained has some statistical significance.

So far as the radon alone is concerned, some national and international definitions and/or criteria have also been proposed previously. For example, ICRP has expressed a "radon-prone area" as one in which more than 1% of dwellings have a radon concentration of more than 10 times the national average value [16]. The National Radiological Protection Board (NRPB) in UK has used the term "affected area" rather than "radon-prone area" and has expressed it as an area within which a certain percentage of present and future dwellings might exceed a predetermined action level and from which certain consequences would follow [17]. Also Scott [18], selecting one of the three concepts for radon-prone areas that he has proposed, defines a radon-prone area as an area where the average risk (radon concentration) to the population is high enough to justify an action programme. The derived criterion is arithmetic mean (AM) radon concentration greater than some value; e.g., AM $\geq$ Y Bq m^{-3}.

The above criteria only separate the areas into two groups as well as they are based on levels of different measured environmental parameters. This author based on the existing experiences on HLNRAs and bearing in mind to establish criteria more dependent on a system of limitation of annual effective equivalent dose received by public rather than on environmental parameters and based on the definition given in the introduction above proposes the following criteria for classification of natural radiation areas by dividing them into four groups [19]:

1. A "Low Level Natural Radiation Area" (LLNRA) or a "Normal Level Natural Radiation Area (NLNRA)"; an area or a complex of dwellings where soil, outdoor air, indoor air, water, food, etc. have exposures and/or radioactivity levels to lead to an internal and/or external public exposure, which falls below or is equal to two times the average global annual effective equivalent dose from natural sources given for example by UNSCEAR [14]; i.e. 2 x 2.4 = 4.8 or $\approx$ 5 mSv y^{-1} or a dose level $\leq$ 5 mSv y^{-1}. No remedial action is recommended for such LLNRAs although some simple measures can always mitigate the national average annual effective equivalent dose.
2. A "Medium Level Natural Radiation Area" (MLNRA); an area or a complex of

dwellings where soil, outdoor air, indoor air, water, food, etc. have exposures and/or radioactivity levels to lead to an internal and/or external public exposure which is higher than the upper limit for LLNRA, or 5 mSv y^{-1}, but it falls below or is equal to a pre-established level or limit; for example the dose limit of 20 mSv y^{-1} for radiation workers [7]. A remedial action is required to be implemented within a time frame to be determined; for example within 5 years.

3. A "High Level Natural Radiation Area" (HLNRA); an area or a complex of dwellings where soil, outdoor air, indoor air, water, food, etc. have exposures and/or radioactivity levels to lead to an internal and/or external public exposure which is higher than 20 mSv y^{-1}, the upper limit for a MLNRA, but it falls below or is equal to, for example, 50 mSv y^{-1}, the former ICRP dose limit for radiation workers. A remedial action should be implemented subject to a regulatory control within a time frame to be calculated; for example within one year.

4. A "Very High Level Natural Area" (VHLNRA); an area or a complex of dwellings where soil, outdoor air, indoor air, water, food, etc. have exposures and/or radioactivity levels to lead to an internal and /or external public exposures higher than the upper limit for a HLNRA; i.e. 50 mSv y^{-1}. Evacuation is recommended as the first step in the remedial action programme to be enforced by a regulatory authority.

The criteria proposed above are also summarized in table 1. They are believed to be easily applied as a regulatory framework and as a "system of dose limitation", compatible with that of the ICRP, for natural radiation indoors and outdoors and they can be also directly applied to define radon-prone areas and to classify them accordingly. Based on the criteria proposed, remedial actions could be also implemented for regulatory purposes; Yet the above criteria are open for criticism, improvement, and/or approval to be used as an international criteria for radiation-prone areas. It should also be mentioned that the term "HLNRA" is used in this paper as a general term commonly applied in practice. To take into account factors such as the population size, the area and/or number of dwellings involved, the criteria can be also based on a "collective effective equivalent dose concept", still being under investigation by the present author.

III. HIGH LEVEL NATURAL RADIATION AREAS WORLDWIDE

Detailed studies on some HLNRAs and hot springs used as spas in Austria [1], Brazil [2], China [3], India [4,5], Iran [6] and Japan [20] as well as in some other countries have led to some interesting results in particular from epidemiological point of views. Some areas such as Badgestein in Austria [1], Ramsar in Iran [6,15], Slovenia [21], Tuwa in India [22] and Rudas in Budapest, Hungary [23] have thermal waters used as spas. They are usually rich in radium and thus having high radon concentrations indoors, which deliver high doses to staff as well as to the visitors, some of which have been introduced long ago [24].

In Brazil, due to high thorium contents and traces of uranium in minerals, gamma exposure levels of up to 2 mR h^{-1} have been detected registering a dose from 1 to 32 mGy y^{-1} with an average dose rate to public of 6.4 mGy y^{-1} in Guarapari, built on a peninsula over monazite sand deposits, and Araxa over one volcanic intrusive area [2].

On the SW coast of India, in an area with 140,000 inhabitants, a sizable proportion of the population receives exposures exceeding 10 mGy y^{-1} with the highest personal dose rate of

Table 1
The criteria proposed in this paper for natural radiation area as a system of dose limitation.

Classification	Criteria	Remedial Action (RA)
LLNRA; Low (or Normal) Level Natural Radiation Area	Potential public exposure ≤ e.g. two times natural average global effective dose of UNSCEAR ($\approx$ 5 mSv y^{-1})	None
MLNRA; Medium Level Natural Radiation Area	Potential public exposure > 5 mSv y^{-1} and ≤ e.g. 20 mSv y^{-1}; the present ICRP limit for workers	Within 5 years
HLNRA; High Level Natural Radiation Area	Potential public exposure > 20 mSv y^{-1} and ≤ 50 mSv y^{-1}; the former ICRP limit for workers	Subject to regulatory control (RC) and RA within 1 year
VHLNRA; Very High Level Natural Radiation Area	Potential public exposure > 50 mSv y^{-1}	Subject to RC, evacuation and urgent RA with the assistance of the government

32.6 mGy y^{-1} belonging to a resident of a house that registered 38.4 mGy y^{-1} [5]. The average radiation level was estimated to be 15.7 mGy y^{-1}, in contrast to an overall mean of 2.08 mGy y^{-1} in the nearby control area. Although the dose levels are significant, the results of the demographic survey for dose-genetical effect correlation, and epidemiological studies as well as studies on chromosomal anomalies of human blood cells and plants have shown no conclusive statistically significant biological effects on the population of HLNRAs in India in comparison to control groups [5].

In the HLNRAs of Ramsar, Talesh Mahalleh has the highest radiation levels with personal doses registered by personal dosimeters up to 132 mGy y^{-1} having a mean value of 10.2 mSv y^{-1} with a maximum personal dose detected by inhabitants of a house [6,15]. The potential gamma exposures indoors and outdoors measured by survey meters range form 0.05 to 9 mR h^{-1}, with annual potential exposures in houses ranging from 0.6 to 360 mGy y^{-1}. One school, in particular, showed potential gamma exposures ranging from 5.6 to 15 mGy y^{-1}. The dose received by the inhabitants in the area from gamma exposures ranged up to 132 mSv y^{-1} [6, 15]. The gamma exposures in Brazil and India seem to be almost equal but the level in Ramsar is a few times higher than these two areas. On the other hand, the HLNRAs of India are densely populated, where epidemiological study data can be of interest, unlike areas in Brazil and in Ramsar, Iran having lower population densities.

A new elevated natural radiation area was recently studied in hot spring region of Mahallat called Abegarm-e-Mahallat in central Iran where 5 hot springs with a mean temperature of 46 ± 1 °C are used as spas [25]. The isodose curves of gamma exposures have divided the region into five zones with increasing exposures of ≤0.8, 0.8-1.7, 1.7-2.6, 2.6-3.5 and 3.5-4 μGy h^{-1} in a 4 km^2 area. A ^{222}Rn concentration of up to 2731 kBq m^{-3} has been detected in the springs. The region in under current study with some results presented at this conference.

In Yangjiang, China, which has been under study since 1972, the average effective dose is 5.4 mSv y^{-1} which is about three times higher than its mean background of 2 mSv y^{-1}. The results of epidemiological studies over a population of 80,000 have led to a firm conclusion; i.e. no difference was found in mortality rates from all types of cancer or due to leukaemia in the HLNRAs of Yangjiang in comparison with the control group studied [3].

Some areas in Japan, with exposure rates; (i) below 7.6 μGy h^{-1}, (ii) between 7.6 to 10.5 μGy h^{-1} and (iii) above 10.5 μGy h^{-1} respectively with population sizes of (i) 2,230,300; (ii) 2,885,787; and (iii) 2,790,818 from 39 areas including 28 cities and 11 towns and villages have been under epidemiological investigations, as reported by Iwasaki and co-workers [20]. The results of cancer mortality rates of public exposed to natural radiation levels in different geological zones in Japan no detectable increase in cancer mortality.

Another area which has been recently introduced is Miri Lake Area in Nuba Mountains in Sudan [26]. The preliminary studies have shown an average population exposure of 38.4 mS y^{-1}. The HLNRAs discussed above have been tentatively classified in table 2, based on

Table 2
Tentative classification of some areas, commonly known HLNRAs, based the criteria given in table 1.

Location	External Exposure Level (mSv y^{-1})	Internal Exposure Level (mSv y^{-1})	Annual Effective Dose (mSv y^{-1})	Type of NRA
Guarapari Brazil [2]	-	-	6.4	MLNRA
Yangjiang China [3]	2.1	3.3	5.4	MLNRA
SW Coast India [5]	15.7	15.7*	31.4	HLNRA
Japan [20]	0.96	0.96*	1.92	LLNRA
Ramsar Iran [6,15]	10.2	10.2*	20.4	HLRA
Mahallat Iran [25]	7	7*	14	MLNRA
Lake Miri Sudan [27]	-	-	38.4	HLNRA

* Between 50-60% of total annual effective dose is from radon inhalation in normal as well as in elevated areas [14].

the available data. As it can be seen in table 2, the elevated natural radiation areas, commonly known as HLNRAs, have been classified simply into LLNRAs, MLNRAs and HLNRAs.

This classification is very important for studies in such areas as well as psychologically for researchers and the public. Of course for a more precise classification, necessary data for potential internal and external exposures are required.

IV. HIGH LEVEL RADON AREAS IN HOT SPRING REGIONS AND IN DWELLINGS

The effective dose to the public in some areas in the world, either elevated or enhanced, may be higher than even 10 times the average value of 40 Bq m^{-3} indoors and 10 Bq m^{-3} outdoors given by UNSCEAR [14]. As also stated above, this is due to the combination of local geological formation, construction materials, and types and ventilation of some houses. The outdoor radon comes mainly from the soil, the level of which depends on the type of rock and the radon flux density from the earth, height above the ground and dispersion in the atmosphere: which are all affected by the meteorological conditions. The radon indoors in houses built directly on the ground is largely due to soil gas that enters any cracks and openings in the structure of the house and especially its floor. Radon also emanates from construction materials, tap water, etc. where the type of soil or rock plays an important role. For example, rocks pertaining to uranium deposits, phosphates and in particular alum shale from Upper Cambrian or Lower Ordovician alume shale have ^{226}Ra activity concentrations ranging from 600 to 4500 Bq kg^{-1}; while those from diorite, sandstone and limestone vary from 1 to 60 Bq kg^{-1} [14]. Very high soil gas concentrations of 700 kBq m^{-3} in alum-shale soil in Sweden have led to very high outdoor and indoor radon concentrations in the region [27]. Also, high outdoor radon concentrations have been measured in the provinces of Manitoba and Saskatchewan in Canada having respectively 59 and 61 Bq m^{-3} which are about 6 times higher than the global average [28]. Therefore, depending on the type of rock, the radon level outdoors and indoors varies significantly from place to place leading to some high levels in some areas and dwellings. Some high-level radon dwellings as taken from UNSCEAR [14] or from other reports will be reviewed below.

One very high-level and important historic area is Saxony and the neighbor state Thuringen, as reported by Becker [29], where 40 underground and four open-pit mines have been in operation over the years. More than 200,000 underground miners have been exposed with annual exposures of 30-300 WLM corresponding to an annual effective equivalent dose of 0.3 to 3 Sv y^{-1}. These levels were reduced later to about 10 to 40 mSv y^{-1} by the implementation of remedial actions. The dumps from the mines have covered an area of 10,000 km^2 with radon levels of 1 to 3 kBq m^{-3} in open area. The houses using the old and new tailings or waste rocks as building materials and constructions above shallow shafts and adits have indoor concentrations more than 100 kBq m^{-3} in storage rooms and 30 kBq m^{-3} in the living rooms (with a current figure showing around 150 kBq m^{-3}) [29].

In Schneeberg and Schlema [30,31], the houses built on the mining waste materials have median radon concentrations of about 300 Bq m^{-3} with a 7 fold increase compared with the median value in dwellings of the former West Germany and the global average. Further, in 60 houses in Schneeberg, radon concentrations of more than 10 kBq m^{-3}, with maximum values reaching even 80 kBq m^{-3}, have been measured with maximum values in cellars of about 200 kBq m^{-3} [30]. A similar area is Joachimsthal in Czech Republic where houses built of prefabricated blocks of slag concrete with radium concentrations up to 300 kBq kg^{-1}, have accumulated unacceptably high radiation and radon levels indoors [32].

Another interesting area is Badgastein in Austria with a population of 6500 where 20 thermal springs originate in the center of town. The major part of 5 million liters of water from the springs with a mean concentration of about 1.5 kBq m^{-3} of ^{222}Rn, flows to 120 hotels and treatment centers distributed over the town [1]. The air activities in the periphery, which range from 37 to 185 Bq m^{-3} indoors and 3.7 to 55 Bq m^{-3} outdoors, are lower than in the center where the springs originate and concentrations range from 74 to 555 Bq m^{-3} indoors and 18 to 130 Bq m^{-3} outdoors. The mean level is even higher in the bathrooms, with a concentration of about 3.3 kBq m^{-3}; and it is the highest in the "Thermal Gallery". This is a former gold mine near Badgastein, used as a natural inhalation facility for therapeutic purposes with a mean ^{222}Rn concentration of over 100 kBq m^{-3}, where the short lived products are 70 to 80% in equilibrium. The mine is a treatment house with medical examination room in the entrance, followed by therapy and recreation areas. The employees work in the house in an environment with a mean radon concentration of about 0.3 to 5.5 kBq m^{-3}. The scientific development of the former gold mine to the facility (Thermal Gallery) has been recently reported [33].

In Ramsar where 9 hot springs exist with ^{226}Ra contents from 1 to 146 kBq m^{-3} [6, 15], ^{222}Rn levels measured in about 473 rooms of about 350 houses in different villages as well as in the town of Ramsar showed arithmetic mean values of 615 Bq m^{-3} in Talesh Mahalleh, 326 Bq m^{-3} in Chaparsar, 258 Bq m^{-3} in Schools of Ramsar; 246 Bq m^{-3} in Ramak, 111 Bq m^{-3} in the town of Ramsar, 50 Bq m^{-3} in Sadat Mahalleh and Katalom, 49 Bq m^{-3} in Tonekabon and 27 Bq m^{-3} in Talesh Mahalleh of Katalom and 90 Bq m^{-3} in old and 50 Bq m^{-3} in new Ramsar Hotels [34]. Maximum values of 3070 Bq m^{-3} were observed in Talesh Mahalleh. In one house, the effective equivalent dose due to external and internal exposures including radon has been measured to be 128, 161, 167 and 11.2 mSv y^{-1} respectively for the father (53 y), housewife (47 y), son (17 y) and daughter (22 y) [6,15,34]. The results have been easily justified by their residence times indoors and outdoors.

There are also some mineral and thermal springs used as spas for therapeutic purposes in Hungary some of which are located at the foot of Mount Gellért in Budapest and have been the subject of some studies. High radon concentrations of up to 7.15 kBq m^{-3} have been detected [23]. The effective equivalent doses were determined to be 14 to 41 mSv y^{-1} for staff and 4 mSv y^{-1} for visitors with 2-3 visits per week (104 hours per year) each lasting 1.5 hour. Also there are thermal spas in Tuwa, India with ^{226}Ra concentrations ranging from 400 to 900 Bq m^{-3} leading to radon concentrations indoors in thermal water up to 40 kBq m^{-3} and in dwellings up to 420 kBq m^{-3} [22] and in Slovenia with ^{226}Ra concentrations in thermal drinking water of about 140 Bq m^{-3} while having ^{222}Rn concentration of about 16.7 kBq m^{-3} and relatively low indoor radon concentrations of up to 190 Bq m^{-3} in the air [21]. Table 3 summarizes some data obtained in some hot springs known worldwide.

In a village of 2,600 inhabitants in an Alpine region of Western Tyrol, Austria, very high levels of radon have been reported indoors [35]. The measurements made, for example during the period January - April 1992, showed radon concentrations of 20 to 88,000 Bq m^{-3} with a median value of 1,180 Bq m^{-3} on the ground floor of 346 houses and 21 to 210,000 Bq m^{-3} with a median value of 3,750 Bq m^{-3} in the basement. In 208 houses (63.5%), the mean annual radon concentration exceeded the Austrian action level of 400 Bq m^{-3} [35].

There are many other areas around the world where a certain percentage of houses screened also shows high or very high level radon concentrations some examples of which

Table 3
^{226}Ra and ^{222}Rn levels in some hot springs uses as spas with mean values in parenthesis.

Location	No. of Springs	Radioactivity Range (kBq.m^{-3})	
		^{226}Ra	^{222}Rn
Badgestein Austria [1,33]	20	0.04-4.9 (1)	20-4500 (555)
Ramsar Iran [6,15]	9	1-146 (31)	1-160 (64.3)
Mahallat Iran [25]	5	0.48-1.35 (1.02)	145-2731 (710)
Slovenia [21]	-	0.01-0.6	1.0-63
Tuwa (Gujarat State) India [22]	25	0.4-0.9 -	4-40 -
Rudas (Budapest) Hugary [23]	9	-	≤7.15 -

will be reviewed below. Measurements were made in 55,000 randomly selected houses in 38 states divided in 225 regions in the Unites States to identify houses with screening level of radon [36]; twenty four regions were selected as having the highest concentrations, with 78.4% above 74 Bq m^{-3}, 57.3% above 148 Bq m^{-3}, 31.7% above 296 Bq m^{-3}, and 8.6% above 740 Bq m^{-3}. However, an extremely high radon level exceeding 410 kBq m^{-3}, has been measured in the basement of a house in Prescott, in the state of Arizona, which had a well opening in the basement with the extremely high radon concentration of 3.5 MBq m^{-3} in the well water [37].

In a survey made in Belgium, it has been estimated that about 100,000 houses have radon concentrations above 150 Bq m^{-3}, 10,000 above 400 Bq m^{-3}, and 1,000 above 4,000 Bq m^{-3} [38]. Also, in a survey made in the massif of Visé in Belgium, known to geologists for its medical and geological features, of several hundred houses surveyed, 160 houses (about 2% of them) showed radon concentrations above 400 Bq m^{-3}, and three houses showed concentrations above 3000 Bq m^{-3} [39].

In a survey made in Sweden, about 1% of houses have shown concentrations above 800 Bq m^{-3}, with the highest being 40,000 Bq m^{-3}. The studies made by Akerblom et al. [27] on the effects of soil gas radon concentrations in 105 Swedish houses showed indoor radon variations from 20 to 20,000 Bq m^{-3}, and soil gas radon levels from 5,000 Bq m^{-3} in sandy soil to 700,000 Bq m^{-3} in alum-shale rich soils. Also radon surveys in other countries show high radon levels in certain dwellings; for example the highest level of 5920 Bq m^{-3} in Nova Scotia in Canada; 20,000 Bq m^{-3} in former Czechoslovakia; 4687 Bq m^{-3} in France; 3070 Bq m^{-3} in Iran; 15400 Bq m^{-3} in Spain; 10,000 Bq m^{-3} in UK, etc. as taken from table 1 (quoted from UNSCEAR [14].

Some of the above radon-prone areas are given in table 4. They are classified according to the criteria proposed based on the available data. It seems that the criteria can also be simply be applied to radon-prone areas. However, more research is being carried out by the author to further justify the above criteria.

Table 4
Tentative classification of some radon-prone areas.

Location	Radon Concentration (kBq.m^{-3})		Effective Dose (Sv.y^{-1})	Type of NRA (Based on maximum)
	Outdoor	Indoor		
Germany: Saxony [29]	1-3	30: living room 100: storage	1.51 5.05	VHLNRA
Germany: Schneeberg Schlema [30,31]	- -	0.30 10-80: in 60 houses	0.015 0.50-4.04	MLNRA VHLNRA
Austria: Badgestein periphery Center	0.004-0.055 0.018-0.130	0.037-0.185 0.074-0.555	0.002-0.01 0.003-0.028	HLNRA
Austria: Western Tyrol [35]	-	0.020-88	0.001-4.44 0.06	VHLNRA
Iran: Ramsar [15,34]	-	0.027-0.615	0.001-0.031	HLNRA
India: Tuwa [22]	0.004-0.4	420	21.21	VHLNR

V. CONCLUDING REMARKS

Establishment of the criteria for elevated natural radiation areas is highly necessary for radiological studies. The criteria proposed are believed to provide a reasonable base for this purpose. The literature shows lack of detailed information to reach any firm conclusion in HLNRAs. More extensive researches are invited in particular on epidemiology and radiobiology as well as on internal dose assessment which assist in a more precise classification of such areas as well as to establish radiation risk. Conducting a systematic approach with the assistance of the international agencies such as IAEA, WHO, FAO, UNSCEAR, etc. will assist in obtaining necessary data for risk assessment. Also establishment of an International Committee will assist in harmonization and standardization of any activities regarding studies in natural radiation areas worldwide.

REFERENCES

1. J. Pohl-Rüling, F. Steinhäusler and E. Pohl, Procs. 2nd Special Symp. on Natural Radiation Environment, 19-23 Jan. (1981), K.G. Vohra, U.C. Mishra, K.C. Pillai and S. Sadasivan (eds.), Wiley Eastern Limited, India (1982) 107.
2. T. L. Cullen and E. Penna-Franca (eds.), Procs. Int. Symp. High Natural Radioactivity, Pocos de Caldas, Brazil, 16-20 June (1975), Academia Brasileria de Ciencias, Rj (1977). Also see E. Penna-Franca, the same conference (1977) 29.
3. L. Wei, Y. Zha, Z. Tao, W. He, D. Chen and Y. Yuan, Procs. Int. Conf. on High Levels of Natural Radiation, M. Sohrabi, J.U. Ahmed and S.A. Durrani (eds.), Ramsar, Iran, 3-7 Nov. (1990), IAEA Publication Series, Vienna (1993) 523.
4. U.C. Mishra, Procs. Int. Conf. on High Levels of Natural Radiation, M. Sohrabi, J.U. Ahmed and S.A. Durrani (eds.), Ramsar, Iran, 3-7 Nov. (1990), IAEA Publication Series, Vienna (1993) 29.
5. C.M. Sunta, Procs. Int. Conf. on High Levels of Natural Radiation, M. Sohrabi, J.U. Ahmed and S.A. Durrani (eds.), Ramsar, Iran, 3-7 Nov. (1990), IAEA Publication Series, Vienna (1993) 71.
6. M. Sohrabi, Procs. Int. Conf. on High Levels of Natural Radiation, M. Sohrabi, J.U. Ahmed and S.A. Durrani (eds.), Ramsar, Iran, 3-7 Nov. (1990), IAEA Publication Series, Vienna (1993) 39.
7. International Commission on Radiological Protection, ICRP Publication 60, Oxford, Pergamon Press, Oxford (1991).
8. K.G. Vohra, U.C. Mishra, K.C. Pillai and S. Sadasivan (eds.), Procs. 2nd Special Symp. on Natural Radiation Environment, 19-23 Jan. (1981), Wiley Eastern Limited, India (1982).
9. M. Sohrabi, J.U. Ahmed and S.A. Durrani (eds.), Procs. of the Int. Conf. on High Levels of Natural Radiation, Ramsar, Iran, 3-7 Nov. (1990), IAEA Publication Series, IAEA, Vienna (1993).
10. L. Tommasino, G. Furlan, H. A. Khan and M. Monnin, Procs. 1st Workshop on Radon Monitoring in Radioprotection, Environmental Radioactivity and Earth Sciences ICTP, Trieste, Italy, April 3-14, (1989), World Scientific Publ. Co., Singapore (1990).
11. G. Furlan and L. Tommasino (eds.), Procs. 2nd Workshop on Radon Monitoring in Radioprotection, Environmental and/or Earth Sciences ICTP, Trieste, Italy, 25 Nov.-6 Dec. (1991), World Scientific Publ. Co., Singapore (1993).
12. M.C. O'Riordan and J.C. Miles (eds.), Procs. of a Conf., London, 26-27 March (1992); Radiat. Prot. Dosim. 42 (1992).
13. G. Campos-Venuti, A. Janssens, M. Olast, S. Peirmattei, J. Sinnaeve and L. Tommasino (eds.) Procs. 1st Int. Workshop on Indoor Radon Remedial Action, Rimini, Italy, 27 June-2 July (1993), Radiat. Prot. Dos. 56 (1994) 1.
14. United Nations Scientific Committee on the Effects of Atomic Radiations, United Nations, New York (1993).
15. M. Sohrabi, Procs. 2nd Workshop on Radon Monitoring in Radioprotection, Environmental, and/or Earth Sciences; ICTP, Trieste, Italy, 25 Nov.-6 Dec. (1991), G. Furlan and L. Tommasino (eds.), World Scientific, Singapore (1993) 98.
16. International Commission on Radiological Protection, ICRP Publication 65, Oxford, Pergamon Press, Oxford (1993).

17 G.A.M. Webb, Radiat. Prot. Dosim. 42 (1993) 191.
18 A.G. Scott, Health Phys. 64 (1993) 435.
19 M. Sohrabi, Chapter 3.5., Radon Measurements by Etched Track Detectors; Applications in Radiation Protection, Earth Sciences and Environment, Book in Press, R. Ilic and S.A. Durrani (Eds.), World Scientific Publ. Co., Singapore (1996).
20 T. Iwasaki, M. Minowa, S. Hashimoto, N. Hayashi and M. Murata, Procs. Int. Conf. on High Levels of Natural Radiation, Ramsar, Iran, 3-7 Nov. (1990), M. Sohrabi, J.U. Ahmed and S.A. Durrani (eds.), IAEA Publication Series, IAEA, Vienna (1993) 503.
21 I. Kobal and A. Renier, Health Phys. 53 (1987) 308.
22 L.U. Joshi and U.C. Mishra, J. Radioanalytical Chem., 59 (1980) 672.
23 P. Szerbin, Gy. Köteles and D. Stùr, Rad. Prot. Dosim. 56 (1994) 319.
24 M. Eisenbud, Environmental Radioactivity, 3rd. Edition, Academic Press, San Diego (1987).
25 M. Sohrabi, M. M. Beitollahi, Y. Lasemi and E. Amin Sobhani, 4th ICHLNR, Beijing China, 21-25 Oct. (1996).
26 O.M. Mukhtar and F. Raouf Elkhing, Book of Abstract, Int. Conf. on High Levels of Natural Radiation, Ramsar, Iran 3-7 Nov. (1990) 15.
27 G. Akerblom, P. Anderson and B. Clavens, Radiat. Prot. Dosim. 7 (1984) 49.
28 R.L. Grasty, Health Phys. 66 (1994) 185.
29 K. Becker, Procs. Int. Conf. on High Levels of Natural Radiation, Ramsar, Iran, 3-7 Nov. (1990), M. Sohrabi, J.U. Ahmed and S.A. Durrani (eds.), IAEA Publication Series, Vienna (1993) 281.
30 G. Keller, Environment International 19 (1993) 449.
31 R. Lehman and R. Czarwinski, Radiat. Prot. Dos. 56 (1995) 41.
32 J. Thomas and L. Moucka, Procs. Int. Conf. on High Levels of Natural Radiation, Ramsar, Iran, 3-7 Nov. (1990), M. Sohrabi, J.U. Ahmed and S.A. Durrani (eds.) IAEA, Vienna (1993) 183.
33 J. Pohl-Rüling, Environment International 19 (1993) 455.
34 M. Sohrabi, H. Zainali, Sh. Mahdi, A.R. Solaymanian and M. Salehi, Proc. Int. Conf. on High Levels of Natural Radiation, Ramsar, Iran, 3-7 Nov. (1990), M. Sohrabi, J.U. Ahmed and S.A. Durrani (eds.), IAEA publication Series, Vienna (1993) 365.
35 O. Ennemoser, W.W. Oberaigner, P. Brunner, P. Schneider, F. Purtscheller, V. Stingl an V. W. Ambach, Health Phys. 69 (1995
36 B. Alexander, N. Rodman, S.B. White and J. Phillips, Health Phys. 66 (1994) 50.
37 K.J. Kearfort, Health Phys. 56 (1989) 169.
38 G. Eggermont, 1st Int. Seminar on Managing the Indoor Radon Problem, Mol, Belgium 14-16 May (1990).
39 A. Poffijn, G. Eggermont, S. Hallez and P. Cohilis, Radiat. Prot. Dosim. 56 (1994) 77.

Origin of a new high level natural radiation area in the hot spring region on Mahalat, central Iran

M. Sohrabi[1], M. M. Beitollahi[1], Y. Lasemi[2] and E. Amin Sobhani[2]

[1]National Radiation Protection Department, Atomic Energy Organization of Iran, P.O.Box 14155-4494, Tehran, I.R. Iran

[2]Department of Geology, Tarbiat Moalem University Mofateh Avenue, No. 49, Tehran, I.R. Iran

1. INTRODUCTION

Areas with high levels of natural radiation are found around the world namely in Austria [1], Brazil [2], China [3], India [4], Iran [5], etc., the studies of which have led to some interesting results especially in epidemiology [3,6,7]. The origins of such areas and the behavior of natural radionuclides can be of topical importance in environmental radiological investigations. One area of concern has been Ramsar in Iran, with nine mineral springs used for medicinal purposes, which has been under an intensive study [8]. Such studies have been also extended in this research to other areas in Iran namely "Abegarm-e-Mahallat". This region also has high levels of natural radiation originating from hot springs. The water from such springs has deposited vast amounts of calcium carbonate in the surroundings, leading to the formation of huge travertine mines, among the largest in Iran under routine exploitation. In this research, the origin of natural radioactivity in the region of Abegarm-e-Mahallat and its levels in five existing hot springs, and in various igneous and sedimentary rocks, especially in travertine, were investigated. In particular, ground radiation exposure levels were determined with the aim of drawing an isodose map of the region as well as for estimating potential public radiation exposure. A detailed article on the formation of travertine in this hot spring region has also been submitted [9]. Some preliminary results of such investigations are presented and discussed.

2. FIELD AND LABORATORY EXPERIMENTS

Abegarm-e-Mahallat is a region to the north-east of Mahallat, a town in the central part of Iran, where five hot springs namely Shafa, Solaymani, Donbeh, Soda and Romatism, with a mean temperature of 46 ± 1 °C, are used by visitors as spas. To obtain a natural radiation exposure map of the area and its geographical boundaries, a Mini-Instruments "Smart Ion" ion chamber survey meter was used. The data obtained was transferred onto a 1:20,000 scale map.

Water samples were taken from every spring orifice at various periods during a year to determine Radium-226 (^{226}Ra), using the emanation method [10, 11]. Polyethylene bottles were used for sampling water and nitric acid was added to prevent the wall adsorption of the radionuclides.

Radon-222 (^{222}Rn) concentrations in the hot springs were measured using a liquid scintillation counting method. Samples from springs were taken in glass vials by injecting 10 ml of water under Dupont mineral-oil-base scintillator. The measurements were made using a Beckman S6500 Liquid Scintillation Counter. To cross check the results, a Studsvik 2413-A portable radon sampler was used.

Samples of Quaternary tufa and travertine were also taken from the Solaymani, Donbeh, Soda and Shafa springs and from a travertine mine. Samples from different geological formations in the area such as post-lower Miocene granodiorite; Eocene shale, marl, and conglomerates; Cretaceous limestone; and Jurassic shale and sandstone were also taken for the determination of radioactivity and iron content. Potassium-40 (^{40}K), Thorium-232 (^{232}Th), Radium-226 (^{226}Ra) and Uranium-238 (^{238}U) levels in various rock samples, after necessary preparations, were measured using a Canberra GC4020 High Purity Germanium (HPGe) Gamma Spectrometer with a relative efficiency of 40%. Iron concentrations in the travertine samples were measured using the Ortec 6110 X-Ray Fluorescence (XRF) System.

3. RESULTS AND DISCUSSION

The radiation exposure map of the Abegarm-e-Mahallat region, covering an area of about 4 km^2, is shown in Fig. 1. Different radiation areas are separated from each other by isodose curves, dividing the region into five zones of increasing radiation exposure levels : <0.8, 0.8-1.7, 1.7-2.6, 2.6-3.5 and 3.5-4 μGy h^{-1}. Radiation above normal background levels ($\approx$ 0.1 μGy h^{-1}) are mainly limited to the Quaternary travertine in the vicinity of the hot springs. Maximum radiation exposure of 3.5-4 μGy h^{-1} occurred in the Solaymani hot spring zone near a limonite deposit and bedded travertine containing iron in its superficial layers.

The radionuclides present in the region were identified and quantified by gamma spectrometry of the travertine samples, the results of which are shown in table 1. The results indicate that the concentration of ^{232}Th present is below the detection limit of the system (2 Bq kg^{-1}); that of ^{40}K is at normal levels with a mean value of 227$\pm$88 Bq Kg^{-1}; and that of ^{238}U has a mean value of 19.4$\pm$8.4 Bq Kg^{-1}. It can also be concluded that there exists only one instance of secular equilibrium between ^{238}U and ^{226}Ra, located at a site 1 Km north-east of Soda station outside the radiation zones of Fig. 1. At this site, as in some parts of the region, the rocks are older because the ^{226}Ra has been initially deposited in them in high concentrations and it has decayed with a half-life of 1620 years to the present state of secular equilibrium with ^{238}U. Table 1 also shows that ^{226}Ra is the predominant radionuclide, transported from its source by the hot springs. Considering the iron concentration in these samples, one can infer a positive correlation between ^{226}Ra and iron.

Mean seasonal levels of ^{226}Ra and ^{222}Rn in the travertine forming hot springs, spanning the period from autumn 1993 to summer 1994, are listed in table 2. Although ^{226}Ra concentration in the hot spring waters is relatively high (max. 1.35 kBq m^{-3}), it is still insignificant compared to that in sulphurous hot springs (a reductive milieu) such as in the Ramsar high level natural radiation areas (HLNRA) which reaches up to a value of 146 kBq m^{-3} [12].

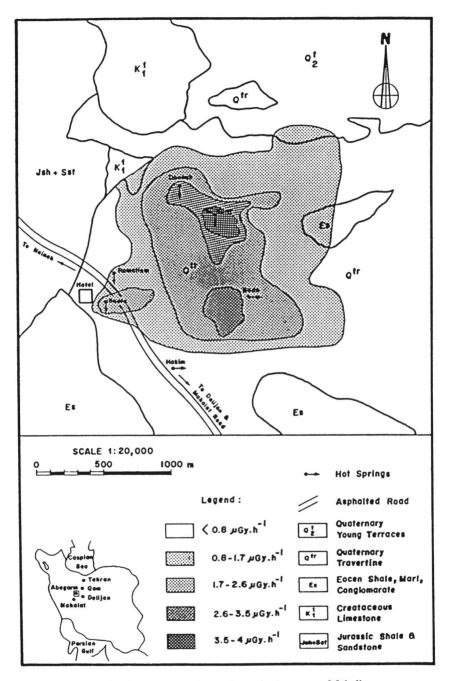

Fig 1. Map of natural radiation of Abegarm-e-Mahallat.

Table 1
Mean concentration of natural radionuclides and iron in travertine and tufa of Abegarm-e-Mahallat region.

Type of Sample	No. of Samples	Mean ± SD (Bq kg^{-1})				Fe (ppm)
		^{238}U	^{232}Th	^{226}Ra	^{40}K	
Travertine (yellow) Solaymani Station	3	18±3	<2	5220±224	219±45	20277±1850
Travertine (yellow) Donbeh Station	3	17±2	<2	4310±190	261±52	15352±3121
Tufa (yellow) Soda Station	3	23±5	<2	712±159	224±28	2588±236
Travertine (white) Shafa Station	3	19±3	<2	71±15	196±23	1493±377
Travertine (white) 1 Km NE of Soda Station (mine)	7	20±5	<2	21±3	235±41	1260±133

Furthermore, since ^{226}Ra is soluble in reductive and insoluble in oxidative conditions [13], and given that the Abegarm-e-Mahallat springs are classified as sulphated (oxidative milieu), it appears that ^{226}Ra levels are due to the alpha recoil phenomenon at source.

In general ^{222}Rn, due to its short half life of 3.8 days, can only migrate a short distance in water. Thus at sites where the concentration of ^{222}Rn is high, it can be assumed that radon is close to its parent radionuclides, i.e. ^{238}U and ^{226}Ra. But the water current velocities in this region are high due to the water pressure and the presence of CO_2. Accordingly, ^{222}Rn can migrate longer distances in these hot springs; i.e. this migration depends mostly on the physical properties of the stream bed (e.g. structure and porosity of the bedrock). The ^{222}Rn to ^{226}Ra ratio ranges from a minimum of 132 at the Soda station to a maximum of 5690 at the Romatism site.

Table 2
Mean concentration of ^{226}Ra and ^{222}Rn in hot springs of Abegarm-e-Mahallat.

Mean ± SD (kBq m^{-3})	Sampling Site				
	Shafa	Solaymani	Donbeh	Soda	Romatism
^{226}Ra	1.35±0.13	1.09±0.11	1.07±0.12	1.10±0.13	0.48±0.05
^{222}Rn	208±5	318±17	150±55	145±37	2731±98

In order to determine the origin of the radioactivity in the Abegarm-e-Mahallat region, it was necessary to measure levels of natural radioisotopes in rock samples across and outside the region. To this end, a variety of rock samples underwent gamma-spectrometric analysis, the results of which are given in table 3. They indicate that the highest radioactivity was observed in samples of post-lower Miocene granodiorite in which ^{226}Ra and ^{238}U are in secular equilibrium.

Table 3
Mean concentration of natural radionuclides in different rocks of the Abegarm region.

Type of Sample		No. of Samples	Mean ± SD (Bq kg^{-1})			
			^{238}U	^{232}Th	^{226}Ra	^{40}K
Post Lower Miocene	Granodiorite	3	179±28	16±5	151±13	935±170
Eocene	Shale	3	46±7	12±2	49±4	890±95
	Marl	3	29±6	<2	31±6	799±113
	Conglomarate	3	16±5	<2	15±3	612±74
Cretaceous	Limestone	3	31±5	<2	37±6	220±31
Jurassic	Shale	3	57±9	23±5	53±7	971±107
	Sandstone	3	20±3	<2	22±5	935±117

A body of Post-lower Miocene granodiorite is found at Dodehak, approximately 1 Km north of the Abegarm-e-Mahallat travertine. Sphene and Apatite are among the accessory uranium-containing minerals in that body. Considering the results obtained, it seems clear that the body of granodiorite is the source of the high levels of natural radiation in the region. The radioactive materials such as ^{238}U and ^{226}Ra from these accessory minerals are carried to the surface by the magmatic waters.

4. CONCLUSION

It can be concluded that the hot springs in Abegarm-e-Mahallat are responsible for the distribution of radionulides (especially ^{226}Ra) in the region, in turn leading to the creation of an HLNRA. The higher ^{226}Ra levels in some travertine are an indication of their younger age compared to travertine with normal or baseline ^{226}Ra activity. These younger travertine are best classed as belonging to the late Quaternary (Holocene) period.

The presence of high ^{222}Rn levels in the hot springs leads to population exposure. The travertine used for building and decorative purposes, particularly indoors, can expose an even greater number of people both internally and externally. The area is under further investigation from a radiation protection point of view especially as regards population exposure.

ACKNOWLEDGMENT

The authors would like to extend their gratitude to Mrs. M. Assefi and Mr. Alirezazadeh for their valuable assistance in the laboratory.

REFERENCES

1. J. Pohl-Rüling, F. Steinhäusler and E. Pohl, In: Procs. 2nd Special Symp. on Natural Radiation Environment, Bombay, India, 19-23 Jan. (1981), K. G. Vohra, U. C. Mishra, K. C. Pillai and S. Sadasivan (eds.), Wiley Eastern Limited (1982) 107.
2. T. L. Cullen and E. Penna Franca (eds.), Pocos da Calgdes, Brazil, Procs. of the Int. Symp. on Areas of High Natural Radioactivity, 16-20 June (1975), Academia Brasileira de Ciencias, RJ (1977).
3. L. Wei, Y. Zha, Z. Tao, W. He, D. Chen and Y. Yuan, In: Procs. Int. Conf. on High Levels of Natural Radiation, Ramsar, IR Iran, 3-7 Nov. (1990), M. Sohrabi, J. U. Ahmed and S. A. Durrani (eds.), IAEA Publication Series (1993) 523.
4. U. C. Mishra, In: Procs. Int. Conf. on High Levels of Natural Radiation, Ramsar, IR Iran, 3-7 Nov. (1990), M. Sohrabi, J. U. Ahmed and S. A. Durrani (eds.), IAEA Publication Series (1993) 29.
5. M. Sohrabi, In: Procs. Int. Conf. on High Levels of Natural Radiation, Ramsar, IR Iran, 3-7 Nov. (1990), M. Sohrabi, J. U. Ahmed and S. A. Durrani (eds.), IAEA Publication Series (1993) 39.
6. T. Iwasaki, M. Minowa, S. Hashimoto, N. Hayashi and M. Murata, In: Procs. Int. Conf. on High Levels of Natural Radiation, Ramsar, IR Iran, 3-7 Nov. (1990), M. Sohrabi, J. U. Ahmed and S. A. Durrani (eds.), IAEA Publication Series (1993) 503.
7. M. Sohrabi, J. U. Ahmed and S. A. Durrani (eds.) Procs. of Int. Conf. on High Levels of Natural Radiation, Ramsar, IR Iran, 3-7 Nov. (1990), IAEA Publication Series (1993).
8. M. Sohrabi, In: Procs. 2nd Workshop on Radon Monitoring in Radioprotection, Environmental, and/or Earth Sciences, Trieste, Italy, 25 Nov.- 6 Dec. (1991) G. Furlan and L. Tommasino (eds.), World Scientific Publishing Co. (1993) 98.
9. Y. Lasemi, M. M. Beitolahi, M. Sohrabi and Z. Lasemi (in press).
10. M. Sohrabi, H. Mirzaee, M. M. Beitollahi and S. Hafezi, In: Procs. of an Int. Conf. on High Levels of Natural Radiation, Ramsar, IR Iran, 3-7 Nov. (1990), M. Sohrabi, J. U. Ahmad and S. A. Durrani (eds.), IAEA Publication Series (1993) 425.
11. D. R. Rushing, W. J. Garcia and D. A. Clark, In: Procs. Symp. on Radiological Health and Safety, IAEA SM/412-44, Vienna, Austria (1963) 189.
12. H. Mirzaee, and M. M. Beitollahi, AEOI Scientific Bulletin, Nos. 11 and 12, (1993) 97.
13. R. B. Wanty, and L. C. S. Gundersen, In: Procs. of the Georad Conference Special Publication, M. A. Marikos and R. H. Hausman (eds.) No.4 (1988) 147.

Levels of ^{222}Rn and its daughters in the environment of Xinglong district in Hainan province

Zeng Qingxing, Cheng Xuesheng, Li Yan, Huang Zhaohui, Deng Zhihong and Liu Aihua

Wuhan Sanitary-Epidemiological Station, No. 24 Jianghan Bei Lu, Wuhan 430022, P. R. China

1. INTRODUCTION

Hot spring has been widely used for bath. There is high content of Rn in the hot spring water. Rn may induce lung cancer. It is of common concern whether the health of hot spring employees may be affected by the Rn that they were exposed to. It was reported that Rn in water is the major source of indoor Rn. We used solid nuclear track detector and continuous Rn meter to measure the Rn in hot spring to find out the radiation dose of Rn in the utilization of hot spring and to appraise its harmful effect.

2. MATERIAL AND METHOD

2.1. Objects of measurement

Hotels and restaurants in Xinglong District in Hainan Province use hot spring for bath. The sampled area were the kitchen, sleeping rooms and offices for duty.

2.2. Measurement instrument

In order to get representative annual average, CR-39 was used to measure Rn in air. It should be placed somewhere other than corners, doors, windows and heights with air disturbed by breath. The sampling duration was 3 months. In order to determine the equilibrium factors of Rn and its daughters, double layer filter method and 3-step method were used to measure Rn and its daughters in air. Rn in water was measured by scintillation method, the sensitivity being 0.11 Bq·L^{-1}

3. RESULTS AND ANALYSIS

3.1. Contents of Ra and Rn in hot spring water

We have measured the Rn and Ra in hot spring water for bath in the holiday hotels in Xinglong District (Table 1). It is shown that the contents of ^{226}Ra in hot spring water is about 80 times as many as that in surface water. The concentration of Rn is 1210.0 Bq·m^{-3}, about 34.47 times as many as that in the surface water.

Table 1
Concentration of radium and radon in the water

	Hot spring	Ground water
^{226}Ra(Bq l^{-1})	1.03±0.22	0.012±0.01
^{222}Rn(Bq l^{-1})	1210.0±.82.0	35.01±3.0

3.2. The results

Potential energies of Rn and its daughters are shown in Tables 2,3. It is shown that Rn indoors and outdoors in Xinglong District are higher than that in other areas in Hainan Province. Rn indoors is generally contributed by the base of house, the soil and stone around house, building material, air outdoors, water supply and fuel as arranged in decreasing order. 30%-94% of Rn in water are released to air according to the ways in which the water is used. Water rich in Rn water is the major source of indoor Rn. That is the reason why the indoor Rn in Xinglong is higher than that in other areas. In order to find out the contribution of building materials to Rn, we made measurements (Table 4). The results show that National Standard for hot spring is not surpassed.

3.3. Exposure of employees to Rn in a holiday hotel

We placed some personal Rn meters on staffs working in different rooms. The duration is one month. The result is shown in Table 5. It is shown that the highest value is 208.2 Bq·m^{-3}, the lowest is 46.5 Bq·m^{-3}, and the average is 90.03 ± 51.19 Bqm^{-3}.

Table 2
Concentration of radon and its daughters in the air

Location	Radon(Bq.m^{-3})		Radon progeny (MWL)	
	Indoors	Outdoors	Indoors	Outdoors
Hotel A	118.75	45.66	11.30	4.03
Hotel B	54.16	29.42	5.28	3.2
Hotel C	33.71	15.92	3.06	1.49
Hotel D	141.40	91.18	12.18	7.87
Hotel E	88.93	11.69	7.71	1.129
Mean	87.39 ± 44.41	38.77±32.16	7.90±3.88	3.54 ± 2.70

Table 3
Radon measured with CR-39

Location	Radon (Bq.m^{-3})
Hotel A	96.54
Hotel B-1	43.7
Hotel B-2	40.9
Hotel C-1	26.8
Hotel C-2	47.5
Hotel D	51.7
Hotel E-1	22.3
Hotel E-2	28.3
Mean	44.65±23.43

Table 4
Concentration of radium in building material

Location	Sample	^{226}Ra(Bqkg^{-1})
Hotel A	Brick	155.07
Hotel A	Soil	107.79
Hotel B	Soil	166.42
Hotel B	Stone	164.53

Table 5
Exposure of staffs in a holiday hotel

Number	Radon(Bqm^{-3})
1-2	46.5
2-2	71.0
3-2	80.0
3-1	134.6
4-1	159.0
4-2	75.8
5-2	75.8
6-1	208.2
6-2	56.4
7-1	51.3
7-2	95.5
8-2	66.2
Mean	95.03±51.19

3.4. Variation of indoor Rn

We measured Rn in 24 hours continuously in holiday hotels in Xinglong. The results are as follows. That night 22-h: 110 Bq m^{-3}; 23-h: 114 Bq m^{-3}; the next day 1-h: 121 Bq m^{-3}; 3-h: 129 Bq m^{-3}; 6-h: 132 Bq m^{-3}; 9-h: 139 Bq m^{-3}; 12-h: 143 Bqm^{-3}; 16-h: 153 Bq m^{-3}; 17-h: 150 Bq m^{-3}; 20-h: 143 Bq m^{-3}; and 22-h: 139 Bq m^{-3}.

It is concluded that the variation in 24 hours is not pronounced. The peak emerges at hours 16:00 and 17:00. It may be attributed to the Rn released from bath water.

3.5. Estimation of exposure of residents

Assuming that the average breath rate of adult is 0.79 m^{3}h^{-1} indoors, and 1.0 mm^{3}h^{-1} outdoors, the range of FP indoors is 0.01-0.05, the activity median diameter of gaseous aerosols is 0.1-0.2 μm, the average occupancy factor is 0.8 indoors and 0.2 outdoors, then the Rn and its daughters absorbed by adult residents can be calculated according to the dose conversion factor given in the UNSCEAR Report 1988. The average effective dose for persons in Xinglong District is 3.3 mSv.

REFERENCES

1. UNSCEAR Source, effects and risks of ionizing radiation. New York: UN, 1998.
2. Smith H. Lung cancer risk from indoor exposures to radon daughters. Radiat Prot Dosim,20 (1987) 195.
3. Man Yin W Tso et al. Indoor and outdoor ^{222}Rn and ^{220}Rn daughters in Hong Kong. Health Phys. ,53 (1987) 175.
4. Ren Tianshan et al. Indoor 222Rn measurements in the region of Beijing, P. R. China. Ibid , 53 (1987) 219.
5. Zeng Qingxiang et al. A sample study of the concentration of indoor radon and radon progeny and its influencing factors in Wuhan City. Indoor Air '90, Toronto, Canada ,113, 1990.
6. UNSCEAR. Ionizing radiation: sources and biological effects. New York: UN,1983.

Radon concentration and radiation dose to the population exposed in the high elevation area of Sichuan

Liu Yigang and Mao Yahong

Sichuan Institute of Radiation Protection, 19, Dianxin Road, Chengdu 610041, China

ABSTRACT

We measured ^{222}Rn and its daughters concentration by scintillation chamber and Makov method in the high elevation areas of Sichuan. We measured more than 1,500 places where the average elevation is about 2,500 m. The results suggest that the ^{222}Rn and its daughter concentrations do not tend to change with the elevation (1,000m - 3,500m). But there is a trend that radon and its daughter concentrations change markedly with the landform (compared between prairie and other landforms). And the change of the indoor ^{222}Rn concentration is contrary to that of its daughters. The arithmetic average ^{222}Rn concentrations are 22.7 ± 22.5 Bq m^{-3} for indoors and 17.6 ± 14.1 Bq m^{-3} for outdoors. The arithmetic averages of the ^{222}Rn daughter concentration are (12.5 ± 14.2) x 10^{-8} J m^{-3} for indoors and (4.9 ± 3.3) x 10^{-8} J m^{-3} for outdoors. According to the model and parameters of estimation recommended by UNSCEAR 1993 Report, the assessed radiation effective dose from ^{222}Rn to the population in the areas is 0.87 mSv/(a·man).

1. INTRODUCTION

Radon is the major cause leading to high lung cancer incidence of people. And radon dose contribution occupies 54% of total natural radioactive dose. So scientists are concerned with the levels and distribution of radon in different geographical areas. For this reason, we measured radon and its daughter concentrations in different geographical areas (in range of elevation 1.0km - 3.5 km) in Sichuan Province of China and obtained about 3,000 data. Now we analyze these data as follows.

2. MEASUREMENT

We adopted scintillation counter and filter membrane methods to survey the radon and its daughters. In order to ensure the reliability of all the data, we had the radon measurement instrument calibrated and we collected two simultaneous samples at every location to reduce random error. Since the study was carried out over a long period of time, we provided the two instruments with ^{241}Am surface source for self-calibration whose area equals that of the detectors.

We did our best to collect and measure the samples at almost the same time of every day and kept the houses at the condition as they were (windows and doors opened or closed). The samples were all collected 1.5m above ground.

3. RESULTS AND DOSE ESTIMATES

We analyzed the 3,000 measured data, which are shown in the following tables(Table 1 and 2):

Table 1
Radon and radon daughter concentrations in different elevation areas

Elevation (m)		^{222}Rn concentration (Bq m^{-3})			^{222}Rn daughters concentration ($\times 10^{-8}$ J m^{-3})		
		Sample	Range	Average	Sample	Range	Average
<2000	Outdoor	80	2.6~38.0	14.8±5.3	78	2.8~11.4	6.4±1.1
	Indoor	160	1.7~70.5	19.1±6.3	144	1.0~182.0	15.9±14.8
2000~3000	Outdoor	99	1.8~82.1	14.0±12.4	36	1.8~19.5	8.5±7.2
	Indoor	295	0.5~374.1	28.8±21.6	278	0.4~83.8	11.7±8.2
>3000	Outdoor	59	0.6~237.0	9.6±3.4	65	0.7~6.7	6.9±5.8
	Indoor	66	2.1~155.5	29.4±29.1	196	0.7~81.5	9.3±4.8

Table 2
Radon and its daughter concentrations in areas of different landform

Landform		^{222}Rn concentration (Bq m^{-3})			^{222}Rn daughter concentration ($\times 10^{-8}$ J m^{-3})		
		Sample	Range	Average	Sample	Range	Average
Plateau	Outdoor	156	2.6 ~ 82.1	15.5±8.7	95	1.8 ~ 11.4	5.6 ± 1.9
	Indoor	327	0.5 ~ 374.1	20.6±13.0	207	0.4 ~ 182.1	13.3±12.3
Canyon	Outdoor	25	1.8 ~ 43.4	12.4±10.0	25	2.0 ~ 20.4	9.9±7.4
	Indoor	133	1.7 ~ 165.1	26.6±21.6	253	1.0 ~ 83.8	13.0±8.6
Prairie	Outdoor	57	0.6 ~ 237.0	7.6±1.3	59	1.0 ~ 83.8	2.9±0.8
	Indoor	61	2.1 ~ 155.5	35.1±32.8	158	0.7 ~ 5.4	9.2±5.8
Total	Outdoor	238	0.6 ~ 237.0	17.6±14.1	179	0.7 ~ 81.5	4.9±3.3
	Indoor	521	0.5 ~ 374.1	22.7±22.5	618	0.4 ~ 182.1	12.5±14.2

According to the UNSECEAR 1993 Report, the indoor occupancy factor of 0.8 is adopted, indoor and outdoor equilibrium equivalent factors are 0.4 and 0.8, respectively. The estimated annual effective doses from radon are 0.60 mSv a^{-1} for indoors and 0.27 mSv a^{-1} for outdoors.

4. CONCLUSION

In the high elevation areas of Sichuan, the indoor and outdoor radon concentrations are in the range of 0.5 - 374.1 Bq m^{-3} and 0.6 - 237.0 Bq m^{-3}, respectively; the corresponding average values are 22.7(22.5 Bq m^{-3} and 17.6 ± 14.1 Bq m^{-3}, respectively. The indoor and outdoor radon daughter concentrations are in the range of (0.4 ± 182.1)(10^{-8} J m^{-3} and (0.7 - 20.4) x 10^{-8} Jm^{-3}, respectively; the corresponding average values are (12.5 ± 14.2) x 10^{-8} J m^{-3} and (4.9 - 3.3) x 10^{-8} J m^{-3}, respectively. Compared with the average concentration of whole Sichuan (indoor radon and its daughter concentrations are 17.8 Bq m^{-3} and 6.0 x 10^{-8} J m^{-3}, respectively, while outdoor values are 12.8 Bq m^{-3} and 5.5 x 10^{-8} J m^{-3} respectively.), the indoor radon concentrations in high elevation areas are higher, but the outdoor values are lower. This is caused by the high wind frequency and fast wind speed in high elevation areas.

The two tables show that radon and its daughter indoor concentrations are higher than those outdoor in the whole high elevation areas. There is no evidence indicating that radon and its daughter levels vary with the elevation, but radon and its daughter concentrations do vary with the landform. From Table 2 we can see that the indoor radon concentrations in prairie areas are higher than those in the plateau and canyon with very significant difference ($P < 0.01$). We think this is because in prairie the floors of houses are almost all wooden, which makes the radon in soil under the floors easy to release. Moreover, the residents in prairie are accustomed to their doors and windows closed against wind.

The outdoor radon concentrations in prairie are lower than those in the other high elevation areas. This is because the wind in prairie is stronger and the wind frequency higher than that in the other high elevation areas.

In the high elevation areas, the annual effective dose to the population exposed to indoor and outdoor radon is 0.87 mSv a^{-1}, which is lower than the world's average value of 1.2mSv a^{-1}.

REFERENCE

1. David Bodansky et al. Indoor radon and its hazards, University of Washington Press 1987.
2. United Nations Scientific Committee on the Effects of Atomic Radiation, Ionizing Radiation: Sources and biological effects. New York. United Nations; 1993.
3. Crameri et al., Indoor Rn levels in different geological areas in Switzerland, Health Physics, 1 (1989)29.
4. Bernard L. Cohen. ,A national survey of ^{222}Rn in U.S. homes and correlating factors, Health Physics. ,2 (1986)175.
5. Yibin Cheng, Indoor concentrations in homes of Sichuan residents and the dose to the population exposed to radon and its daughters, Indoor Air, 3 (1993), 327.

^{222}Rn concentration in different types of cave dwellings in Gansu Province

Gao Pingyin[1], Sun Wei[1], Tian Deyuan[2], Huo Junlan[1], Tao Yufang[1], Zhang Shouzhi[2], Wang Yin[1], Xu Birong[1] and Guo Shanxiang[1]

[1]Gansu Institute of Radiation Protection, Lanzhou 730000, Chin

[2]Laboratory of Industrial Hygiene, Ministry of Health, Beijing 100088, China

1. INTRODUCTION

A survey of indoor and outdoor ^{222}Rn and its daughter levels in Gansu Province conducted in 1986-1988 has discovered that most residents in Pingliang and Qingyang Prefectures live in underground dwellings (cave dwellings) and the ^{222}Rn level in cave dwellings is far higher than that in common dwellings (^{222}Rn concentration ranges from 35.5-347 Bq/m^{-3}, mean 101 Bq/m^{-3}) [1]. Therefore, cave dwellings may provide an excellent choice for studying indoor radon and risk of lung cancer.

In order to demonstrate the high radon level in cave dwellings and to ascertain the exact ^{222}Rn levels in different types of caves dwellings, radon level measurements were conducted with activated-carbon detectors in the suburbs of Xifeng City in 1991. This paper presents the results of those measurements.

2. TYPES OF CAVE DWELLINGS

In Pingliang and Qingyang Prefectures cave dwellings ("*yao-dong*" in Chinese) is a unique style of residence constructed entirely underground. The caves are generally constructed around a courtyard. All caves are almost of the same shape and construction. There are four types of cave dwellings, according to their positions compared with ground level and the types of construction: (1) Ground *kang* are caves entirely below ground level; (2) half-dim and half-light caves are those partly below ground and built into the side of road; (3) light caves are built into the side of the a hill; and (4) hooped *yao* are built on the ground level, but with a thick-walled construction, an interior room design similar to those of types (1)-(3) and a appearance similar to common house.

3. SAMPLING AND MEASUREMENT

The sampled houses were selected randomly. The samples were collected from the four types of cave dwellings, common dwellings and outdoors.

The activated carbon detector [2] is a cylinder aluminum box, inner diameter 10.5cm, high 5.0cm, containing about 160 g of activated carbon, made from high quality coconut shell. After sampling is started, with the box open, the carbon bed is exposed to the air for about three days. The carbon absorbs radon from the air in house. Close and seal the box immediately after sampling and send to laboratory for measurement.

The samples were measured with Ge (Li) gamma-ray spectrometer. After calibration with radon source, relating to count rates of the full energy peaks of 294 keV, 350 keV and 609 keV gamma-rays of radon daughters, ^{222}Rn concentration can be expressed approximately as follows:

$$C_{294} = 186.0 R_{294} e^{\lambda t_1} t_2^{-0.564}$$
$$C_{352} = 138.9 R_{352} e^{\lambda t_1} t_2^{-0.611} \quad (1)$$
$$C_{609} = 180.5 R_{609} e^{\lambda t_1} t_2^{-0.593}$$

where C_i denotes the ^{222}Rn concentration in the house air (Bq/m^{-3}); R_{294}, R_{352}, R_{609} are count rates of the full energy peaks of 294 keV, 352 keV and 609 keV gamma-rays, respectively (cps); t_1 is time interval from sampling center to measuring center (hours); t_2 is time exposure time of activated carbon (hours); λ is ^{222}Rn decay constant.

^{222}Rn concentration of sampled house is given by

$$C = \frac{1}{3}(C_{294} + C_{352} + C_{609}) \quad (2)$$

where C is ^{222}Rn concentration of sampled houses; C_{294}, C_{352}, C_{609} are the same as in equation (1).

4. RESULTS AND DISCUSSION

The results of ^{222}Rn concentration in the different dwellings are shown in Table 1. The results show that in cave dwellings the ^{222}Rn levels are far higher than that in the common dwellings.

Ground *kang*, light caves and half-dim and half-light cave are main types of cave dwellings. Hooped *yao* is a new and uncommon type of cave dwelling discovered in recent years. The results show that the ^{222}Rn concentrations in ground *kang*, light caves and half-dim and half-light caves are consistent. This is

Table 1
^{222}Rn levels in different type of dwellings (Bq/m^{-3})

Dwellings	Number	Range	Mean
Ground *kang*	15	56.6-202	131
Light caves	24	53.3-389	172
Half-dim and half-light caves	24	52.3-419	164
Hooped *yao*	14	21.7-150	70.7
Common dwellings	4	24.7-53.6	39.4
Outdoors	1	21.4	21.4

because the construction and ventilation are similar in these cave dwellings. It is shown that ^{222}Rn level is not relative to cave types. Pingliang and Qingyang Prefectures seem to be excellent places for studying indoor radon and risk of lung cancer.

We shall survey further the variation of ^{222}Rn levels and influencing factors in cave dwellings.

REFERENCES
1. Gao Pingyin, J. Radiol. Med. Prot., 11 (1991) 32.
2. George A.C., Health Phys., 46 (1984) 867.

Concentration of radon and its daughters in environment and internal doses to population in Qinghai Province

Ni Jianqing, Cao Zhiyou, Ma Qinglu and Liang Zhuqi

Qinghai Occupational Disease Prevention and Treatment Center, 69 Nanchuan West Road, Xining 810012, China

There are many nuclides that emanate radon on to the earth surface. As it is estimated, a half of natural radiation come from radon and its daughters. In recent years it has been found that the concentration of radon and its daughters are quite high in some resident environment. Radon is a primary factor to induce lung cancer in human being. In order to find out and evaluate the levels of radon and its daughters in the population dwelling in Qinghai plateau, we carried out an investigation from June 1986 to October 1989.

1. INVESTIGATION METHOD

1.1 Item

Rn-222 and its daughter concentration and distribution feature in air indoors and outdoors were investigated. The doses to population were estimated.

1.2 Method

Radon concentrations were determined by scintillation method using ZYW8501 meter. The volume of scintillation flask is 0.7L. The background dose level of instrument is 0.04 cpm. Its lowest detection limit is 1.1Bq m^{-3}. This meter was calibrated according to national regulations (k=13.05 Bqm^{-3}cpm^{-3}, relative error less than 30%).

Radon daughters were determined by Thomas method using FD-3005 low background meter. The radon samples were gathered by filter paper (No.1). The sampling time was 5 minutes. The mean exhausting speed was 48.6 L min. The total filter efficiency was 0.29. The samples 2-5, 6-20 and 21-30 minutes after filtering were determined for counts, respectively.

1.3 Site distribution:

The determined sites were selected according to density of population (every twenty-five thousand people for site). We chose 162 points indoors and 161 points outdoors. The indoor points are chiefly dwellings with one door and one window for natural ventilation. The areas of rooms differ in size. The dwellings in cities and towns are storied house built of brick and concrete and heated by central heating, while the dwellings in rural and pastoral areas are one-store houses made of mud and wood and tents heated with coal, wood and animal dung. The outdoors points were chosen on open field (every point was 5 meters apart from buildings and 1.5 meters above ground). See Figure 1.

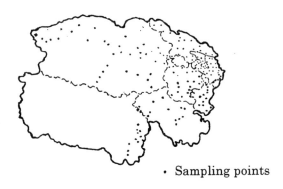

· Sampling points

Figure1. Distribution of points for determining radon and its daughter

2. RESULTS AND DISCUSSION

2.1. Time variation

The concentrations of radon outdoors were highest in December (16.72 Bqm^{-3}), lowest in summer (10.48 Bqm^{-3}) and intermediate in spring and autumn (13.80 Bqm^{-3} and 11.44 Bqm^{-3}). The radon concentration peaked at 2-6 o'clock, descended at 6-12 o'clock and was steady at 12-20 o'clock in a day. It gradually rose after 20 o'clock. The maximum values was three times as many as the minimum.

The sampling time was at 8-13 o'clock(80%) in this investigation. The determined values more approximate to the average values in a day.

2.2. Distribution of radon concentrations in indoors and outdoors

The values indoors was higher than those outdoors with statistically significant difference ($P<0.05$). The frequency of concentrations followed log-nomal distribution.

Table 1
Radon concentration and potential energy distribution in whole province.

District	No. of Samples	Rn-222(Bqm^{-3})			α potential(mWL)		
		Range	$\overline{X}$	s	Range	$\overline{X}$	s
Xining							
Indoor	22	5.56-54.81	31.43	13.24	3.17-6.42	4.77	0.96
		(28.05)	(1.71)		(4.67)	(1.24)	
Outdoor	22	2.18-26.39	14.74	6.65	1.82-6.56	3.80	1.30
		(12.45)	(2.03)		(3.57)	(1.45)	
Haidong							
Indoor	37	10.64-105.27	33.07	23.60	1.08-9.12	3.45	2.32
		(27.23)	(1.83)		(2.98)	(1.90)	
Outdoor	35	2.83-43.11	11.10	8.36	0.53-6.99	1.91	1.17
		(8.87)	(1.91)		(1.71)	(1.82)	
Hainan							
Indoor	17	7.01-55.35	21.08	15.18	0.22-2.58	1.54	0.39
		(14.79)	(2.12)		(1.16)	(2.19)	
Outdoor	17	1.96-22.79	4.87	1.84	0.30-1.35	0.80	0.37
		(4.52)	(1.50)		(0.68)	(1.56)	
Huangnan							
Indoor	19	8.70-44.55	17.86	9.06	0.59-6.55	2.29	1.39
		(16.11)	(1.57)		(1.96)	(1.77)	
Outdoor	19	2.39-10.92	5.84	2.34	0.59-1.88	1.05	0.44
		(5.41)	(1.51)		(0.97)	(1.49)	
Haixi							
Indoor	30	4.35-26.06	12.50	5.93	0.16-3.53	1.26	0.90
		(11.22)	(1.61)		(0.98)	(2.08)	
Outdoor	30	2.29-15.90	5.46	2.57	0.12-2.52	0.76	0.45
		(5.04)	(1.48)		(0.64)	(1.82)	
Haibei							
Indoor	15	9.73-41.37	14.16	9.67	0.55-4.45	1.60	1.09
		(12.19)	(1.68)		(1.33)	(1.83)	
Outdoor	15	2.38-10.08	5.59	2.41	0.32-1.53	0.71	0.37
		(5.13)	(1.55)		(0.64)	(1.61)	
Yushu							
Indoor	12	1.78-25.48	10.27	6.54	0.24-3.45	1.39	0.88
		(8.59)	(1.97)		(1.15)	(1.95)	
Outdoor	12	1.80-17.06	6.71	4.21	0.29-2.77	1.09	0.68
		(5.70)	(1.80)		(0.93)	(1.82)	
Guoluo							
Indoor	10	3.39-14.84	8.18	3.58	0.46-2.01	1.11	0.48
		(7.39)	(1.63)		(1.00)	(1.63)	
Outdoor	11	1.80-8.05	5.00	1.61	0.29-1.31	0.81	0.26
		(4.71)	(1.48)		(0.76)	(1.48)	
Total							
Indoor	162	1.78-105.27	20.94	16.79	0.16-9.12	2.41	1.90
		(17.60)	(3.49)		(1.77)	(2.23)	
Outdoor	161	1.80-43.11	8.01	6.09	0.12-6.99	1.40	1.23
		(6.49)	(1.88)		(1.05)	(2.21)	

Note: Figure in brackets are geometric means and standard deviations

The ratios of radon and its daughter concentrations indoors to outdoors were 2.61 : 1 and 1.72 : 1, respectively. The population weighted averages were similar to geometric means.

Table 2
Population weighted averages of radon and its daughter concentrations

Sampling point	Radon-222 (Bqm^{-3})		α potential energy (mWL)	
	Geometeric mean	Population weighted averages	Geometeric mean	Population weighted averages
Indoor	17.60	16.73	1.77	1.63
Outdoor	5.17	6.49	1.00	1.01

Radon daughter equilibrium factors in air (F) were 0.43±0.15 for indoors and 0.63±0.25 for outdoors. These levels were close to the values in UNSCEAR Report (0.5 indoors, 0.6 outdoors).

The concentrations of radon were on the high side in Xining and eastern region owing to lower altitude, flourishing industry, dense population and air pollution (radon levels were 10 Bqm^{-3} outdoors and 25 Bqm^{-3} indoors).
In contrast, the radon concentrations were on the low side in pastoral areas owing to higher altitude, smooth air current (radon levels were 5 Bqm^{-3} outdoors and 15 Bqm^{-3} indoors).

2.3. Radon and its daughter concentrations in dwellings of different structure and heating

The levels raried with structure dwellings and mode of heating. The values from high to low were those in storeyed houses, bungalows with brick ground, bungalows with concrete ground, and tents. The radon concentration values for different mode of heating from high to low were those with coal, central heating, animal dung and firewood. The (potential energy values from high to low were those with central heating, coal, animal dung and firewood. There was statistically significant difference between those with animal dung and coal and central heating (P<0.05). However, because there were obvious differnces in ventilation condition, area and height of dwellings, the factors influencing on radon concentration remain to be throughly investigated and studies.

2.4. Radon concentrations at different altitudes

Table 3
Radon and its daughter concentrations in dwellings of various structures

Type	No. of samples	Rn-222 (Bqm^{-3})			α potential energe(mWL)		
		Range	$\overline{X}$	s	Range	$\overline{X}$	s
Storeyed houses	39	4.35-55.35	21.37	13.69	0.16-6.42	2.83	1.92
Bungalous	61	6.32-105.27	22.60	18.20	0.22-7.81	2.48	1.86
Brick ground							
Soil ground	21	5.58-60.27	19.46	15.49	0.54-8.86	2.09	1.91
Concrete ground	24	5.56-96.01	20.71	20.55	0.55-9.12	2.15	2.88
Tents	6	3.39-8.28	5.82	2.02	0.46-0.99	0.79	0.27

Table 4
Radon and its daughter concentrations in dwelling with different heating modes

Mode of heating	No. of samples	Rn-222 (Bqm^{-3})			α potential energe(mWL)		
		Range	$\overline{X}$	s	Range	$\overline{X}$	s
Central heating	36	4.35-55.35	22.42	13.69	0.16-6.00	2.83	1.92
Coal	80	5.56-105.3	22.61	20.12	0.22-9.12	2.36	2.01
Animal dung	22	1.79-44.55	13.24	10.23	0.24-6.55	1.82	1.46
Firewood	4	6.53-18.62	11.36	5.78	0.96-2.46	1.42	0.70

The levels from high to low were those in group 1 (<2,500 meters), group 2 (2,500-3,000 meters) and group 3 (>3,000 metres). The ratio between them was 1:0.55:0.44. The values between group 1 and groups 2,3 had statistically significant difference (P<0.01). Thus it can be seen that radon concentrations outdoors decline with rising altitude (Table 5).

Table 5
Radon concentrations outdoors at different altitudes

Group	No. of samples	Altitude (meter)	Rn-222(Bqm^{-3})		
			$\overline{X}$	s	Range
1	43	<2500	12.71	8.20	2.41-43.11
2	54	2500-3000	6.95	4.91	1.96-25.40
3	62	>3000	5.57	2.67	1.80-17.06

Table 6
Per capita effective dose equivalent

District	Indoor(mSv)	Outdoor(mSv)	Indoor+Outdoor(mSv)
Xining	0.81(0.72)	0.74(0.63)	1.55(1.35)
Haidong	0.85(0.70)	0.56(0.45)	1.41(1.15)
Huangnan	0.46(0.41)	0.30(0.27)	0.70(0.68)
Hainan	0.54(0.38)	0.25(0.23)	0.79(0.61)
Haixi	0.32(0.29)	0.28(0.25)	0.60(0.54)
Haibei	0.36(0.31)	0.28(0.26)	0.64(0.57)
Yushu	0.26(0.22)	0.34(0.29)	0.60(0.51)
Guoluo	0.21(0.19)	0.25(0.24)	0.46(0.43)
Total	0.54(0.45)	0.40(0.33)	0.94(0.78)

Note: Figures in brackets are geometric means.

The low radon concentration high altitude area is attributed to low atmospheric pressure and humid air.

Because these values did not come from the same selected sites, they could not be further evaluated; they are provided here for reference only.

3. DOSES TO POPULATION

3.1. Doses estimation

Assuming the adult breath rates of 0.79 m^{-3} h^{-1} indoors and 1 m^{-3} h^{-1} outdoors and occupancy factors of 0.8 indoors and 0.2 outdoors.and in accordance with the recommendation in UNSCEAR 1982 Report, taking radon daughter covnersion factors of 0.016mSvBqm^{-3} indoors, 0.031mSvBqm^{-3} outdoors, the population doses were estimated (Table 6). Radon daughter effective dose equivalent to population in whole province was 0.94mSv. This level was similar to the whole world mean but was higher than those in Beijing and England (Table 7).

Table 7
The dose equivalent resulting from radon daughters

Area	Dose equivalent(mSv)
England	0.7
Beijing	0.79
Whole world	0.80
Qinghai	0.94

The size distribution of atmospheric radioactive aerosols and its measurement

Jin Yihe[a], M.Shimo[b], Zhuo Weihai[a], Xu Liya[a], Fang Guoqiu[a]

[a]Fujian Institute of Radiation Health Protection, Fuzhou 350001,China

[b]Gifu College of Medical Technology, Japan

1.INTRODUCTION

Various aerosols exist in atmosphere . Some are natural, and others arise from mankind activities. Aerosol are important study objects for researchers in many fields. Studied contents include particle concentration, chemical composition, size distribution, mass concentration, radioactivity, optical character, etc. Up to now, because there is no perfect method to observe human lung dose caused by radioactive aerosols. Estimation are usually based on the present dosimetric model and measured radiation energy of radioactive aerosols. Owing to the selection and variation of different parameters (trachea shape, target cell location, breathing model, and size distribution) for dosimetric model, the calculated doses may fluctuate. Among them, the activity median diameter (AMD) of aerosol particles attached with radon progeny is an important influent factor. As the AMD ranges from 0.05 to 0.17μm, the lung dose may fluctuate from 100% to -20% (NCRP Report No.78). So researchers devote more and more attention to the study of the size distribution of radioactive aerosols.

The diameters of radioactive aerosols range from 0.001 to 100μm. At present, it is still impossible to measure the aerosols in diameters of whole range by using the same method. From dosimetric point of view, natural radioactive aerosols can be regarded as those attached and unattached fraction of radon progeny, their sizes approximately range from 0.001mm to 0.5mm. So the wire screen type diffusion battery which is easily controlled and small volume was adopted. We present here a simple measuring method and results for the attached fraction of radon progeny.

2. MEASURING METHOD

2.1. Measuring principle

When original radioactive aerosols (radioactive concentration Z0) with uniform size pass through the wire screens, their concentrations reduce to Z due to the diffusion effect. Once the wire screen structure (effective volume, wire diameter and screen size, etc.) and environment conditions (temperature, pressure, etc.) were fixed, the particle penetration ratio ($p = Z / Z_0$) will be the function of airflow rate, particle size (r) and screen number (n). As the airflow rate can be fixed in actual measurement, p is only the function of r and n, i.e. $p = R(n,r)$. For different size particles attached with radon progeny, p can be expressed as the following formula:

$$p(n) = \int R(n,r) f(r) dr / \int f(r) dr \qquad (1)$$

f(r) is the function of size distribution. According to the formula(1), once airflow rate is fixed, the penetration curve can be measured by changing the number n of screen. Having been mathematically treated, size distribution can be calculated out.

2.2. Sampling and measurement

Sampling equipment is shown in Fig.1. Airflow enters the SB from the left to the right. In each SB 43 chips of wire screens can be installed. When more than 43 chip are needed in the diffusion process, each SB should be series connected. The effective screen diameter is 40 cm, and the screen is made of stainless metal wire with 165 meshes adhered to a brass ring which is 5mm thick, 40cm inner diameter and 60cm outer diameter. A groove is made on the ring, and then the screens set with air tighted sticky gaskets is connected in order to let the entire airflow penetrate the screen set. According to the study made by K.Sugiyama et al., the size distributions measured by using different screens with 165~250 meshes were nearly the same, and the results were also similar to those by using other kinds of diffusion battery, so we adopted the cheaper wire screen with 165 meshes.

Particles having passed SB were collected by filters(TOYO, TM-80, 0.8 μm). The joint between filter and filter holder should be air tighted. Airflow rate for sampling was fixed to 4 L / min. Flowmeters and their regulators were connected to the air-way of each SB in order to keep the flowrate highly consistent and stable during the whole sampling process. This feature is important to the measuring results.

Because of the different number of screens in each branch of SB and the different activity of particles collected in the filters, a penetration curve can be drawn out. To establish a penetration curve, at least 9 data (p values) obtained from 9 different numbers of screens and blank SB are needed. That is to say, 10 branches of SB and 10 α counters are needed in the mean time. Owing to the limited availability, only 4 branches of SB and 4 αcounters were used, and sampling for 3 times in each process was needed. The process in detail is scheduled as shown in Fig.3.

During the 9 different diffusion processes in this study, the numbers of screens used in SB were 10, 15, 25, 40, 55, 80, 110, 159 and 269, respectively.

2.3. Method for data analysis

Based on the penetration curve, Ikebe analysis method (Ikebe.Y. 1972) was used to determine the size distribution. It is to determine the size distribution of polydisperse submicron aerosols by a response matrix method. But negative value for formula(1) may exist some time, so a progressively approaching was used to find the approximate solution for $f(r)$. In order to lower the error($< \pm 5\%$), more than 16-time progressively approaching was needed. Due to a large amount of calculation, a specialized software developed by Nagoya University of Japan was used for calculating and drawing. Typical size distributions are shown in Fig.4.

2.4. Sampling location

In order to avoid the interference of artificial aerosols, a piece of quiet lawn in our institute far away from roads was selected for outdoor sampling, and a laboratory in which few staffs worked was selected for indoor sampling.

2.5. Measuring period and items

After one and a half years of preparation, formal measurement were carried out once a week from July 1992 to December 1993. Sampling time was from 8 am to 12 am. In the case of failure in sampling, another measurement was made in the same week. In the process of sampling, concentrations of radon, radon daughters and meteorological factors (temperature, atmospheric pressure, humidity, etc.) were also measured.

3. RESULTS AND DISCUSSION

Sixty-two radioactive aerosol size distribution curves (11 for indoors, 51 for outdoor) were measured. The size distribution for outdoors showed a geometric normal distribution(Fig.3). The peak diameter for outdoor size distribution ranged from 0.10 to 0.56μm, their frequency distribution is shown in Fig.4. As shown in Fig.4, the diameters mainly distribute in the range of 0.15 - 0.40μm, accounting for over 70% of total results. The peak diameters for indoor size distribution mainly gathered in range of 0.15 - 0.40μm. Its peak diameter distribution was not analyzed for few number of samples.

The arithmetic mean peak diameter for indoor and outdoor size distribution were (0.20 ± 0.09) μm and (0.27 ± 0.11) μm, respectively. As shown in Table1, the measuring result for indoors is the same as the recommended data by

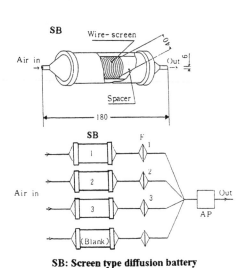

SB: Screen type diffusion battery
F: Filter   AP: Air pump

Fig. 1 Schematic diagram for measuring the size distribution of radioactive aerosols by screen type diffusion battery.

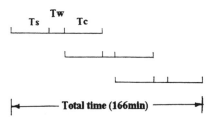

Fig. 2 A time table for measurement.

Ts: Sampling time (40 min)
Tw: Wait time (2 min)
Tc: Count time (40 min)

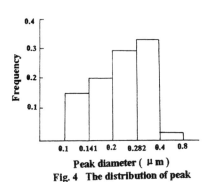

Fig. 4 The distribution of peak diameter in ourdoor

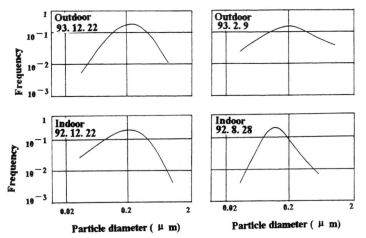

Fig. 3 Some examples of size distribution of radioactive earosols

UNSCEAR (1988); the result for outdoors is slightly higher than the recommended data, but is similar to that reported by Jacobi (W. Jacobi and K. Eisfeld, 1980).

Information on the size distribution of radioactive aerosols in China is scanty. So authors usually quote the recommended data by UNSCEAR (1988) to estimate the lung dose, i.e. the dose transfer coefficients are based on the information that the AMD are 0.2 µm for indoor and outdoor aerosols. Based

Table 1
Comparison of results in this study and other report

Source	AMD(mm)	
	indoors	outdoors
UNSCEAR(82)	0.2	0.1
UNSCEAR(88)	0.2	0.2
Jacobi-Eist	0.25*	
NCRP No. 78	0.05-0.17*	
This study	0.20	0.27

*not separated for indoors and outdoors

on the result in this study, the lung dose caused by outdoor radon progeny was over estimated in in the previous studies for Fujian Province ; but for indoors, the estimated value is the same. About 90% of lung dose is contributed by indoor radon daughters, so it is basically feasible for Fujin Province to estimate lung dose using the recommended data by UNSCEAR (1988). Whether this conclusion can be extended nationwide needs further study.

Futhermore, correlation between the size distribution and the concentrations of radon, radon daughters and meteorological factors were also analyzed, but no significant correlation was found among the present data.

REFERENCES

1. NCRP Report No.78, 1, 1984.
2. Cheng. Y. S. and Yeh. H. C., Aerosol Sci. 11(1980)313.
3. K. Sugiyama, M. Shimo and Y. Ikebe, Research letters on atmospheric electricity, 6 (1986) 67.
4. Ikebe. Y., Pure Appl. Geophys, 98 (1972) 197.
5. W. Jacobi and K. Eisfeld, GSF Report, S-626, 1980.

A new sub-nationwide survey of outdoor and indoor ^{222}Rn concentrations in China

Yihe Jin[a], Iida Takao[b], Zuoyuan Wang[c] and Abe Shirou[d]

[a]Fujian Institute of Radiation Health Protection, Fuzhou 350001, China

[b]Nagoya University, Nagoya 464, Japan

[c]Laboratory of Industrial Hygiene, Ministry of Health, Beijing 100088, China

[d]National Institute of Radiological Sciences, Chiba 263, Japan

1. INTRODUCTION

From November 1988 to January 1990 and from February 1991 to March 1993, international cooperative surveys of outdoor and indoor ^{222}Rn concentrations in mainland of China were carried out by China and Japan. There were two purposes. One was to study the distribution and alteration of radon concentration and to estimate the contribution of radon and its daughters to the population dose in China. The other was to study the interaction of air masses over mainland of China and offshore islands. The latter can provide basic experimental data for the theories of aerodynamicics and dissemination of atmosphere pollution because radon is an optimal tracer element for the study of long distance movement of air.

2. INSTRUMENT AND METHODS

2.1. Instrument
The measurements were made using electrostatic integrating ^{222}Rn monitor with cellulose nitrate (Aloka, GS-201B) developed by Takao Iida et.al. (Fig.1). According to Currie's definition, the detection limit is found to be 0.4 Bqm^{-3} for an exposure time of two months.

2.2. Monitoring duration

The monitoring lasted From November 1988 to January 1990 and from February 1991 to March 1993. The exposure time was 2 months for each period.

2.3. Arrangement of monitoring

Fuzhou, Guiyang, Wuhang, Shanghai, Nanjing, Nantong, Xi'an, Beijing, Hohhot, and Changchun, the 10 central cities in six different directions of China, were selected as surveyed locations. Gaoxiong of Taiwan and Nagoya of Japan were selected as control cities (Fig.2). In the first survey the measured points were 102 in the 6 cities (Fuzhou, Guiyang, Shanghai, Nanjing, Xi'an and Beijing) and in the second survey they were 96 in the 7 cities (Fuzhou, Guiyang, Wuhang, Shanghai, Nantong, Hohhot and Changchun). In principle, the measurement was made indoors and outdoors simultaneously at each point. The outdoor monitors were placed from 1 to 1.5 m above the ground. The indoor monitors were placed in bedrooms and offices of houses constructed with concrete and brick, which are typical buildings in the mainland of China.

3. RESULTS AND DISCUSSION

3.1. Outdoor ^{222}Rn concentrations
3.1.1. Distribution of outdoor ^{222}Rn concentrations

The average outdoor ^{222}Rn concentrations at 10 cities of China are shown in Table.1. As shown in Table.1, Fuzhou, Guiyang and Shanghai received both surveys, and their results of the first and second surveys agreed well for outdoor ^{222}Rn concentrations, but did not for indoor ^{222}Rn concentrations, owing to the different buildings for the first and second surveys.

The annual mean of the 10 cities was 8.8 Bq·m^{-3}. This is higher than the worldwide mean (5Bq·m^{-3}), given in UNSCEAR Report 1988. Among the 10 cities, Wuhang ranked the highest, 13.5(7.9 - 24.3)Bq·m^{-3}, and Nantong the lowest, 4.6(2.2 - 5.2)Bq·m^{-3}. Compared with Gaoxiong and Nagoya, ^{222}Rn concentrations in Fuzhou, Guiyang, Wuhang, Xi'an, Beijing, Hohhaot and Changchun were higher, but those in Shanahai, Nanjing and Nantong were nearly at the same level. From the viewpoint of geographic location, the distribution of outdoor ^{222}Rn concentrations in the mainland revealed a tendency: the concentrations were higher in southern parts than those in northern parts, and they were lower in offshore areas than those in inland areas.

The seasonal variations of outdoor ^{222}Rn concentrations for 2 months are shown as black points in Fig.3; the dotted lines represent annual average outdoors. Since the first survey there were much more samples, and the results from the first survey were similar to those from the second survey, Fig.3 is based on the results from the first survey in the 6 cities. The outdoor ^{222}Rn concentrations ratio between the maximum and the minimum ranged from 1.5 to 2.2, with an average of 1.8 (those for the 10 cities ranged 1.2 - 2.8 with an average

Fig. 2 The locations of survey of radon concentrations in China.

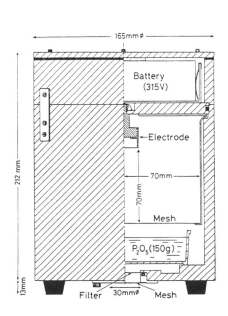

Fig. 1 Electrostatic intergrating ^{222}Rn monitor.

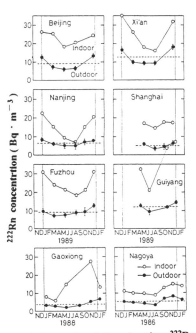

Fig. 3 Seasonal variation of outdoor ^{222}Rn concentrations at 6 cities in China and control cities.

Table 1
Average outdoor and indoor ^{222}Rn concentration in China(Bq・m^{-3})

Location	Monitoring time	Outdoor N[1]	Outdoor Mean	Outdoor S	Indoor N[1]	Indoor Mean	Indoor S	indoor/outdoor Summer	indoor/outdoor Winter
Fuzhou	88/11~ 89/11	10	9.5	1.9	10	13.9	3.0	1.5	2.4
Guiyang	88/11~ 89/11	4	12.2	2.0	4	38.0	13.9	2.2	3.6
Shanghai	88/11~ 89/11	5	4.8	1.6	5	16.5	4.8	4.3	2.7
Nangjing	88/11~ 89/11	10	6.4	1.3	10	13.7	5.3	1.8	2.8
Xi'an	88/11~ 89/11	14	11.2	3.0	9	23.9	7.7	1.7	2.2
Beijing	88/11~ 89/11	10	8.0	2.5	11	22.6	3.4	3.0	2.1
Fuzhou	91/07~ 92/07	9	11.8	3.1	9	18.8	3.4	2.1	1.7
Fuzhou[2]			10.7			16.4		1.8	2.1
Guiyang	91/07~ 92/07	1	11.6	1.1	1	17.5	6.1	1.1	2.1
Guiyan[2]			12.1			33.9		2.0	3.3
Shanghai	91/07~ 92/07	5	5.2	1.0	5	29.3	7.3	3.5	4.0
Shanghai[2]			4.9			18.6		4.2	2.9
Wuhang	91/07~ 92/07	11	13.5	6.2	15	17.5	8.4	1.2	1.3
Nangtong	91/07~ 92/07	2	4.6	1.1	1	11.8	7.2	2.4	4.6
Hohhot	91/07~ 92/07	4	8.4	1.6	4	10.8	2.0	1.1	1.3
Changchun	91/07~ 92/07	7	8.0	1.1	4	17.2	6.2	2.6	1.6
Mean	First survey[3]		8.7	2.6		21.4	8.4	2.4	2.6
Mean	Second survey[4]		9.0	3.5		17.6	5.9	2.0	2.5
Mean	Average[5]		8.8	2.9		19.5	6.5	2.1	2.6
Gaoxiong	85/11~ 86/11	1	4.0	2.0	1	15.5	9.1	3.9	
Nagoya	88/11~ 89/11	6	5.5	1.8	6	11.3	2.3	2.1	

1) Numbers of monitors;
2) The average of the first and second surveys;
3) The average of the first survey(88/11~ 89/11) in the 6 cities;
4) The average of second survey(91/07~ 92/07) in the 7 cities;
5) The mean of the first and second survey in the 10 cities.

of 1.8, as shown in Table2). The parterres of seasonal variations in the 10 cities were similar to those in control cities and most area in the world, i.e., the outdoor ^{222}Rn concentrations show a clear seasonal pattern of a summer minimum and a winter maximum.

The variation of outdoor ^{222}Rn concentrations in different geographic sites and seasons can be explained as the due to differences in ^{226}Ra contents and moisture in soil, landform, atmospheric stability and meteorological condition.

Table 2
The outdoor ^{222}Rn concentration ratio between the winter and summer

Location	Ratio	Location	Ratio
Fuzhou	1.7	Guaoxiong	3.5
Guiyang	1.5	Nagoya	2.6
Shanghai	1.9		
Wuhang	2.8		
Nangjing	1.6		
Nangtong	1.9		
Xi'an	1.9		
Beijing	2.2		
huhehaote	1.6		
Changchen	1.2		
Mean	1.8		

Table 3
Induced annual effective dose equivalet

Location	H_{eff} (mSv·a^{-1})		
	Indoor	outdoor	Total
Fuzhou[a]	0.46	0.15	0.61
Guiyang[a]	0.95	0.17	1.12
Shanghai[a]	0.52	0.07	0.59
Wuhang	0.49	0.19	0.68
Nangjing	0.38	0.09	0.47
Nangtong	0.33	0.06	0.39
Xi'an	0.67	0.16	0.83
Beijing	0.63	0.11	0.74
Huhehaote	0.30	0.12	0.42
Changchen	0.48	0.11	0.59
Mean	0.51	0.12	0.63
Gaoxiong	0.43	0.06	0.49
UNSCEAR(1988)	1.0	0.07	1.1

a. Weighted average of first and second survey

3.1.2. Comparisons between the two sides of the straits

The average width of Taiwan strait is 200km. The nearest distance between the mainland and the archipelagos of Japan is 1,000km. Because the half-life of ^{222}Rn is 3.8d, in the northwest wind season lasting from autumn to winter, it is imaginable that high outdoor ^{222}Rn concentrations of mainland may affect that of Taiwan and Japan. Though qualitative surmise had early been made by scholars of Taiwan and Japan, quantitative estimation has rarely been reported.

As shown in Table.2, the ratios of ^{222}Rn concentrations between the maximum (in winter) and the minimum (in summer) in every two months ranged from 1.2 to 2.8 with an average of 1.8. However, the ratio in Gaoxiong was up to 3.5. This implies demonstrate that ^{222}Rn in the airflow over Taiwan was largely affected by the airflow over the mainland. ^{226}Ra concentration(40.0 Bq·kg^{-1}) in the soil of Taiwan is similar to the average (42.0 Bq·kg^{-1}) of the 10 cities in the mainland. It is supposed that without the influence of the airflow in the mainland, the ratio of outdoor ^{222}Rn concentrations between winter and summer should also be 1.8, but the value in fact was 3.5. Thus, it can be considered that the excessive 1.7 resulted from the airflow in the mainland. In other words, in winter, about 50% of outdoor ^{222}Rn concentrations in Gaoxiong was contributed by the airflow of the mainland. Estimated by the same way, about 35% of outdoor ^{222}Rn concentrations in Japan derived from that in mainland of China in winter. The variation in extent of influence between Taiwan and Japan was due to the different distances of airflow movement.

3.1.3 Height variation

The height of typical dwelling in mainland of China ranged from 1 to 6 stories. In order to find out the variation of outdoor radon concentration with height in living environment, monitors were set at the balconies of 1 to 6 stories in Fuzhou and Xi'an. There were no prominent variations in the annual means of ^{222}Rn concentrations in each story from the height of 1 to 30 m in Xi'an. In Fuzhou, there were no prominent change in different stories either, except for the height of 1 m, where it was little higher.

3.2 Indoor ^{222}Rn concentrations

There are several factors which may affect indoor ^{222}Rn concentration. Besides the factors of natural environment, there are the factors of living environment. Even in the same city and with the same nationalities, the dwellers born in various places have quite different living habits in China. This is quite different from the unique nationality of Japan. These differences could make the survey of indoor ^{222}Rn concentration much more complicated. The buildings selected in this survey were mainly typical bedrooms and offices in mainland of China, so the results may basically reflect the level of indoor ^{222}Rn concentrations in the mainland.

The averages and seasonal variation of indoor ^{222}Rn concentrations are shown in Table1 and Fig.3. The highest (33.9 Bq·m^{-3}) was in Guiyang, and the lowest (10.8 Bq·m^{-3}) in Hohhot. The annual mean of the 10 cities was 18.6 Bq·m^{-3}. The seasonal pattern of indoor ^{222}Rn concentrations was the same as that outdoors, i.e., minimum in summer and maximum in winter. As seen the right column of Table.1, the ratios between outdoor and indoor ^{222}Rn concentrations in the mainland range from 1.1 to 4.2 with an average of 3.7, really the same as those of Gaoxiong.

3.3 Estimation of population dose

We used the estimating model recommended in UNSCEAR Report (1988), which involves equilibrium factors of 0.8 and 0.4, occupancy factors of 0.2 and 0.8 for outdoors and indoors,respectively and dose conversion factor of 1×10^{-5}[mSv·h^{-1}(Bq·m^{-3})$^{-1}$] for indoors and outdoors. The final estimated results are shown in Table4. The average annual population effective dose equivalent was 0.63 mSv·a^{-1}. Among the 10 cities, the highest value(1.12 mSv·a^{-1}) was in Guiyang and the lowest (0.39 mSv·a^{-1}) in Nantong. The cause has been explained as that the annual average outdoor ^{222}Rn concentrations in mainland was higher than that worldwide, whereas the population dose from radon and its daughter products was lower than that worldwide as shown in Table3. This is mainly because of the high ventilation rate in Chinese buildings and living habits, which caused the indoor average annual population effective dose equivalent (0.51 mSv·a^{-1}) far less than that worldwide (1.0 mSv·a^{-1}).

4. CONCLUTION

The survey of outdoor and indoor ^{222}Rn concentrations at 10 cities of China was carried out cooperatively by China and Japan, lasting from November 1988 to January 1990 and from February 1991 to March 1993. The annual mean of outdoor ^{222}Rn concentrations in the 10 cities was 8.8 Bq·m^{-3}. The highest (13.5 Bq·m^{-3}) was in Wuhang, and the lowest (4.6 Bq·m^{-3}) in Nantong. Seasonal pattern of outdoor ^{222}Rn concentrations in China was similar to those in the world. The concentration was maximum in winter and minimum in summer. In winter about 50% of outdoor ^{222}Rn concentrations in Taiwan was contributed by the airflow in the mainland. Outdoor ^{222}Rn concentrations did not change prominently in the height from 1 m to 30 m in typical living environment of China.

The annual mean of indoor ^{222}Rn concentrations in the 10 cities was 18.6 Bq·m-3. The highest (33.9 Bq·m^{-3}) was in Guiyang, and the lowest (10.8 Bq·m^{-3}) in Hohhot. The seasonal pattern of indoor ^{222}Rn concentrations was the same as that outdoors, i.e., that was minimum in summer and maximum in winter. The ratios between indoor and outdoor ^{222}Rn concentrations in the mainland range from 1.1 to 4.2 with an average of 3.7.

The annual mean of population effective dose equivalent resulting from radon and its daughters was 0.63 mSv·a^{-1}. The highest was in Guiyang and the lowest in Nantong, 1.12 and 0.39 mSv·a^{-1}, respectively.

The varying ^{222}Rn concentrations in the mainland can be analyzed by the factors of geological conditions, geographic sites, meteorological conditions, etc. To quantitatively analyze the radon concentration in more detail, the study of radon exhalation of soil and continuous monitoring of radon concentration should be conducted in future.

REFERENCES

1. Iida T., Ikebe Y., Hattori T., et al. Health Phys., 54 (1988)139.
2. Michiko Abe and Siro Abe, Proceeding of the 15th NIRS Seminar,79, 1987.
3. Ikebe Y., Kojima S., and Shimo M, Res. Lett. Atomosph. Electr., 51 (1983) 3.
4. Lin Y.M. and Chen C.J. , Hoken Butsuri, 385 (1985) 20.
5. UNSCEAR Report, 487, 1988.

III Natural radionuclides

Indian research on high levels of natural radiation: pertinent observations for further studies

P.C. Kesavan

Biosciences Group, Bhabha Atomic Research Centre, Trombay, Bombay - 400 085, India

1. INTRODUCTION

The High Natural Background Radiation Areas (HBRAS) of the Thekkumbhagam- Purakkade Coastal Strip (T-P strip) in the Kerala State and Manavalakuruchi of Tamil Nadu of India are of specific interest as the inhabitants in these regions receive the highest collective dose. Over 140,000 inhabitants receive an average dose of 15 to 25 mGy per year in Kerala; about 2000 inhabitants of Chinnavilai village in the Manavalakuruchi of Tamil Nadu receive radiation doses ranging from 20 to 40 mGy per year. While some records put the human settlement in the strip since the 7th century, one report (1) appears to be rather conservative that most of the present inhabitants, Arayas, have been there just for over 30 generations.

The studies so far undertaken have been designed to look for harmful genetic effects in plants, animals and human beings in accordance with the current "radiation paradigm". Thus it is expected that chronic radiation doses would enhance "genetic load". The present "radiation paradigm" that very low and high doses produce the same type (i.e. not the opposite type) of biological effects provides the foundation for the "linear, no-threshold" relationship between the effects and the dose. In short, the studies so far carried out could be useful only to verify the tenability of the "linear, no-threshold" hypothesis with reference to a given effect and not for detecting adaptive advantages that the organisms including the human inhabitants of HBRAS might have "gained" from low chronic exposures to high level background radiations throughout their lifespan. It is therefore, expected that once the laboratory-based experimental studies establish the consistent occurrence and the scientific basis of "radiation hormesis" the epidemiological studies would design innovative approaches to spefiecy the "positive gains" in the health status and longevity of the HBRA inhabitants. This is a likely possibility.

There is also an urgent need to draw a distinction between the low LET radiobiological effects of chronic low dose, low LET external radiation exposures and the internal radiation due to inhaled/ingested radionuclides. These have obvious implications to our achieving a better perspective for peaceful applieations of nuclear energy agriculture. The point is that there is a qualitative difference between external exposure to a for industry, medicine and dose of gamma radiation, and ingesting cobalt-60 in equivalent amount to obtain the same dose. The present review briefly deals with certain aspects which need to be highlighted from the Indian studies on rats, plants and humans in the HBRAS. These are as follows:

2. GENETIC STUDIES ON RATS (RATTUS RATTUS) IN THE T-P STRIP

The black rats (*Rattus rattus*) then representing the descendants of the 800th to 1000th generation in the HBRAs were chosen for detailed investigations of six dental and fifteen skeletal measurements [2]. The metrical variables are usually approximately normally distributed around a mean value which is characteristic for a given population. Populations may differ both in their means and in their variances, that is, the degree of scatter around the mean. If the genetic variance is increased as a result of a higher mutation rate, a higher proportion of individuals with more extreme deviations from the mean value and correspondingly fewer near the middle of the distribution would be expected to result. The rat populations in the HBRAs received about 7.5 times greater radiation dose than those in the control areas. The author concluded, "whereas our findings thus give no positive indication of genetic damage to the rats living on the strip, they do not rule out the possibility of induced mutations lurking beyond the reach of our method". In this regard , Rose [1] points out that based on extensive work with laboratory mice [3], the minor skeletal variants studied in Kerala rats are overwhlemingly influenced by nongenetic factors in their manifestation. In nutshell, an explanation for the lack of positive evidence for possible enhancement of mutation frequencies by low chronic exposures is the choice of inappropriate parameter. This argument is not supported by experimental mutagenesis in crop improvement. Polygenic mutations with epigenetic interactions are more readily induced by ionizing radiations. It is believed that the statistical methods employed by Prof. H. Gruneberg are good enough to detect the polygenic mutations if these are really induced by chronic low level exposures.

3. EPIDEMIOLOGICAL STUDIES IN THE HUMAN INHABITANTS

During its first phase of epidemiological studies in the I 970s, BARC could **not** find any significant difference between the groups exposed to different radiation levels. The indicators of genetic change employed by BARC were: sex ratio among offspring; fertility index; infant mortality; pregnancy terminations; multiple births; gross abnormalities [4].

4. BARC STUDIES ON THE INFLUENCE OF INTERNAL AND EXTERNAL RADIATION ON CYTOGENETIC ABNORMALITIES IN PLANTS GROWING IN THE NORMAL AND HBRAS OF KERALA

Considering that cytogenetic damage would be a much greater sensitive parameter, Dr. A.R. Gopal-Ayengar and his associates at the BARC, Bombay started a series of studies initially with plant species growing in the normal and high background radiation areas. Some of these regions have had background radiation levels upto 250 times greater than the normal background levels. That is, that a radiation level of 5.0 mR/hr is compared with normal background radiation levels of 0.02 mR/hr [5]. There were a few morphologically aberrant plants and accumulation of radionuclides of thorium series in plants has been reported [6]. However, the most significant aspect of their studies is regarding relationships between external radiation level, internal radionuclide content and the cytological effects. In the HBRAS, alpha emitting radionuclides account for about 60 per cent of the total internal dose received by the plants. The contribution from gamma emitters is nearly 40 per cent, while that of beta emitters is only about 0.01 percent. A significant linear relationships were observed between total (internal + external) dose and the observed cytological damage. It was also noted that the slope value for the regression of internal dose on cytological damage is higher by an order of magnitude than the corresponding value for the regression of external dose on cytological damage. These findings are in agreement with another study [7] on *Tradescantia*, clone 02, plants grown in monazite-containing sand cultures which revealed that the contribution of absorbed radionuclides to biological damage as measured in terms of the frequency of somatic mutation rates of staminal hairs, is significantly greater than that of external radiation exposure.

A significant observation was that there was no difference in the frequency of somatic mutations in the stamens following exposures at dose-rates of 0.03 mR/hr (control), 0.08 mR/hr, 0.16 mR/hr, 0.32 mR/hr, 0.65 mR/hr and I .30 mRJhr for 280 days. The total exposure at the end of 280 days was 0.201R (0.201 cGy) for the control and 8.736 R (8.736 cGy) for the maximally exposed groups maximally exposed groups were 2.77 ± 0.706 and 3.63 i 0.819 respectively. It is therefore .The somatic mutation frequencies for the control and the evident that a "linear, no-threshold" relationship between the frequency of somatic mutations in the stamen hairs of Tradescantia and chronically accumulated total dose of 8.74 cGy does not exist. Mutations in the stamen hairs form a sensitive system. There was however, a demonstrable mutagenic activity associated with the content of alpha-emitting radionuclides in the plant tissue.

The fact that per capita daily intake of gross alpha, gross beta, 85 keV gamma, radium-228 and potassium-40, activity are estimated to be 215, 3648, 96, 162 and 3551 pCi, respectively [8] and that urinary excretion of thorium and Ra-228 in male adults residing at Chinnavilai village (HBRA) are atleast an order of magnitude higher than those collected from subjects residing at normal background radiation level areas [9] emphasizes the urgent need to assess the body burden of the radionuclides in the individuals of the HBRA inhabitants who have significantly greater incidence of dicentrics and rings than the normal frequency expected in relation to their age.

5. INDIAN STUDIES IN CYTOGENETICS OF HBRA POPULATION

At the outset, it is pointed out that Verma et al. [10] reported a substantially increased frequency of chromosomal aberrations in the HBRA populations as compared to that in the normal (control) populations. The most common types of chromosomal aberrations observed by them were deletions and acentric fragments. They had however, noted that the frequency of dicentrics and rings was considerably lower. These authors from the All India Institute of Medical Sciences, New Delhi had also published [11] a paper claiming a higher incidence of Down syndrome births in the HBRA of Kerala; however this had met with a great deal of criticisrn including the total absence of any Down syndrome case in the control population.Today, their own data base for Down syndrome collected from various parts of India (Normal background levels of radiation) render their earlier report in Nature of greater incidence of Down syndrome in the HBRA population invalid. The point is that the reported incidence of Down syndrome in the HBRA population is just not different from its incidence elsewhere in India.

So far as the human cytogenetic data of BARC is concerned, the studies at this point (meaning that these studies are still going on) suggest the following:

(i) With reference to the unstable aberrations (dicentrics and rings), an increased frequency becomes somewhat discernible only in the adult populations. However, the data are quite scattered and no pecise conclusions could be drawn at this juncture. Therefore, the studies are still in progress to clarify the situation.

(ii) In the newborns, there is no significant difference in the incidence of chromosomal aberrations (dicentrics + rings) and micronuclei in peripheral lymphocytes between the control and HBRA populations (Table 1). While the BARC data on the adult populations do not yet allow an unequivocal conclusion that dicentrics + rings are significantly greater among the inhabitants of the HBRAS, the Chinese data [12] are significant ($P < 0.05$) in this regard. If further studies establish that the inhabitants above a certain age in the HBRAs have indeed significantly higher frequencies of dicentrics and rings and/or translocations, then the role of body burden of radionuclides would have to be considered. It has been reported [8] that the mean daily intake of Ra-228 by the population living in the monazite area of Kerala is about 162 pCi which is about 40 to 50 times the normal daily intake of this radionuclide.

Table 1
Cytogenetic studies on the newborns in the high background radiation areas of Kerala coast

Area	Chromosomal changes*		Micronuclei**
	Dicentrics	Fragments	
	per 1000 cells		
NBR	0.194	0.255	1.36
	(6283)	(6283)	(90)
HBR	0.179	0.230	1.14
	(3444)	(3444)	(90)

NBR - normal background and HBR - high background radiation areas
Figures within parenthesis denote the number of newborns; dicentrics include rings; fragments are largely acentric and include minutes.
Unpublished data from: *V.D. Cherian, C.J. Kurian, Birajalaxmi Das, C.V.Karrupasamy, E.N. Ramachandran, M.V.Thampi, K.P.George and P.S.Chauhan.
** Birajalaxmi Das, C.V.Karrupasamy, M..Thampi and P.S.Chauhan

The observation that the plants which accumulate large amounts of radionuclides have considerable meiotic abnormalities, but the seeds, which they set give rise to almost normal plants suggests that certain mechanisms to sieve out genetically imbalanced gametes/zygotes must be operating. In this regard, the occurrence of apoptosis in plant systems (13,14) is of relevance for further studies in the plant communities ofHBRA. Similarly, the significance of apoptosis as a part of biological defense mechanism of mammalian systems is being increasingly appreciated.

6. CONCLUSIONS

1. The HBRA studies so far carried out, within the scope oftheir resolving power, have not revealed any adverse biological effects. These include the search for dental and skeletal mutations in rats, and cytogenetic studies ofthe newborns and adult populations.
2. Cytogenetic studies with plants which accumulate considerable quantities of radionuclides and which do not have such mobility and excretory processes as animals implicate internal radiation to be largely responsible for meiotic

abnormalities. In fact, external radiation doses upto 8.7 cGy do not induce somatic mutations in the sensitive stamen hair system of *Tradescantia* .

3 . More recently, the data base for Down syndrome for different parts of India has become more reliable and acceptable. These are based on about 90,000 births during the past three years. It turns out that the incidence of Down syndrome in the HBRAs of Kerala earlier reported by N. Kochupillai et al. [11] is not different from that elsewhere.

4. The lack of discernible support from the HBRA studies for the "linear, no-threshold" relationship between dose and biological effects has important implications for further studies in basic radiobiology and cellular defense mechanisms. In particular, the role of apoptosis needs to be studied. The other basic question is regarding the "paradigm shift" and the "linear, no-threshold" relationship between biological effects and low level chronic exposures to ionizing radiations.

5 . There have been a few suggestions that we should investigate for mutations in the minisatellite regions in the normal and HBRA populations as has been done by Kodeira et al. [15] for the gametes of the 'control' and of the parents exposed to A-bombs. These studies also failed to show that exposure to A-bomb radiation enhanced the frequency of mutations in the germ cells. However, at present the phenotypic impact of mutation at the minisatellite regions are unknown and are probably none.

REFERENCES

1 . K.S.B. Rose, Nucl. Energy, 21 (1982) 395.
2. H.Gruneberg, Nature, 204 (1 964) 222.
3. U. Ehling, Genetics, 54 (1966) 1381.
5. A.R.Gopal-Ayengar, G.G.Nayar, K.P.George and K.B. Mistry, Indian J. Exptl. Biol., 8 (1970) 3 13.
6. K.B. Mistry, A.R.Gopal-Ayengar and K.G. Bharathan, Health Physics, I I (1965) 1457.
7. G.G. Nayar, K.P.George and A.R. Gopal-Ayengar, Radiation Botany, 10 (1970) 287.
8. K.B. Mistry, K.G. Bharathan and A.R.Gopal-Ayengar, Health Physics, 19 (1970) 535.
9. A.C Paul, T.Velayudhan, P.N.B. Pillai, C.C. Maniyan and K.C. Pillai, Environ. Radioactivity, 22 (1994) 243.
10. I.C. Verma, N. Kochupillai, M.S. Grewal, G.R. Mallick and V. Ramalingaswami, in : International Symposium on Areas of High Natural Radioactivity, p 185, Pecos de Caldas, Brazil, 1975.
1 I . N. Kochupillai N, I.C. Verma, M.S.Grewal and V. Ramalingaswami, Nature, 262 (1976) 60.
12. DeQing Chen and Wei Luxin, J. Radiat. Res., supplement 2 (1991) 46.

13. J.T. Greenberg, A.Guo, D.F. Klessig and F.M. Ausubel, Cell, 77 (1994) 551.
14. R. Mittler and E. Lam, Plant Physiology, 108 (1995) 489.
15. M. Kodeira, C. Satoh, K. Hiyama, K. Toyama, Am. J. Hum. Genet., 57 (1995) 1275.

Brazilian research in areas of high natural radioactivity

A. R. Oliveira[a], C. E. V. de Almeida[a], H. E. da Silva[b], J. H. Javaroni, A. M. Castanho, M. J. Coelho and R. N. Alves[c]

[a]Radiation Medicine Division, Brazilian Nuclear Industries, Rua Mena Barreast Brazil

[b]Laboratory of Radiological Sciences, Rio de Janeiro State University, Rio de Janeiro, Brazil

[c]Department of Nuclear Engineering, Military Institute of Engineering, Pça. General Tibúrcio 80, Rio de Janeiro, 22290-270 Brazil

Studies in areas of high natural radioactivity in Brazil have been conducted since the late 1950's. These areas have essentially two features: occurrence of monazite sands, mainly located along the Atlantic Coast (Guarapari, Meaipe) and zones of volcanic intrusives, in the interior of the state of Minas Gerais (Poços de Caldas, Tapira and Araxá). These studies aimed basically at assessing the level of external radiation resulting from exposure to ground and building materials and evaluating the intake of selected radioactive elements (Ra^{228} and its daughters) derived from consumption of food produced in these regions. Recently, attention has been paid to an area located in the Amazon Rainforest (Pitinga, Amazon state) where a village was settled close to a tin mine whose ore is rich in radioactive materials. The houses in the village were built with material extracted from the ore. Approximately 2000 individuals have been submitted periodically to medical evaluation and laboratory investigations in order to evaluate the potential hazardous effects of low doses of ionizing radiation. Occupational and environmental research work are still in progress. This paper reviews briefly the areas of high natural radioactivity in Brazil and outlines the research work performed in Pitinga. Preliminary results are discussed and compared with findings provided by former studies.

1. INTRODUCTION

One of the most efficient tools to estimate the risks to which individuals exposed to low radiation dose are subject is by means of the follow up of populations residing in areas of high natural radioactivity. The concern demonstrated by public health authorities, nuclear regulatory agencies and environmental organizations on this subject has risen over the last years, particularly due to the fact that risk estimates based on extrapolations from exposure to high radiation doses (atomic bomb survivors and irradiated patients for cancer treatment) offer no accurate estimation of dose-effect relationship for exposure to low radiation doses, taking into account the uncertainties due to typical linear, infra and supra-linear extrapolation models.

The natural occurrence of radionuclides in nature is almost negligible. Under different circumstances, radionuclides belonging to the uranium and thorium series may "contaminate" the environment, representing a potential concern from a public health standpoint. The study of populations living close to or in areas of high natural radiation, i. e. with appreciable concentrations of $K40$, $U238$ and $Th232$, thus becomes indispensable. Over the past forty years, studies have been - and are still being - carried out in areas of high natural radiation. Four of these areas have been extensively reported by many authors: the coastal regions of Southeast Brazil (Guarapari, Meaipe); inland in the state of Minas Gerais (Poços de Caldas, Araxá and Tapira) [1]; the coastal regions of Southwest India (Kerala State) [2]; parts of Dong-anling and Tongyon, in the Guandong province, set on the South China sea, in China [3], and finally, Ramsar in Iran, on the Caspian sea.

2. RESEARCH IN AREAS OF HIGH NATURAL RADIOACTIVITY IN BRAZIL

Studies have basically focused on two regions of high natural radioactivity in Brazil: Monazite sand regions running along the Atlantic coastline, and the volcanic intrusions located in the Minas Gerais state. In the Guarapari and Meaipe regions, the occurrence of heavy minerals (monazite, ilmenite, zircon, and rutile) is due to heavy weathering and the decomposition of the massifs located parallel to the coastline. After reduction in size, and carried away as alluvia by the various rivers running through the region to the ocean, they were stratified and deposited in river-beds and sea-floors. The result of this phenomenon is the formation of long strips of deposits of radioactive materials, intercalated with inactive layers of ordinary sand [4]. These deposits have been mined over the past forty five years. This entire coastline region, which runs some 500 miles length has attracted heavy tourism, that has prompted deep-rooted alterations in the radiological characteristics of the region.

Poços de Caldas is an outstanding example of tectonic process, where an alkaline plug measuring over 1,000 sq. km (30 x 35 km) was raised to a height of some 400-500 meters above the adjacent granite. Heavy weathering and erosion lowered the central part of this area, leaving in its edge the harder rock intact, which conferred the area the appearance of a huge volcanic crater, although no true explosion ever took place here. Uranium and thorium are found separately in this alkaline province, uranium is associated with zirconium, and thorium with iron. As the outcome of this geological process, some seventy radioactive anomalies were formed, constituting three or four areas where either uranium or thorium predominate [5].

3. DOSIMETRIC STUDIES

Some twenty measurements were taken in Guarapari, whose results varied from 1,3 to 12 mSv/year. In Meaipe, the sister-town of Guarapari, doses ranged from 2 to 42 mSv/year. As mentioned above, changes due to progress and the transformation of these towns into beach resorts, with paved streets, new highways and intense urbanization has brought about sweeping changes in their radiological characteristics. The external doses are due principally to the obvious and typical presence of monazite sands in the region, which are characterized by a high Th232/U238 ratio [4].

In Poços de Caldas, a larger number of points were monitored [4]. Sixty sites measured showed doses varying from 1 to 10 mSv/year. In addition to measurements taken in areas belonging to the uranium mine there, which showed a major contribution of U238 and its series in the total annual dose rate, measurements taken some distance away from the mine showed a contribution of thorium, due to its high presence in the soil or in association with uranium. In general, doses in the mineralized region of Poços de Caldas are far higher than in other parts of Brazil.

Another important study analyzed the ingestion of radioactive materials contained in food. These foods (beans, maize, vegetables and fruits) were consumed regularly by the inhabitants of regions with a high concentration of radioactive materials. In the Tapira region, they represented the main, if not the only source of subsistence. Preliminary studies carried out in 1964 ([6] showed high concentrations of Ra isotopes in some foods produced locally. Additional studies confirmed that local communities featured a high intake of Ra isotopes. Concentrated in the Araxá - Tapira and Barreiro regions, these studies continued under way for three years, involving some thirty families. The highest concentrations of Ra226/228 were found in roots and tubers, as well as in manioc flour, and to a lesser extent in fruit and vegetables. In other foods, radioactive content was negligible. These figures were equivalent to around 192 and 16 times the standard figures noted in the USA and UK respectively. The problem of internal contamination due to consumption of foods by the local population was initially over-estimated, since from a population of some 2,000 people, only a small fraction featured any ingestion of radioactive materials higher than five times the normal average.

4. STUDIES CARRIED OUT IN A NEW AREA OF HIGH NATURAL RADIATION IN BRAZIL

In 1983 a tin deposit was discovered in the Amazon region, close to the border of Amazon state with the Roraima state. In view of its specific characteristics - remote location, size and strategic importance - it was decided to build townships, totaling homes to house approximately 5000 inhabitants. Construction of these townships, as well as the access roads made use of materials coming from the mineralized regions, which was characterized by an association of tin with uranium and thorium. Field measurements began in 1990, involving both indoor and outdoor monitoring, in order to establish the radiation dose to which people living in the three townships (A, B and C - southeast, northeast and center) were subject. Other studies analyzed the daily habits of the inhabitants, including their food intake, and the average amount of time spent at home.

The initial surveys lead to the conclusion that, from the start-up of construction of the townships through to 1992, there were sweeping alterations in the radiological characteristics of the area. Building materials came from areas located some distance away from the mine, and only later was building material from the mineralized areas used. In township A the external measurements found figures of around 0.25 µSv/h. In the B and C townships, southeast, local measurements reached some 0.30 µSv/h. Township C, in the northeast, posted the highest outdoor exposure rates, due to the widespread dissemination of radioactive materials, in view of its construction and urbanization. In C township, Center, the average figure was around 0.30 µSv/h. Indoor results reflected the contribution of radiation coming from the walls and floors of the houses. In a simplistic manner, in order to calculate the annual doses due to indoor exposure, it was calculated the product of the indoor measurements by occupation factors and total annual hours of exposure, which was expressed in µSv/h.

The figures found for these total doses in individuals among the population of Pitinga were uniform, in relation to townships A, B and C. Although township A presented a lower background and occupancy rate for its dwellers, the indoor exposure rate was higher.

The distribution of doses in Pitinga does not feature the same pattern as that noted in Guarapari, corresponding only to a sub-set of the exposure levels found in this beach resort [4]. These rates are situated around the same level as those for Poços de Caldas, but are lower than those of Meaipe and the Araxá-Tapira region, which are characterized by spots of high natural radioactivity.

Indoor exposure is associated with the presence of radioactive materials in the walls and floors of homes, and is due to gamma radiation of the thorium and uranium series. With regard to corrective actions undertook to minimize exposure in certain homes, the same criteria were adopted as those used by other U.S. investigators for the surveys carried out at Grand Junction, Colorado [7], meaning that there was no need to structurally modify these homes. Although the doses inside the houses are not high, it was recommended that certain steps be taken to minimize the radiation exposure to residents, such as ensuring better air circulation by keeping windows open, encouraging people to carry out tasks outside the house whenever possible, urbanization of the township to reduce the level of radioactive airborne dust, and control of the radionuclides found in the water and locally-produced biomaterials.

5. MEDICAL AND EPIDEMIOLOGICAL CONSIDERATIONS

Neither the Guarapari Project nor the campaigns performed by Brazilian researchers to Poços de Caldas and the Araxá-Tapira-Barreiro region were much concerned with medical or epidemiological surveys. They were conceived basically to assess the effects of exposure to low radiation doses. The main objectives of these studies were focused on dosimetric estimation, both indoor and outdoor, as well as the to survey the sites studied from geological point of view. Cytogenetic analysis using the modified Moorhead technique based on 72-hour lymphocyte cultures were also performed and documented.

Although the main objective of the studies in Pitinga was to estimate the external dose received by workers and local community, it was considered equally important to assess, at a certain moment, the potential impact of this exposure on the health of local residents, as well as to follow up the inhabitants medically, in order to detect any changes that could be attributed to the effects of ionizing radiation. It became quite clear that the performance of epidemiological studies would constitute a formidable venture, with very few chances of success. The population density in the Amazon region is extremely low. Pitinga in fact only came into existence due to the discovery of a commercially valuable mineral deposit, when it was decided to launch not only the mining and milling activities, but also to develop the townships to house workers and their families. During its most productive or active phase, Pitinga housed some 3,500 people from all over Brazil. Its largely unskilled labor-intensive operations were mostly handled by workers from poorer parts of Brazil where jobs were in short supply, particularly in the North and Northeast region, characterized by its low social, economic and cultural levels. The technical and administrative, as well as the management staff came from almost all over the country, more particularly the South and Southeast states, which are characterized by striking social, economic, "ethnic" and geographical differences, as well as variations in customs, compared to the origin of the operation workers.

As we did not take into account the small size of the population, the difficulty of selecting an adequate control group, the limited stability of the population in the region, this fact alone, in our view, would already hamper any attempt to carry out comparative population studies in a prospective manner. Instead, we opted for a program that could offer an instant overview of the medical situation at both the individual and group levels, through a concentrated effort to examine a few thousand people during the first months, followed by regular medical check-ups of this same population, when it would be possible to obtain data and information on health, with special attention to social habits, previous occupational history (exposure to clastogenic agents, for example), eating habits, incidence of degenerative diseases in the family, particularly cancer, health of offspring, pregnancies and the occurrence of congenital abnormalities. This task was not difficult, as all staff are examined regularly once a year, while family members enjoy free medical aid offered by the only hospital in the area, which is supported by the venture. Despite an understanding of the drawbacks of this type of medical, occupational and social survey, it was concluded that there was no other way of rapidly and objectively assessing any possible effects linked to exposure to ionizing radiation. It is important to stress that the workers were exposed to other occupational hazards which could cause "noise" regarding interpretation of the data, and that the doses to which population was exposed were so low that any exceptionally altered number of cancers would be in single figures, if so, according to the risk predicted by the international scientific community [8], which cases would probably never be detected. Despite alarming lurid news stories on the occurrence of bizarre diseases in newborns in the region, this has never been confirmed clinically. A good information program and correct diagnosis of each case eased tensions and calmed the population. Special attention was given to the occurrence of any type of malignancies. Pediatricians were requested to notify the occurrence of any severe congenital malformations, particularly Down's Syndrome. No cytogenetic studies were carried out to detect chromosome aberrations.

The medical studies were launched in September 1990. Interviews with the hospital manager, medical staff, particularly pediatricians and obstetricians, as well as X-ray technicians and laboratory staff were carried out. A detailed review of the health indicators and vital statistics was carried out, with nothing special being noted during the five years prior to the start of the work. During the ten years under analysis (1985-1994) of which the first five were based on medical records, no indication was noted of any changes in the nosocomial profile in the Pitinga region, nor any alteration in cancer incidence, whose occurrence in the region is extremely low, and practically limited to a few cases of cervix and breast tumors. The vast majority of the population consists of workers and their families (wife and young children) as well as young people, who have no more than a high school education. Mention should be made of the appearance of two cases of congenital tumors during that period: one retinoblastoma and one Wilms' tumor. Nor was any exceptionally large number of cases of Down's Syndrome noted, normally associated with pregnancy in elder women, finding which is compatible with the relatively young age of pregnant women in the region. Trivial congenital deformities were detected, within the expected frequency.

Lectures offered to all health care professionals, senior staff and managers were devoted to explain the efforts of the multidisciplinary team to carry out the radiometric, biological, medical and occupational studies. Visits were made to the mines and the facilities. General advice was given to workers on how to avoid unnecessary exposure to chemical and physical agents in the work-place, as well as on avoiding eating, drinking or smoking in industrial and mining areas. A brief questionnaire was drawn up for completion by the physicians or social workers trained to do so, when each person was examined, covering identification, social and job history, pathological antecedents, occurrence of degenerative diseases, and, for women, history of pregnancies and health conditions of offspring. Drills were carried out to train the staff responsible for completing the questionnaire.

CONCLUDING REMARKS

As in any study carried out in an area of high natural radioactivity, drawing conclusions regarding the possible effects of low radiation doses or dose rates is a difficult and complex task that is often surrounded by traps and dead-ends. The studies launched in China in 1972 already offer a consistent amount of guidance. The first conclusions drawn from the comparison between the incidence of cancer in the high background radiation areas (Dong-anling and Tongyon) and in the control areas (Taishan and Enping) clearly and unmistakably indicate that there was no increase in the cancer incidence, except for leukemias in the areas of high natural radiation. The number of people included in these surveys is in the hundreds of thousands range, of which a sample of 1,000 individuals was taken in each region, which is large enough to endow the study with statistical validity. The first conclusions are encouraging, and should strengthen the opinion of some researchers that doses as low as 5 - 10 mSv/year are not able to increase cancer incidence in the population exposed thereto. Naturally, these surveys would be consistent to challenge the linear model of dose-effect relationship for cancer causation without threshold for exposure to low doses of radiation.

Also praiseworthy are the efforts of researchers carrying out a demographic surveys of the entire region where monazite occurs in the state of Kerala, Southern India, which runs 55 km along the coast between Thekkumbhagun and Purakkadu. They are also assessing external irradiation and the intake of food containing high levels of radioactive materials, in addition to surveys designed to profile the health conditions and habits of the studied population. Despite the huge size of this task and the challenge faced by the researchers, it is expected that the first results could contribute for a better understanding of the effects of low radiation doses.

Brazilian research in areas of high natural radioactivity has much to learn in the natural and unexpected laboratory of Pitinga. The programs to be coped with are different from those performed in China, India, Iran and other countries with areas of high natural radiation, particularly taking into account the population universes involved in each study. In Pitinga, the population is small, and represented mainly by medium-term inhabitants, subject to fluctuations in number in the course of the process, with social and economic characteristics that do not allow it to be characterized as uniform and stable. In counterpart, it is perfectly characterized and controlled medically, in addition to offering a good idea of the dose to which it is exposed. It is very early to state that any failure to note changes in the cancer incidence among the population or congenital malformations is coherent with the findings observed in China, for instance. However, taking into account the dose rate in the region of some 3 mSv/year, which is lower than that in most of the areas currently under study, it is not very probable that there will be any increase in the incidence of cancer that could be attributed to ionizing radiation.

The sanitary problem derived to the presence of Th and U in the building materials used in the houses, which even boosted indoor exposure due to radon was resolved in a fast and creative way by transforming these homes into "sanctuaries" dedicated to environmental and radiation protection activities. In homes that still have some type of indoor exposure, residents are recommended to keep their windows opened as much as possible, in order to ensure good ventilation and removal of the radon, as well as reducing the amount of time spent in the most heavily affected rooms. The problem caused by the presence of a large quantity of airborne radioactive particles in residential areas was minimized by urbanization and road-surfacing, as well as spraying water along the access roads leading to the mines and facilities.

Medical supervision of the entire population is handled on a regular basis, following the schedule implemented by the entrepreneur. The medical and dose records have expanded, incorporating new data and information brought in through the various campaigns. Despite its remoteness from the more developed parts of Brazil, the Pitinga Mining, Industrial and Residential Complex will continue to exist at least over the next few decades, in view of the potential of its mineral reserves. We believe that this natural laboratory may join the list of areas of high natural radioactivity known worldwide, such as Guarapari and Poços de Caldas. Medical findings in this case will certainly contribute to better knowledge and understanding of the effects of low doses of radiation.

REFERENCES

1. E. Penna Franca, J.C. Almeida and J. Becker. Health Physics No 11 (1965) 699.
2. A. R. Gopal-Ayenger et al. Proceedings of the 4th Internal Conference on the Peaceful Uses of Atomic Energy (1972) 31.

3. High Background Radiation Research Group. Science No 209 (1980) 877.
4. T.L. Cullen. Academia Brasileira de Ciçncias (eds). International Symposium on Areas of Natural Radioactivity, Rio de Janeiro, Brazil 1977.
5. F. X. Roser and T. L. Cullen. J. A. S. Adams and W. M. Lowdes (eds), University of Chicago Press, 1964.
6. M. Eisenbud, H. Petrow, R. T. Drew, F. X. Roser, G. Kegel and T. L. Cullen. J. A. S. Adams and W. M. Lowdes (eds). University of Chicago Press, 1964.
7. M. V. J. Culot, K. J. Schiager and H. G. Olso. Health Physics No 30, 1976.
8. United Nationseto 161, Rio de Janeiro 22271-100, Brazil Scientific Committee on the Effects of Atomic Radiation -UNSCEAR, Sources and Effects of Ionizing Radiation, 1993.

Natural radioactivity of soil samples in some high level natural radiation areas of Iran

M.Sohrabi, M. Bolourchi, M. M. Beitollahi, and J. Amidi

National Radiation Protection Department
Atomic Energy Organization of Iran
P.O. Box 14155-4494, Tehran
Islamic Republic of Iran

1. INTRODUCTION

Man has always been exposed to external and internal radiations from primordial radionuclides, principally from the uranium (^{235}U, ^{238}U) and thorium (^{232}Th) series and potassium-40 (^{40}K) present in the earth's crust. Soil, as the cradle of human life, is a major component of building materials such as bricks, cement, concrete, etc., as well as being the nutrient medium for plants and vegetables and the carrier of surface water. So, it is the main environmental component of human exposure, the radioactivity analysis of which is of prime importance for internal and external dose assessment [1].

To make a natural radioactivity analysis of soil in Iran, measurements of the ^{226}Ra, ^{232}Th and ^{40}K contents of soil samples, taken from different parts of the country in particular from the high level natural radiation areas (HLNARs) of Ramsar (a northern coastal city) and Mahallat (a village located at the central part of Iran), were carried out by gamma spectrometry. Based on the radioactivity levels determined, the gamma absorbed dose rates in air at one meter above the ground were calculated using the procedure applied by Yu and co-workers [2]. The project is being continued in parallel with other activities on environmental radiation measurements such as ^{226}Ra measurements in water [3], determination of public exposure from different man-made and natural sources [3,4], etc. Some preliminary results are presented and discussed in this paper.

2. MATERIALS AND METHODS

A total of 357 soil samples were taken from 85 locations in different provinces of Iran such as Tehran, Gilan, Semnan, etc. and in particular Ramsar and Mahallat regions. All the samples were air dried and sieved to grain sizes of less than 0.63 mm. Samples of about 300 grams were prepared and placed in cylindrical gas-tight containers with the same geometry as the sample containers used for efficiency calibration. They were kept for at least three weeks before their measurements to reach a radioactive equilibrium between the radium and its short-lived decay products [5].

The ^{226}Ra, ^{232}Th and ^{40}K concentrations in the samples were determined directly using a high purity germanium (HPGe) detector with an energy resolution of 2.0 keV at an energy of 1332 keV, and relative efficiency of 20% coupled to a CANBERRA MCA-Series 100. The counting time for each sample was 16 hours except for higher radioactivity samples for which shorter counting times were applied.

The radioactivity concentrations were determined by the 609 keV photopeak of ^{214}Bi for ^{226}Ra, the 583 and 911 keV photopeaks of ^{208}Tl and ^{228}Ac respectively for ^{232}Th and the 1460 keV for ^{40}K [5,6]. Based the levels of radionuclide contents of soil, the absorbed gamma dose rates were determined using the procedures applied in the literature [2].

3. RESULTS AND DISCUSSION

The radioactivity concentrations of ^{226}Ra, ^{232}Th and ^{40}K and calculated absorbed gamma dose rates in air at one meter above the ground are given in table 1. More than 40 locations have selected in Tehran since over 20% of the population of Iran belong to this city. Table 1 shows range of values of several measurements from different locations in each province.

Table 1
Natural radionuclide contents of soil samples and calculated gamma absorbed dose rate in air.

Province	No. of Samples	Specific Activity (Bq kg^{-1})			Dose Rate (nGy h^{-1})
		^{226}Ra	^{232}Th	^{40}K	
Ardebil	12	15-45	7-38	709-940	53-102
Gilan	17	14-33	9-23	555-770	44-68
Khoozestan	20	8-37	6-17	250-773	20-72
Markazi	17	15-33	9-30	528-840	42-73
Mazandaran	64	10-55	8-42	298-970	32-107
Semnan	19	20-46	7-25	634-897	46-88
Tehran	129	11-40	10-42	261-980	33-94

Considering table 1, the ^{226}Ra concentration ranges from 8 to 55 Bq kg^{-1}, ^{232}Th concentration from 6 to 42 Bq kg^{-1} and ^{40}K concentration from 250 to 980 Bq kg^{-1}. The concentration levels have a wide range of values due to the heterogeneous nature of the soil which is due to the parent rock composition and relevant processes such as erosion. The average calculated absorbed dose rates from terrestrial gamma rays in normal background areas is 64 ± 15 nGy h^{-1}. This is higher than the estimated world average [7], mainly due to the high concentration of ^{40}K in the soils of Iran.

In order to survey the above radionuclide levels in the soils in HLNRAs of two parts of Iran namely Ramsar and Abegarm-e-Mahallat, soil samples were taken from different locations the results of which are shown in table 2.

Table 2
Natural radionuclide contents in soil samples and calculated gamma absorbed dose rates in air from HLNRAs in Iran.

Locations	No. of Sample	Specific Activity (Bq kg^{-1})			Dose Rate (μGy h^{-1})
		^{226}Ra	^{232}Th	^{40}K	
RAMSAR					
Ab-e-Siah	4	(12.0-15.0)x10^3	17-30	763-945	6.2-7.8
Chapar-Sar	7	(4.3-12.6)x10^3	25-41	369-720	2.3-6.5
Dasht-e-Jalam	4	(0.1-0.2)x10^3	27-37	373-444	0.1-0.3
Katalum	5	(0.2-0.4)x10^3	30-39	333-483	0.1-0.2
Khak-e-Sefid	9	(23.9-26.5)x10^3	20-47	750-911	12.3-13.6
Ramak	4	(17.2-23.6)x10^3	15-35	637-829	8.8-12.1
Sadat-Mahalleh	4	(0.3-0.5)x10^3	29-40	605-819	0.2-0.4
Talesh-Mahalleh	11	(11.5-16.7)x10^3	23-45	426-758	5.9-8.6
Vazirgarma	5	(0.6-0.9)x10^3	19-38	300-531	0.3-0.5
MAHALLAT					
Donbeh	5	(4.3-6.1)x10^3	15-36	516-721	2.2-3.2
Hakim	4	(0.5-1.1)x10^3	21-39	449-599	0.3-0.6
Romatism	4	(1.3-2.0)x10^3	21-35	440-561	0.7-1.1
Shafa	4	(4.2-4.8)x10^3	30-41	471-543	2.2-2.4
Soda	4	(4.1-5.3)x10^3	23-38	364-539	2.1-2.8
Solaymani	5	(6.7-7.3)x10^3	17-29	439-873	3.5-3.8

The predominant natural radionuclide in these soil samples given in table 2 is ^{226}Ra. One may conclude that it is the hot springs in these areas that are the source of the distribution of ^{226}Ra and the cause of high levels of natural radiation [3].

The maximum mean concentration of ^{226}Ra in the Ramsar soil samples was 25.2 kBq kg^{-1}, taken from the Khak-e-Sefid areas. The corresponding value for the Mahallat soil was 7.0 kBq kg^{-1}, from the Solaymani hot spring area. Concentration ratios of ^{226}Ra in soil samples from these two regions compared to the normal background areas were 900 and 250 respectively. In some soil samples from the HLNRAs of Mahallat (Solaymani, Soda and Donbeh) and Ramsar (Ab-e-Siah), the high concentration levels of ^{226}Ra are associated with the yellow color of the soils (iron oxides). According to the results obtained, gamma absorbed dose rates at 1 meter above the ground level of the Mahallat and Ramsar regions are approximately 55 to 200 times higher than normal background areas. ^{226}Ra levels in the hot spring water of Mahallat and Ramsar, which are the source of high natural radioactivity of the soil samples, are from 0.48 to 1.35 kBq m^{-3} in the Mahallat and from 1.06 to 146.54 kBq m^{-3} in Ramsar [3,8].

4. CONCLUSION

According to the results obtained especially in the HLNRAs of Ramsar and Mahallat, it can be concluded that radiological monitoring of such areas is needed to estimate the external and internal doses of the inhabitants for epidemiological and biological studies. Such studies have been partially carried out in Ramsar and complementary studies are underway. The new HLNRAs of Abegarm-e-Mahallat, will need more extensive radioactivity monitoring studies especially for radon gas in air and radium and radon in hot springs and other water sources for dose assessment [3]. A program has been underway to reach the above goals by the National Radiation Protection Department of the Atomic Energy Organization of Iran the results of which are being reported elsewhere.

REFERENCES

1. M. Sohrabi, In Procs. 2nd Workshop on Radon Monitoring in Radioprotection, Environmental, and/or Earth Sciences, Trieste, Italy, 25 Nov.- 6 Dec. (1991), G. Furlan and L. Tommasino (eds.), World Scientific Publ. Ltd., Singapore (1993) 98.
2. K.N. Yu, Z. J. Guan, M.J. Stokes, and E. C. M. Young, J. of Environmental Radioactivity, 17, (1992) 31.
3. M. Sohrabi, M. M. Beitollahi, Y. Lasemi, and E. Amin Sobhani, 4th ICHLNRA, Beijing, China, Oct. 21-25 (1996).
4. M. Sohrabi, J. Roozi-Talab and J. Mohammadi, AEOI-NRPD Report (1996).
5. A. Savidou, C. Raptis and P. Kritidis, Radiat. Prot. Dos., 59, No. 4, (1995) 309.
6. H. Sorantin, and F. Steger, Radiat. Prot. Dos. 7, No. 14, (1984) 59.
7. United Nations Scientific Committee on the Effects of Atomic Radiation, Surveys, Effects and Risks of Ionizing Radiation, United Nations, New York (1988).
8. H. Mirzaee, and M. M. Beitollahi, AEOI Scientific Bulletin, Nos. 11 and 12, (1993) 97.

Investigations about the radioactive disequilibrium among the main radionuclides of the thorium decay chain in foodstuffs

D.C. Lauria, Y. Nouilhetas, L. F. Conti, M. L. P. Godoy, L. Julião and S. Hacon.

Institute for Radiation Protection and Dosimetry, Brazilian Nuclear Energy Commission, P.O.Box 37750-CEP 22780, Rio de Janeiro, Brazil.

This paper addresses an assessment of the activity concentrations of ^{232}Th, ^{228}Th and ^{228}Ra in foodstuffs from a high natural background radiation area. The disequilibrium observed between the activity of the ^{232}Th, ^{228}Ra and ^{228}Th in foodstuffs is discussed here. Due to the high radium bioavailability and the short half lives of the ^{228}Th and the ^{228}Ra, the stored time of the foodstuffs and the growing period of the vegetable have the most important roles in the thorium isotopic disequilibrium.

1. INTRODUCTION

The aim of this research was to assess the contribution of the foodstuffs in the ingestion of long lived members of the thorium chain and identify the main factors that influence the thorium chain radioactive disequilibrium in foodstuffs. The survey was done in the monazite region of Buena, where previous research of thorium metabolic behavior has been made. In the Buena inhabitants feces samples were found concentrations of ^{228}Th more than ten times higher than ^{232}Th [1]. This disequilibrium has already been observed in human feces from inhabitants of Morro do Ferro; a region also characterized by the high concentration of the radionuclides of the thorium chain in soil. In the feces samples, the ^{228}Th/^{232}Th ratio has ranged from two to fifteen. According to the authors, the observed Th isotopic disequilibrium could only be explained by the presence of thorium disequilibrium in the ingested foodstuffs [2]. Thorium-228 is a radionuclide of the thorium-232 chain and product of the radium-228 decay. At first the behavior of the two thorium isotopes should be similar in the environment.

Thus it should not be discrimination between the two radionuclides in the absorption, elimination and other processes in which the radionuclides are involved. However, the radium has a higher mobility and bioavailability than other radionuclides and it can be highly concentrated in the vegetables. Radium uptake into vegetation is 100 - 200 times higher than that of thorium [3]. In the plant, the absorbed ^{228}Ra through its decay produces ^{228}Th. By means of the foodstuff ingestion, ^{228}Th reaches the organism, as it is very little absorbed by the gastrointestinal system (ICRP has adopted a value of 5 x 10^{-4} in its publications 56 [4]). Th is eliminated through the feces. Therefore taking into account, the biological thorium half-life and the half-lives of the radionuclides of Th chain, it can be said that the Th isotopic composition in the human feces should be very similar to the one observed in the ingested foodstuffs. Nevertheless, the information concerning the ^{228}Th and the ^{228}Ra concentrations in foodstuffs were still insufficient to explain the factors influencing the observed disequilibrium. The lack of such information has been attributed: to the difficulties to determine ^{228}Ra, to the complexity of the effect of time on estimating of members of the Th chain [5] and to the difficulties to analyze the isotopic Th spectrum when the ^{228}Th concentration is much higher than the other Th isotopes [6]. Without consistent data, the UNSCEAR 1993 Report has considered the reference values of ^{228}Th concentration the same as the ^{232}Th value for all kinds of foodstuffs [7].

2. MATERIALS AND METHODS

In the first step, a ^{232}Th and a ^{228}Ra survey in some foodstuffs consumed by the inhabitants was made. Subsequently, the main and available region products were chosen in order to determine ^{228}Th too. Soil and vegetable samples were collected in five vegetable gardens and two plantations. Local manioc flour was analyzed as well as two local food samples ready for consumption. The composition of diet[1] was salad (cucumber and tomato), bean, rice, potato and boiled chicken while the diet[2] enclosed salad (cabbage and tomato), manioc flour, bean, rice and roasted meat. Two samples of beans were analyzed: one bean[a] was sampled on the local plantation and another bean[b] was bought at local market.

The foodstuffs were dried and calcimined to be analyzed. The sample ashes were digested with HNO_3, H_2O_2 and HF. Afterwards the coprecipitation with barium sulfate, the dissolution with alkaline ethylenediamine-*tetra*-acetic acid (EDTA) and reprecipitation as $Ba(Ra)SO_4$, the ^{228}Ra was determined by total alpha and beta counting [8]. Preliminary, the ^{232}Th was determined by spectrophotometry with ARZENAZO III. For the second step of the survey the determination of ^{232}Th and ^{228}Th was made by alpha spectrometry by a surface barrier detector. Here it was used the ^{234}Th as tracer, a chromatographic column containing tri-n-octil phosphine oxide, TOPO, to the separation of the elements and a time of counting of sixteen hours [9].

3. RESULTS

Radionuclide concentrations and the ratios between ^{228}Ra and ^{232}Th concentration are presented in table 1. The presented concentration values of Ra

Table 1
Concentrations of ^{232}Th and ^{228}Ra in Buena foodstuffs(Bq/kg$_{fresh}$)

Sample	^{232}Th	^{228}Ra	^{228}Ra/^{232}Th
Kale[3]	1.21±0.02	5.7±0.4	5
Maize	0.018±0.004	0.33±0.04	18
Sweet potato[2]	0.142±0.005	1.07±0.06	7
Manioc	0.025±0.003	3.6±0.3	144
Broccoli	0.043±0.006	3.9±0.6	91
Lettuce	0.21±0.03	0.57±0.06	3
Spinach	0.162±0.006	4.5±0.5	28
Wild chicory	0.036±0.005	19±2	520
Tomato	0.003±0.001	0.30±0.05	101
Mango	<0.06	0.25±0.04	---
Pineapple	<0.05	0.10±0.02	---
Milk	0.012±0.001	0.04±0.01	3
Chicken[3]	<0.009	0.16±0.02	---
Beef	0.009±0.001	0.03±0.01	3

suprscript numbers represent the number of analyzed samples.

and Th are similar to the concentrations found in the regions of high natural radioactivity [10]. It is pointed not only the higher contribution of the leafy vegetables than the other foodstuffs to the human intake of ^{232}Th and ^{228}Ra, but also the high uptake dissimilarity of radionuclides by the different vegetable species. The observed soil-plant concentration ratios (CR) ranged from 1.1×10^{-5} to 1.6×10^{-2} for thorium and from 2.4×10^{-3} to 2.0×10^{-1} for radium. Geometric mean: $CR_{Th}=3.5 \times 10^{-4}$ and $CR_{Ra}=3.3 \times 10^{-2}$ where: CR=Bq.kg^{-1}-dry-vegetable/Bq.kg^{-1}-soil. These values depict the lower bioavailability of thorium in comparison to radium.

On table 2, it is presented the measured values of activity concentrations of ^{232}Th, ^{228}Ra and ^{228}Th in the main foodstuffs. The results show at first a high radioactive disequilibrium between the radionuclides of the decay chain. However, if the decay corrections by considering the time between the sampling and measured are made, (^{228}Ra ($t_{1/2}$ =5.7 years) and ^{228}Th ($t_{1/2}$=1.9 years)) the values of ^{228}Th and ^{232}Th became more similar as it can be seen on table 3.

Table 2
Values of measured activity concentrations of ^{228}Th, ^{232}Th, and ^{228}Ra in foodstuffs(Bq/kg$_{fresh}$)

Sample	^{228}Th	^{228}Ra	^{232}Th
Carrot	0.20±0.01	1.1±0.1	0.017±0.002
Bean^a	0.90±0.03	4.0±0.1	0.43±0.02
Bean^b	0.45±0.01	0.92±0.08	0.075±0.005
Kale	1.0±0.1	3.40±0.03	0.51±0.07
Lettuce	0.24±0.04	0.53±0.06	0.21±0.03
Manioc flour	1.06±0.03	3.8±0.3	0.112±0.005
Sweet potato	0.33±0.01	2.4±0.2	0.118±0.001
Sugar cane	0.34±0.01	0.28±0.02	0.020±0.001
Milk	0.036±0.002	0.04±0.01	0.012±0.001
Beef	0.021±0.001	0.03±0.01	0.009±0.001
Chicken	<0.020	0.16±0.02	<0.009
diet$^{(1)}$	0.103±0.005	0.35±0.04	0.042±0.003
diet$^{(2)}$	0.143±0.005	0.37±0.04	0.015±0.001

In the analyzed vegetables whose period to reach the maturity is short (as carrot, kale, bean^a and lettuce) the concentration of ^{228}Th was approximately the same as that of ^{232}Th. However, in vegetables with long growing periods as sweet potato (growing time six months) and sugar cane (growing time 24 months), it was found values of ^{228}Th concentration over than twice the ^{232}Th. This disequilibrium may be an outcome of the ^{228}Th growing through the decay of the absorbed ^{228}Ra. The obtained results for bean^b and manioc flour both bought at the local market showed that the used time for the decay correction was not sufficient to reach the equilibrium. The stored time has not been considered. Taking into account, the values of ^{228}Th and ^{228}Ra concentrations at the sampling time, it is possible to say that after one year of storage, the ratios ^{228}Th/^{232}Th for bean, maize and manioc flour will be about four, six and thirteen, respectively. The increase of the Th concentration ratios in the manioc flour with the time can be seen on figure 1.

Table 3
Values of concentration of ^{228}Th taking account the ingrowth between the sampling time and analyzed time ($Bq.kg^{-1}_{fresh}$).

Foodstuff	^{232}Th	^{228}Th measured	Tc (days)	^{228}Th due to ^{228}Ra decay	^{228}Th estimated for the sampling day	^{228}Th/^{232}Th
Carrot	0.017±0.002	0.203±0.008	164	0.18±0.02	0.03±0.02	1.7±1.3
Bean[a]	0.43±0.02	0.90±0.03	165	0.6±0.1	0.3±0.1	0.8±0.3
Bean[b]	0.075±0.005	0.45±0.01	164	0.14±0.01	0.36±0.02	4.8±0.4
Kale	0.51±0.07	1.0±0.1	180	0.59±0.01	0.5±0.1	1.0±0.3
Lettuce	0.21±0.03	0.24±0.04	180	0.09±0.01	0.17±0.05	0.8±0.3
Manioc flour	0.112±0.005	1.06±0.03	214	0.76±0.05	0.38±0.08	3.4±0.7
Sweet potato	0.118±0.001	0.33±0.02	30	0.071±0.005	0.26±0.01	2.2±0.1
Sugar cane	0.020±0.001	0.034±0.009	300	0.076±0.005	0.35±0.01	18±1
Milk	0.012±0.001	0.036±0.002	300	0.011±0.003	0.033±0.004	2.9±0.4
Beef	0.009±0.001	0.021±0.001	60	0.002±0.001	0.020±0.001	2.3±0.3
Diet[(1)]	0.042±0.003	0.103±0.005	214	0.071±0.009	0.04±0.01	0.9±0.3
Diet[(2)]	0.015±0.001	0.143±0.005	214	0.079±0.009	0.06±0.01	4.4±0.9

Where: Tc is the elapsed time between the sampling and the separation Ra/Th.

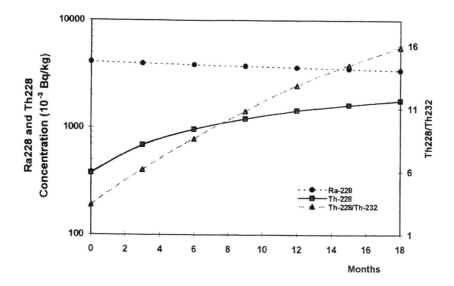

Figure 1. Increase of the thorium concentration ratio in manioc flour, with time after sample collection.

4. CONCLUSION

Among the analyzed samples only the vegetables whose growing period are short, presented equilibrium between ^{228}Th and ^{232}Th. The results suggest an increase of the disequilibrium with the vegetable growing period increase. Due to the higher values of concentration of radium than the thorium in foodstuffs and to the short half-lives of the ^{228}Th and the ^{228}Ra, the storage time of the foodstuffs is a primordial factor for the establishment of the disequilibrium between the ^{228}Th and the ^{232}Th. The disequilibrium should be more accentuated in cereals and industrial products of vegetables. Among the analyzed foodstuffs that can be stored for long time, the manioc and consequently the manioc flour showed the highest ratio ^{228}Ra/^{232}Th tending therefore to present high disequilibrium ^{228}Th - ^{232}Th with the time. The results show the importance of the knowledge of the population dietary habits for the better understanding of the results found in the feces analysis.

REFERENCES

1. Julião L.Q.C., Lipsztein, J.L., Azeredo, A.M.G.F., Dantas, B.M., Dias da Cunha, K. M. A., Radiat. Prot. Dose., 53 (1990) 285.
2. Linsalata P., Penna-Franca E. and Einsenbud M. , Health Phys., 50 (1986)163.
3. Linsalata P., Institute of Environmental Medicine, New York University Medical Center, Report No. DOE / ER60134, New York ,1988.
4. ICRP Publication 56, Annals ICRP, 1990.
5. Fissene I. M., Perry P. M., Decker K. M. and Keller H. W., Health Physics, 53 (1987) 357.
6. Linsalata P. , Rad. Phys. Chem., 34 (1989) 241.
7. UNSCEAR93, Publication E.94. IX.2., United Nations, New York , 1993.
8. Godoy J.M., International Atomic Energy Agency, Technical Reports Series No.310, Vo.1, Vienn , 205, 1990.
9. Lauria D.C. e Godoy J.M., The Science of the Total Environment. , 70 (1988)83.
10. Linsalata P., Morse R.S., Ford H., Einsenburd M., Penna Franca E., Castro M. B., Lob, M, N., Sachett I .and Carlos M. , J. Environmental Radioactivity, 14 (1991) 233.

Transfer of ^{226}Ra, ^{228}Ra, ^{210}Pb and ^{210}Po in the aquatic organism and the foodstuff chain

Yang Xiaotong, Weng Detong, Chen Wenyin, Chen Xiuyun, Chen Jixi and Zhao Shimin

Fujian Institute of Radiation Health Protection, Fuzhou 350001, China

1. INTRODUCTION

Modern science has ascertained the fact that human exposure is mainly contributed by natural radioactive sources. Natural external exposures are originated from cosmic rays, ground and building materials; the internal exposures come from air breathing and food intake. The content of natural radionuclides in the earth crust depends on the geological composition; it is usually high in the igneous rocks. The igneous rock and granite largely distribute in Fujian Province. Former surveys found that the contents of natural radionuclides in soils of Fujian were higher than those in other normal areas throughout China, and so were the contents in water and foodstuffs. Former surveys also showed that among the natural radionuclides except ^{40}K, ^{226}Ra, ^{228}Ra, especially ^{210}Pb, and ^{210}Po made the greatest contribution to human internal exposure. UNSCEAR 1988 Report also indicated that coastal inhabitants took in more ^{210}Pb than others. In order to study the transfer of ^{226}Ra, ^{228}Ra, ^{210}Pb and ^{210}Po in biological organism, especially in foodstuff chain, the authors carried out the survey from 1992 to 1994.

2. MATERIAL AND METHOD

2.1. Sample collection and treatment

According to the dietetic habits of local peoples, the representative aquatic products and the water and sediments in their habitat were collected. For convenience of measurement, the sampling sites were selected at the places with high radioactive background, such as the places with drained in hot spring water. 22 kinds and 55 samples were collected in this study. All samples were collected in the production seasons. According to different experimental requirements, each biological sample was treated by two ways: (1) separation of flesh, bones (or shell), scales, stomach and intestines directly from the raw aquatic organism; (2) after scraping scales and discarding the viscera, rinse and weigh the sample, add tap water and boil, then separate the flesh and bones (or shells). For analysis of ^{210}Po, the sample are treated by wet ashing, while other samples are ashed at 450°C.

2.2. Measuring method and data analysis

^{226}Ra, ^{228}Ra, ^{210}Pb and ^{210}Po were detected in each sample. ^{226}Ra is determined by co-precipitation from the sample with barium sulfate; after an ingrowth period, the ^{222}Rn is removed into an alpha scintillation counting cell for measurement. ^{228}Ra is a weak beta emitter and decays to ^{228}Ac; the ^{228}Ac is then extracted with HDEHP, and is beta counted in a low background counter. ^{210}Pb is then separated by ion exchange; after an ingrowth period of its decay daughter ^{210}Bi, it is beta counted in a low background counter. ^{210}Po is spontaneously deposited on a clean silver disc, and is alpha counted in a low background counter. Their lower limits of detection are 6×10^{-3} Bq/g(ash), 6×10^{-3} Bq/g (ash), 2×10^{-3} Bq/g (ash) and 2×10^{-3} Bq/g (wet), respectively. ^{226}Ra and ^{228}Ra in sediments were measured by gamma spectrometry, the lowest limit of detection was 2.1 Bq/kg (dry). The measuring methods mentioned above had been strictly calibrated.

The measured data of various samples were treated according to statistical requirements. The paired samples were tested for differences between them.

3. RESULT AND DISCUSSION

3.1. ^{226}Ra, ^{228}Ra, ^{210}Pb and ^{210}Po concentration factors of biological organisms

Radionuclides in the water system may be transferred and deposited to aquatic organisms and sediments. The concentration factor (CF) means the ratio of the radioactivity in organism per unit weight to that in water per unit volume. The results of this study are shown in Table 1. As seen in Table 1, organisms have strong concentration capability for ^{226}Ra, ^{228}Ra, ^{210}Pb and ^{210}Po, the CF values ranged from 10^0 to 10^4. ^{226}Ra, ^{228}Ra and ^{210}Pb are similar in nature to calcium, and were mainly deposited in bones (or shells), the CF values ranging from 10^2 to 10^3. ^{210}Po is mainly stored in the soft tissues of organisms, the CF values ranging from 10^2 to 10^4, especially in the flesh of shellfishes and the stomach and intestines of fishes, the CF value reaching 10^4. The CF of freshwater fishes has rarely been reported; the CF values for marine organism in this study are similar to those reported in the foreign and domestic literature.

Radionuclides having entered the water systems for the most part are deposited in the sediments and form a kinetic balance, so called the distribution factor (DF), which means the ratio of radioactivity in sediment per unit weight to that in water per unit volume. In this study, DF value for the 4 kinds of radionuclides ranged from 10^2 to 10^4. The radionuclide absorbed ability in aquatic organisms from sediments was expressed as absorbed factor (AF) in this study, which means the ratio of radioactivity in aquatic organism per unit weight to that of sediment per unit weight. Because radioactive content in sediment is high, the AF value can be usually used as an index for the environmental radioactive pollution, especially for the sediments in mud flats, ponds and lakes.

Table 1
Concentration factors of ^{226}Ra, ^{228}Ra, ^{210}Pb and ^{210}Po in aquatic organisms

Sample	^{226}Ra	^{228}Ra	^{210}Pb	^{210}Po
Freshwater fishes				
Flesh	4.58(0.34-10.1)	3.82(0.83-6.51)	8.03(2.22-23.1)	182(30.9-529)
Bones	206(22.0-513)	232(53.0-509)	119(35.1-242)	649(186-1556)
Scales	168(34.7-406)	230(74.2-464)	132(41.4-268)	644(146-1276)
S&I	-	-	-	3536(2101-11500)
Marine fishes				
Flesh	7.96(3.18-17.4)	16.6(6.11-19.9)	15.9(6.92-24.9)	276(181-363)
Bones	135(58.1-254)	163(76.6-211)	100(43.0-158)	508(-)
Scales	116(115-118)	204(86.0-322)	113(99.1-124)	904(741-1068)
S&I	-	-	-	11848(8090-15600)
Marine shellfish				
Flesh	20.0(11.5-74.0)	24.8(7.23-117)	92.0(33.9-140)	5275(2280-10500)
Shell	204(40.5-368)	398(179-602)	958(525-1727)	295(89.3-690)
Marine Shrimps				
Flesh	17.7(8.77-26.6)	8.50(8.01-9.00)	69.4(25.7-113)	1070(1010-1130)
Shell	81.0(71.6-90.3)	85.8(76.4-95.2)	257(130-384)	5340(5280-5400)
Marine algae	41.8(18.0-84.7)	61.8(13.3-128)	-(-)	212(54.6-545)

Note:* S&I stands for stomach and intestines.

The AF values of the four kinds of radionuclide ranged from 10^{-3} to 10^0 in this study, among which, 10^{-2} to 10^{-1} for ^{226}Ra, ^{228}Ra and ^{210}Pb were absorbed in bones (or shells), and 10^0 for ^{210}Po absorbed in the flesh of shellfishes.

3.2. Transfer of natural radionulides in cooking process of aquatic foodstuffs

The investigation data show that in the nutritious edible parts of aquatic foodstuffs (such as flesh) the concentration of radionuclides except ^{40}K and ^{210}Po is relatively low, but concentration in bones (or shells) is relatively high. During the cooking process, the bones (or shells) usually still adhere to the flesh. In order to study whether the cooking process would increase the radionuclide content in the flesh, collected samples were stewed without spices, the cooked flesh was separated from the cooked bones (or shells) and raw flesh separated from raw bone (or shells), and the contents of ^{226}Ra, ^{228}Ra, ^{210}Pb and ^{210}Po in paired cooked and raw samples were measured for them. Their results (the contents of radionuclides in water were deducted) are listed in Table 2, which shows that the difference in contents of ^{226}Ra, ^{228}Ra and ^{210}Pb between the raw and the cooked aquatic organism is not statistically significant (P>0.05). It suggested that the ^{226}Ra, ^{228}Ra and ^{210}Pb in the bones (or shells) are relatively stable, although their contents are high, and little radionuclides transferred to the flesh during cooking process.

However, quite differently from ^{226}Ra, ^{228}Ra, ^{210}Pb and ^{210}Po does evidently transfer to the flesh from bones (or shells) during the cooking process; especially the difference in contents of ^{210}Po between the raw and the cooked flesh of freshwater fishes and shrimps is very significant (P<0.01). However, there are various methods in Chinese cuisine other than stewing, such as stir-frying, deep-frying, braising and boiling. The conditions of these cooking processes are complicated, and they are not considered in this paper.

3.3. Transfer models of ^{226}Ra, ^{228}Ra, ^{210}Pb and ^{210}Po in aquatic organisms and food chain

The influencing factors on the transfer of natural radionuclides in aquatic organism and food chain are very complicated, such as environmental factors, physical and chemical natures of radionuclides, physiological peculiarity of organisms and cooking methods. These factors interact with, promote or condition each other. Their relationships are summarized in the following diagram. As shown in the diagram, there is a series of complicated steps for natural radionuclides to enter human body from the environment. As regards the four radionuclides, after entering the water system, a large part of them adhere to the sediments, and only several ten thousandths to several hundredths of them are dissolved in the water. Concentrated by biological organism, these radionuclides are stored in the different organs (or tissues) according to their physical and chemical natures. Finally, with the organisms cooked and eaten, these radionuclides does enter the human body. According to these links mentioned above, based on the measuring data, the relationships between the activities of ^{226}Ra, ^{228}Ra, ^{210}Pb and ^{210}Po in foodstuffs and the environment can be evaluated.

Table 2
Comparison of contents of natural radionuclides in the cooked and raw aquatic foodstuffs

Samples	^{226}Ra		^{228}Ra		^{210}Pb		^{210}Po	
Freshwater fish								
Raw flesh	6.71	P>0.05	6.98	P>0.05	14.5	P>0.05	122	p<0.01
Cooked flesh	6.82		6.93		16.6		139	
Raw bone	330	P>0.05	444	P>0.05	325	P>0.05	458	P>0.05
Cooked bone	319		426		307		449	
Marine fish								
Raw flesh	4.18	P>0.05	4.39	P>0.05	32.7	P>0.05	350	P>0.05
Cooked flesh	4.66		7.33		40.4		373	
Raw bone	69.8	P>0.05	221	P>0.05	225	P>0.05	355	P>0.05
Cooked bone	55.1		296		21		394	
Marine shellfish								
Raw flesh	10.1	P>0.05	17.0	P>0.05	156	P>0.05	9827	P>0.05
Cooked flesh	10.4		17.7		161		7750	
Raw shell	196	P>0.05	619	P>0.05	1767	P>0.05	626	P>0.05
Cooked shell	190		534		1288		483	
Marine shrimp								
Raw flesh	12.2	P>0.05	12.6	P>0.05	113	P>0.05	2592	P>0.01
Cooked flesh	11.0		12.9		100		5629	
Raw shell	61.1	P>0.05	132	P>0.05	316	P>0.05	12922	P<0.05
Cooked shell	63.0		152		336		7976	

** Average value unit: $\times 10^{-2}$ Bq/kg.

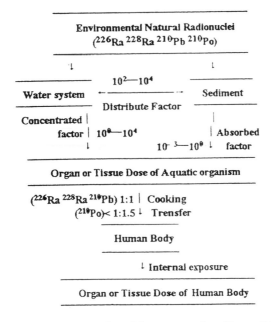

Diagram of transfer of four natural radionuclides in aquatic food chain

4. CONCLUSIONS

4.1. The aquatic organisms have strong concentration capabilities for ^{226}Ra, ^{228}Ra, ^{210}Pb and ^{210}Po in the environment. Similar to Ca, ^{226}Ra, ^{228}Ra and ^{210}Pb, are mainly stored in the bones (or shell), the concentration factors ranging from 10^2 to 10^3 and from 10^0 to 10^2 in the bones (or shells) and in the flesh, respectively. ^{210}Po is mainly stored in the soft tissues, the concentration factors usually ranging from 10^2 to 10^4, of which the value for the stomach and intestines of fishes and the flesh of shellfishes reaching 10^4. The concentration factors of marine organisms are usually higher than those of freshwater organisms.

4.2. Experimental results of stewed aquatic foodstuffs show that: (1) although the radioactivity of bone-seeking nuclides ^{226}Ra, ^{228}Ra and ^{210}Pb in the bones are relatively high, little of them transfer to the edible part during the cooking process, that is to say, cooking process would not significantly increase the radionuclide contents in the food chain; (2) for the freshwater fishes and shrimps, ^{210}Po does transfer to the flesh from the bones (or shells) during cooking process.

REFERENCES

1. UNSCEAR Report, 1988.
2. UNSCEAR Report, 1982.
3. Li Yongqi, Marine Radioactivity, Science Publishing House, 1978.
4. Fujian Province study section of environmental ionizing radiation, Chin. Radiol. Med. Prot., 5, 1985.
5. Yang Xiaotong, et al. , The level of near sea environmental radioactivity in Fujian. In: The level of near sea environmental radioactivity in China. Beijing: China Ocean Press, 1987.
6. Yang Xiaotong, et al. , Chin Radiol Med Prot, 8 (Supplement), 1988.
7. Chen Jixi, et al., Chin. Radiol. Med. Prot. , 8 (Supplement), 1988.
8. Gao Yutang. Statistical method for environmental monitoring. Beijing: Atomic Energy Press, 1981.
9. IAEA, Reference methods for marine radioactivity studies, Vienna, 1970.
10. Woodhead, D.S., Levels of radioactivity in the marine environment and the dose commitment to marine organisms, IAEA-SM-158, 31, 1972.

Intakes of food and internal doses by ingestion in two high radiation areas of China

Zhu Hongda

Institute of Radiation Medicine, Chinese Academy of Medical Sciences and Chinese Union Medical College, P.O.Box 71,Tianjin 300192, China

1. INTRODUCTION

The author presents the results of a nationwide radioactivity survey of foods during 1982, including contents of natural U and Th, ^{226}Ra, ^{228}Ra, ^{210}Pb, ^{210}Po, ^{227}Ac, ^{40}K and ^{87}Rb in food, estimated intakes and resultant committed effective doses (CED) for adults in Yangjiang high background radiation area(HBRA) and an area near U-mining area (ANUA) in comparison with those in normal background radiation area (NA) in the 3rd ICHLNR, held in Ramsar, Iran, 1990 [1].

Along with the development of economic condition and living standard, the diet composition of public in China has changed considerably. Since then the author took part in the first national total diet study in 1990 and organized the second one in 1992. These new surveys have observed the diet composition of Chinese Reference Man(RM), estimated the intakes and CEDs of radionuclides, thereby updating the data on diet radiation in NA in China [2,3]. Based on new diet composition, determined food contents and newly published age-dependent ingestion dose coefficients(IDC) by ICRP [4], this article reports these new data and re-estimates the resulting doses in two high radiation areas.

2. MATERIALS AND METHODS

In the 1982 nationwide radioactivity survey of foods, 14 provinces or autonomous regions were divided into 3 sampling areas. 11-20 kinds of main food were collected from each area and contents of 9 natural radionuclides (or elements) in all of these samples were analyzed by methods described elsewhere [5]. In the 1990 and 1992 total diet studies, 12 provinces or autonomous regions were divided into 4 areas. Average diet compositions were observed in 1080 families for three successive days by means of weighing and recording, 12 categories of food were collected, cooked and homogenized separately. The contents of 6 natural radionuclides were analyzed. The intakes and resulting CEDs were renewed. In the 1992 survey, a pilot study of the age-group difference among 4 age groups, important in radiation protection (2-7 year, 8-12 year and 20-50 year-old males and females) was conducted. The method was described in detail elsewhere [2].

3. UPDATING OF INTAKES AND INTERNAL DOSES FOR THE RADIONUCLIDES BY INGESTION IN NA

3.1. Chinese diet composition and its change

Diet compositions of Chinese RM or adult man, obtained from different surveys, are compared with each other in Table 1.

Table 1
Dietary compositions of Chinese adult man (RM) and their change

Food	Pilot survey		National food survey	
	1982 g/d	1992 g/d(increment in %)	1982 g/d	1990 g/d(increment in %)
Grains	525.8	657.4(+25.0)	517.6	461.4(-10.9)
Beans	10.2	41.6(+307.8)	15.1	39.5(+161.6)
Yam	151.5	49.2(-67.5)	201.4	101.0(-49.9)
Meat	22.8	49.4(+116.7)	27.0	48.6(+81.1)
Egg	7.2	13.3(+80.6)	5.7	17.1(+200.0)
Aquatic product	13.6	23.0(+69.1)	11.6	22.9(+97.4)
Milk	2.5	15.9(+536.0)	2.6	11.0(+323.1)
Vegetable	326.4	306.5(-6.1)	342.7	323.9(-5.5)
Fruit	24.6	34.5(+40.2)	29.3	101.2(+245.4)
Sugar	5.6	2.4(-57.1)	4.4	3.4(-22.7)
Soft drink & water	-	929.6	-	512.1
Alcoholic drink	3.9	16.4(+320.5)	3.0	14.0(+268.4)
Oil & fat	16.8	-	16.1	30.9(+91.9)
Condiment	24.7	-	23.8	23.5(-1.3)

It can be seen from Table 1 that the consumption of plant food remains more than that of animal food, but since 1982 the consumption of animal food, such as milk, egg, fat, aquatic product and meat, and also fruit has increased obviously and that of yam and sugar has decreased. It indicates that the diet composition is tending to be reasonable and of higher quality. The change implies that data on such intake and resultant dose should also be renewed.

3.2. Updating of intakes and resulting CEDs in NA in China and comparison with the world averages

For NA in China, the estimated intakes and CEDs of these 9 natural radionuclides or elements from the past and recent surveys are compared with Table 2, accompanied by comparison with the worldwide averages (WA) [6]. The past reported results are retained in parentheses for comparison. Because U, Th and ^{227}Ac have not been determined in recent surveys, their intakes from the 1982 survey have to be quoted in estimation of CED for recent surveys. Because of its strong homeostatic control, ^{40}K CED has been considered to be independent of its intake and was estimated to be 165 µSv instead of past 180 µSv for NA over the world [6]. In calculation of sum E in Table 2, the average was also adopted.

Table 2
Intakes and CEDs in NA in China and comparison with worldwide averages

Item	IDC µSv/Bq	Annual intake, Bq			CED, µSv/a		
		1982	1990	WA	1982	1990	WA
U	0.0276	10.04		10.01	0.277(0.673)		0.28
^{226}Ra	0.28	22.1	27.1	19	6.188(6.85)	7.588	3.8
^{210}Pb	0.7	69.1	109.1	32	48.37(96.7)	76.37	32
^{210}Po	1.2	59.8	125.5	55	71.76(26.3)	150.60	11
^{227}Ac	2.0	0.2934			0.587(1.115)		
		Total			127.18(131.64)	235.42	47.08
Th	0.4	5.32		1.3	2.128(3.958)		0.52
^{228}Ra	0.66	30.2	67.7	13	19.93(9.97)	44.68	3.9
		Total			22.06(13.93)	48.64	4.42
^{40}K	0.0051	22780	8343		116.18(116.18)	42.55	165
^{87}Rb	0.0013	1279	1118		1.66(1.66)	1.45	6
		Sum A			267.08(263.41)	328.06	
		Sum E			315.90	450.51	222.5

*Sum A comes from actual surveyed result of K and sum E includes quoted worldwide average of ^{40}K

It is shown in Table 2 that owing to renewal of IDC, the estimated CED for ^{210}Pb or ^{226}Ra decreases, but that for ^{210}Po increases. On the other hand, the CED for ^{228}Ra is much higher than the past. The resultant CED of U series and

Th series in the 1990 survey is 85.1% and 120.5% higher than that in the 1982 survey, respectively, mainly attributable to change of diet composition.

From comparison of Chinese CEDs with the world averages, the estimated CEDs from nuclides of Th and U series are higher than the average. The total accounts for about 284μSv, much higher than the average (52μSv), probably being due to their content in environment and diet composition.

Obtained result of age-group difference among the 4 groups and comparison with annual limited intake(ALI) are shown in Table 3. It is indicated that from the point of view of monitoring for radiation protection, ^{210}Pb and ^{210}Po are more important than ^{226}Ra and ^{228}Ra and 8-12 year age-group should be paid more attention to than the others.

Table 3
Annual intakes of radionuclides for different age groups and comparison with the limited value *

Nuclide	2-8 year-old		8-12 year old		20-50 year-old M		20-50 year-old F	
	AI	AI/ALI	AI	AI/ALI	AI	AI/ALI	AI	AI/ALI
^{210}Pb	23.4	0.05	81.0	0.15	144	0.10	110	0.08
^{210}Po	7.9	0.03	31.1	0.08	44.6	0.05	55.2	0.07
^{226}Ra	15.4	0.01	24.4	0.02	40.9	0.01	28.1	0.01
^{228}Ra	13.4	0.05	16.5	0.07	22.9	0.02	31.2	0.02

* The unit of AI is Bq

4. RE-ESTIMATION OF CEDS OF THE RADIONUCLIDES FOR ADULT MAN IN TWO HIGH RADIATION AREAS

Using recently published IDC and the intakes obtained from the 1982 survey, the resulting CEDs for adult man in HBRA and ANUA are re-estimated and shown in Table 4.

It is shown from Table 4 that CEDs from the nuclides of U-series and Th-series in HBRA were 2.5 and 8.0 times as much as in NA respectively, the total accounting for 0.492 mSv/a, while in ANUA only CED from the nuclides of U-series was 6.2 times as much as in NA, but no increment was found in that of Th-series, the total accounting for 0.812 mSv/a. Total CEDs from the two series in HBRA and ANUA were about 3.3 and 5.4 times as much as in NA in 1982. The intakes of ^{40}K and ^{87}Rb in the areas were almost the same as that in NA. It indicates the differences between the two areas: the exposure in HBRA came from nuclides of U-series and Th-series, and increments of Ra and Ac isotopes were more than that of the mother nuclide U or Th, depending on their solubilities; but the exposure in the ANUA came only from U-series, and incre-

Table 4
Re-estimation of CEDs of natural radionuclides for adult man in two high radiation areas *

Nuclide	HBRA			ANUA		
	AI,Bq	CED,µSv	CED/CED'	AI,Bq	CED,µSv	CED/CED'
U	10.66	0.294	1.06	288.15	7.95	28.70
^{226}Ra	146.8	41.10	6.64	518.5	145.18	23.46
^{210}Pb	159.5	111.65	2.31	357.4	250.18	5.17
^{210}Po	131.3	157.56	2.20	311.4	373.68	5.21
^{227}Ac	1.84	3.68	6.27	7.20	14.70	24.53
Total		314.28	2.47		791.39	6.22
Th	7.43	2.97	1.40	5.59	2.24	1.05
^{228}Ra	264.3	174.44	8.75	27.1	17.89	0.90
Total		177.41	8.04		20.13	0.91
Sum		491.69	3.29		811.51	5.44

* CED' represents relevant CED of NA in 1982

ment of U is much more than that of its daughter nuclides because local pollution came from U-abundant liquid waste of nearby U mining and milling. U in the environment was quite soluble.

5. SUGGESTION FOR FURTHER RESEARCH

Owing to large population and quite low and constant dose, high natural radiation area is a good field for chronic low dose-effect research. As one of main entering ways, ingestion should be paid enough attention to. The following aspects are suggested:

5.1. As mentioned above in 3.2, since 1982 the diet compositions in the high radiation areas certainly have changed obviously, so resultant intakes, CEDs and contributions of different foods perhaps also have changed. Therefore, it is necessary to do periodic study by definite procedure and QC measures in order to update the data.

5.2. Considering age-dependent dose and importance to radiation protection, the dietary radiation survey in radiologically important age-groups in the two areas is needed also.

5.3. Seeing that low intake of K is a traditional defect in Chinese diet, the body weight of Chinese RM seems lighter than that of ICRP RM and ^{40}K is main contributor of internal dose, it seems that the internal dose from ^{40}K in China should be verified by means of whole body measurement or analysis of K in organs and tissues.

REFERENCES

1. Zhu H., in: Sohrabi M., Ahmed J.U., Durrani S.A., eds., Proceedings of International Conference on High Levels of Natural Radiation, IAEA, Vienna, 153-160, 1993.
2. Zhu, H., Gao, J., Yin, X. , et al., Radiation Protection Communication , 2 (1996) 1-24 (in Chinese)
3. Zhu, H., Wangm, S., Meng, W. , et al., Radiation Protection, 13 (1993) 85-92 (in Chinese)
4. ICRP Pub.67, Annals of the ICRP, 23(3/4), 1993.
5. Zhu, H. , Zhang, J. , Proc. CAMS and PUMC, 2 (1987) 7-15
6. UNSCEAR, UNSCEAR 1993 Report to the General Assembly, with Annexes. United Nations, New York, 1993.

Biosphere effects of natural radionuclides released from coal-fired power plants

Mao Yahong and Liu Yigang

Sichuan Institute of Radiation Protection, 19 Dianxin Road, Chengdu 610041, China

Abstract: We surveyed eight coal-fired power plants which hold about 90% of firepower generated energy in Sichuan and the corresponding control area. The result suggests that the biosphere effects of natural radionuclides released from the coal-fired power plants come mainly from the fly ashes. The fly ashes released from coal-fired power plants in Sichuan increased markedly the ^{238}U contents of external soil. But the radioactivity depositing on the surface of soil was not found to have been transferred to the crops. The fly ashes caused increase of the natural radionuclides content in the air. The incorporated collective effective dose from fly ashes inhaled by the people living near the plants are in the range of 8.8-91.3 [man·Sv/(Gw·a)] (the average is 37.7), which is contributed mainly by ^{232}Th (occupying about 90% of the total inhaled dose). The coal ash(cinder)water released into the rivers by the coal-fired power plants does not markedly increase the radioactive level of the river water.

1. INTRODUCTION

China is the biggest country producing coal in the world. The greatest part of the produced coal is burned in the coal-fired power plants. Now, about 4,000 million tons of coal are used in coal-fired power plants of China every year. The main ways in which the power plants discharge the radioactive waste into the environment are the release of fly ashes from the stack and the release of ash(cinder) water into the river. In order to gain a clear idea of the biosphere effects caused by these ways, we investigated and measured the eight coal-fired power plants which occupy about 90% of firepower generated energy in Sichuan and the corresponding control area. The results are reported as follows.

2. COLLECTION AND MEASUREMENT OF SAMPLES

2.1. Collection of samples

Since the natural radionuclide concentrations in bottom ashes and fly ashes are approximately identical [1], we replaced fly ashes with bottom ashes. And we collected the fly ash samples in seasons by stages. Every time we collected 0.2 kg by random for 5 days running, then mixed the ashes fully to make a standard sample totaling 1 kg. The soil samples were collected by five-point method. The water samples were collected separately upstream and downstream from the outlet of the discharging sewage. The plant samples collected were chiefly the leafy vegetables around the power plants.

2.2. Analyses of radionuclides

For physical measurement γ-ray spectrometer manufactured by ORTEC Company in U.S.A was used. Few samples were analyzed by radiochemical methods. Finally we intercompared and normalized the two kinds of measurement results.

3. RESULTS AND DISCUSSION

3.1. Atmospheric effects caused by fly ashes

According to the UNSCEAR 1982 Report, Annex C, fly ashes are the major source of the population's additional dose from coal-fired power plants. It occupies 70% of total collective effective dose. For this reason, we put the stress on fly ashes in our study. We measured not only fly ashes but also radon concentration in air at the upwind and downwind directions.

Table 1
Fly ash concentration in air around the power plants(mg·m^{-3})

Plant	Distance from stack(m)	Concentration downwind	Concentration upwind
A	500	0.60	0.36
	1000	0.75	0.39
	1500	0.41	-
B	500	0.44	0.19
	900	0.70	0.14
	1000	0.78	0.21
	1500	0.50	0.19

Table 2
Radon concentration in air around the power plants (Bq·m^{-3})

Plant	No. of samples	Radon downwind	Radon upwind
A	8	10.02 ± 0.50	7.50 ± 0.80
B	14	10.60 ± 0.50	7.50 ± 0.20

Table 1 and Table 2 show that the fly ashes from power plants really affect the air around the power plants. According to the UNSCEAR 1982 Report and the natural radionuclide content we measured in fly ashes, we assessed the normalized collective effective dose from fly ashes inhaled by people living in the range of 80 km around the power plants.

Table 3 shows that the major causes leading to the high normalized collective effective dose are low filter efficiency and high population density. The natural radionuclide contributing the greatest internal exposure dose is ^{232}Th, which occupies about 90% of the total dose. Next is ^{238}U. Then are ^{210}Po, ^{210}Pb and ^{226}Ra. Because homeostasis in human body is maintained by potassium, we did not calculate the additional dose of ^{40}K in fly ashes.

Table 3
Normalized collective effective dose caused by inhaled fly ashes

Plant	Filter efficiency	Population density (per km^{-2})	^{238}U	^{226}Ra	^{232}Th	^{210}Po	^{210}Pb	Collective dose [man·Sv(GW·a)$^{-1}$]
A	0.96	600	7.2	0.6	1.7	1.6	101.1	19.4
B	0.90	400	26.3	454.9	1.9	6.0	5.5	57.6
C	0.92	200	10.3	0.7	151.3	1.6	1.5	8.8
D	0.80	400	35.4	2.5	682.0	8.9	8.3	91.3
E	0.90	300	14.0	1.1	261.3	3.0	2.8	25.8
F	0.90	300	5.4	0.5	140.3	1.7	1.6	13.5
G	0.90	200	18.7	1.2	213.0	2.9	2.7	14.7
H	0.90	800	11.4	1.0	273.6	3.8	3.4	70.4

3.2. Radioactive influence arising from fly ashes depositing on earth

Because the radioactive fallouts remain on the surface of soil in very long time, we collected some soil samples at the downwind direction around the power plants. From the analytical results of these samples we found the radioactivities in all these samples are higher than the background values in the same area. Then we collected and analyzed the surface and the deep(50cm from surface)soil samples. Table 5 shows that the ^{238}U content in the surface soil samples are remarkably higher than that in the deep soil samples ($P<0.01$).

Table 4
Natural radioactivities in soil around the power plants (Bq·kg^{-1})

Place	No. of sample	^{238}U	^{232}Th	^{226}Ra	^{40}K
Plant A:					
Surface	8	48.±2.2	53.5±0.8	38.3±0.5	555.7±4.4
Deep	8	17.5±1.4	47.6±0.8	32.2±0.5	551.7±4.2
Plant B:					
Surface	4	44.3±3.6	75.5±1.4	46.2±0.8	790.6±10.6
Deep	4	27.3±2.2	55.6±1.1	44.3±0.7	619.5±8.0

Then, did the enhanced radioactivities transfer to plant (such as vegetables) proportionately? We compared the radionuclide analysis results of leafy vegetable samples around the power plants with the background values of vegetables in Sichuan Province (Table 5), and found that the radioactivities of U, Th and Ra did not increase remarkably. There are no background values available from the same place. (That is to say, there the quality of soil, climate,

Table 5
Natural radioactivities in leafy vegetables around the power plants (Bq·kg^{-1})

Plant	No. of samples	^{238}U	^{232}Th	^{226}Ra
A	6	8.5±1.3	9.2±0.9	42.2±8.0
B	6	8.2±1.3	7.0±0.7	22.4±4.2
C	4	8.6±1.3	7.8±0.8	32.8±6.5
Background	6	8.5±1.4	7.0±0.3	—

type of vegetables, etc. are all the same and the difference is only that the control vegetables are not polluted by fly ashes from the power plants. We could not find ideal control samples there.) So this conclusion need to be confirmed.

3.3. Radioactive influence arising from ash(cinder) water

During the on-the-spot investigation, we were aware of that most of the power plants discharged the ash (cinder) water into the river nearby. Therefore, we collected and analyzed the ash (cinder) water samples, upstream and downstream river water samples from the outlet of discharging sewage. Table 6 shows the results of these samples. There was no analytical error due to the very low radioactive level in water samples. It is obvious that the natural radioactive contents in the upstream and downstream river water are not remarkably different. Perhaps, it is caused by the vast flow and fast velocity of water in Yangtze River.

Table 6
Natural radioactive contents in water samples ($Bq \cdot kg^{-1}$)

Plant	^{238}U	^{232}Th	^{226}Ra
Plant A:			
Upstream	1.78	0.011	0.32
Downstream	1.73	0.011	0.14
Ash water	6.32	0.11	0.045
Plant B:			
Upstream	0.84	0.011	background
Downstream	0.84	0.016	0.59
Ash water	2.73	0.073	0.14

REFERENCES

1. United Nations Scientific Committee on the Effects of Atomic Radiation : Ionizing Radiation: Sources and Biological Effects, New York: United Nations,1982.
2. United Nations Scientific Committee on the Effects of Atomic Radiation. Ionizing Radiation: Sources and Biological Effects, New York, United Nations,1993.
3. Akira Nakaoka et al., Environmental effects of natural radionuclides from coal-fired power plants, Health Phys., 3 (1984) 407.
4. Papastefanou, C., Radiation impact from lignite burning due to ^{226}Ra in Greek coal-fired power plants. Health Phys., 2 (1996) 187.

IV Fallout from nuclear test

Overview of scope-RADTEST project

René Kirchmann[a], Sir Frederick Warner[a] and Linda Appleby[b]

[a]University of Liége. Belgium

[b]University of Essex, United Kingdom

1.ABSTRACT

The ongoing SCOPE-RADTEST program is examining releases of radio-activity due to nuclear detonations which have occurred at various test sites around the world. for peaceful and military purposes. Aspects relating to both ecological and human effects are being considered. The origin of this program, and a summary of RADTEST'S previous NATO ARW meetings held in Vienna. Austria, and Barnaul. Russia,during 1994 and last year's in Brussels and Liége are described. Program History A previous RADPATH study which was conducted under the auspices of SCOPE-ICSU (Scientific Committee On Problems of the Environment-International Council of Scientific Unions) [1], examined recent developments relating to the biogeochemical pathways of artificial radionuclides. From this program the need to implement the RADTEST (RADioactivity from nuclear TEST explosions) was identified. RADPATH was focused on elucidating the environmental pathways of artificial radionuclides and exploited the data emerging from the Chenobyl accident. The findings were published as "Redioecology after Chernobyl" SCOPE 50 [2-4]. Although RADPATH'S findings are significant, Shapiro and Tsaturov [5] noted that a comparison of releases arising from weapons tests with those due to Chenobyl indicates that the former are between 2 to 3 orders of magnitude greater. In order to address these issues, the RADTEST project was officially established in April 1993. and has an on-going program of workshops scheduled to continue until the project's termination in 1997. The project's Chairman is Sir Frederick Warner (Visiting Professor, University of Essex) and the Vice-Chairmen are Prof. Charles S. Shapiro (San Francisco State University, USA), and Prof. Rene Kirchmaun, University of Liége.

Until the end of 1995 the scientific and administrative organization of the RADTEST project comprised the RADTEST Executive Committee (EC), chaired by Prof.Charles Shapiro, and monitored by the SCOPE EC . In addition a Scientific Advisory Committee (SAC), chaired by Prof. Sir Frederick Warner, provided overall guidance to ensure that the project successfully sought to achieve its scientific, political, economic and social goals. Following a request from the SCOPE Executive Committee meeting (1-3 February 1996, Paris, France) some restructuring of the administration of RADTEST has been necessary. This has been achieved by the amalgamation of the Executive Committee and Scientific Advisory Committee, to form the RADTEST Steering Committee.

RADTEST is focusing on examining the transport. deposition and human health effects of radioactive fallout from nuclear weapons tests. This is being achieved through an intentional collaborative study which has. to date, involved States (Russia. Kazakhstan. USA, China, France, UK), other interested countries and international bodies. An important feature of RADTEST is the involvement of nuclear scientists who were directly or indirectly involved with nuclear testing. Hence this project enables to access restricted work involving data, models and knowledge about the fate of the release of radionuclides and for their possible human health effects to be discussed and considered for the first time.

RADTEST is focusing on the following principal tasks:

(1) Inventory of data on measurements of radionuclides *deposition densities*, and identification of gaps in these data.
(2) Comparison of old and development of new models of *radioactivity transport* for better understanding of the deposition densities of radionuclides *both on and near the nuclear test sites*, including areas downwind where potentially significant episodes of fallout have occurred (such as the Altai region of Russia).
(3) Study of *migration of the radionuclides through the biosphere, including all pathways to humans, and of the effects on other biota that have impacts on humans*. The main focus is to characterize the nature and magnitude of *the dose to humans*. This includes **dose reconstructions from past events**, and also an increased capability for dose prediction from possible future accidental or deliberate explosions.
(4) Analysis of the data on *effects of these doses (including low doses) on human health*.

2.DEVELOPMENTS

Various tests sites worldwide are being considered in the RADTEST study including : the Nevada Test Site (USA); Pacific Islands (USA); Novaya Zemlya (Russia); Semipalatinsk (Kazakhstan); Luc Bu Pu (Lop Nor) (China). Consideration of other sites. such as those of the UK and France, is being included as appropriate. The first RADTEST meeting convened was the North Atlantic Treaty Organization (NATO) Advanced Research Workshop (ARW) held in Vienna, Austria (10-14th January 1994). The purpose of this ARW was to consider the Environmental and Human Consequences of Atmospheric Nuclear Tests, and provided an opportunity to consider and discuss RADTEST'S program, goals, methodology. It involved about 40 experts from 10 countries, including representatives of international organizations (NATO, IIASA, CEC, UIR, IAEA, UNSCEAR and SCOPE).

A total of 48 papers were officially submitted in connection. containing important information and analysis newly available on subjects previously restricted and the meeting considered the results of previous studies which have been undertaken in the Former Soviet Union, the United States, the United Kingdom and China. There was also reappraisal of the RADTEST program, goals and methodology, which led to identification of future areas to be addressed.

Detailed examination of aspects of RADTEST'S Objectives were achieved by the following working groups:

(1) International database (relating to all aspects of nuclear explosion tests including a technical history of all nuclear tests. focusing on data relevant to dose reconstruction);
(2) Dose-reconstruction (fallout modeling, deposition and exposure measurements, pathways to man);
(3) Health effects (epidemiological studies including dose/response and risk factors).

The second RADTEST'S NATO ARW, held in Siberia (5-10th September 1994) was mainly concerned with radioactive fallout in the Altai region of Russia, emanating primarily from nuclear tests undertaken at the Semipalatinsk test site in Kazakhstan. NATO Advanced Study Institute (ASI) publications based upon the findings presented at these ARW's are presently in preparation [6,7].

The third RADTEST workshop, held the following year in Brussels and Liége (27-31 March 1995) was particularly concerned with the environmental consequences of local radioactive fallout from nuclear test explosions and the associated health consequences and epidemiology for the exposed population. This involved some 50 participants from 8 countries and was hosted by the Belgian SCOPE National Committee and the University of Liége, with support from the European Union and NATO. It included the presentation and discussion of 31 written papers, including those contributed reconstruction. medical and epidemiological subjects. Issues relating to the sub-surface in Barnaul relating to dose transport of radioactivity from underground nuclear explosions also provided an input to the Belgium meeting. Information and Database, Dose Reconstruction and Prediction, and Health Effects issues were examined further within RADTEST'S previously identified Working Groups.

3. CONCLUSION

With the ending of the cold war between the West and the former Soviet Union, and an encouraging atmosphere in China for increasing collaboration with the rest of the world's scientific and technical communities, a unique opportunity now exists for an international collaborative study on the fallout from nuclear tests. This subject has never before been jointly studied by the countries that were involved with most of the nuclear test explosions. There is now a new willingness to share data on nuclear tests. much of which unavailable in the open literature, up to now.

RADTEST is enabling the nuclear scientists that were directly or indirectly involved with these tests to share for the first time previously restricted work involving data, models, and knowledge about the fate of the radionuclides released and their possible human health effects. It will support in effect the Comprehensive Test Ban Treaty concluded end September 1996.

4. FUTURE WORK

In the course of the RADTEST programmable. an urgent need to implement a new program to examine radioactivity from Military Installation Sites and effects on population health (RADMIS) has been identified. It was recognized at the SCOPE Executive Committee meeting, February 1996. that all nuclear countries are investing Executive Committee meeting, February 1996. that all nuclear countries are investing considerable efforts in studying these difficult problems, and the network of international expertise developed during the RADTEST project provides an ideal framework within which to develop contacts to address these issues. Consequently, there are ongoing discussions to seek to initiate an international collaborative RADMIS study which will focus on the global problem of disposal of radioactive waste. particularly that arising from military installations, and associated environmental and health effects. The focus of RADMIS will be on the evaluation of risks from the various radioactive wastes resulting from nuclear weapons fabrication and from the equipment containing irradiated fuel, including the following principal tasks:

(1) Study of radioactive waste arising from military installations;
(2) Evaluation of the potential risks for the human populations and the environment from uncontrolled radioactive wastes;
(3) Consideration of remedial actions already taken or proposed for clean-up.

The RADMIS project is expected to provide new, and unique data, which should enable productive advances in this and allied fields. RADMIS will enable experts to share their information about data, models and knowledge about the fate of the release of radionuclides and their possible human health effects.

REFERENCES

1. SCOPE. SCOPE program and directory I 996-98. Scientific Committee on Problems of the Environment. France, 1996.
2. Warner, F.E. and Harrison, R.M. (eds.), Radioecology after Chernobyl (biogeochemical pathways of artificial radionuclides -SCOPE 50), John Wiley & Sons Ltd., 1993.

3. Warner F.E. and Appleby L.J., The post-Chernobyl Environmental Situation, Environmental Management and Health, 7 (1996) 6-10.
4. Warner F.E. and Appleby L.J., New information on Nuclear Tests; Environmental Management and Health, 7 (1996) 27-33.
5. Shapiro, C.S. and Tsaturov, Y., A study of the transport, deposition and human health effects of the radioactive fallout resulting from nuclear test explosions. Symposium on remediation and restoration of radioactive-contaminated sites in Europe, 11-15 October 1993, Antwerp, Belgium, 1993.
6. Shapiro, C.S. (Ed.) NATO ARW Atmospheric Nuclear Tests: Environmental and Human Consequences (in preparation, 1996 a).
7. Shapiro. C.S. (ed.), NATO ARW Long-term Consequences of Nuclear Tests for the Environment and Population Health (Semipalatinsk/Altai Case Study (in preparation, 1996 b).

Assessment of risks from combined exposures to radiation and other agents at environmental levels

Thomas Jung and Werner Burkart

BfS Institute for Radiation Hygiene, D-85762 Oberschleissheim / Munich, Germany

ABSTRACT

Even in areas with elevated levels of natural radiation, ionizing radiation is only one of many natural and anthropogenic agents potentially effecting human health. At the workplace, complex exposures such as radon and smoking or asbestos and smoking, respectively, have been shown to lead to overadditive effects. Hence, efforts to address interactions between different agents are needed to assess and quantify deleterious effects on human health. On the other hand, already the elucidation of possible health risks from single agents, their dependence on exposure levels, exposure rates, age at exposure and their expression in time is a complex endeavour. Therefore in the past and the present the main emphasis in radiation protection, toxicology, and public health is on the study and assessment of single toxicants. The existing data base on combined effect is rudimentary, mainly descriptive and rarely covers exposure ranges large enough to make direct inferences to present day low dose exposure situations.

In view of the multitude of possible interactions between the large number of potentially harmful agents in the human environment, descriptive approaches will have to be substituted by the use of biomechanistic models for critical health endpoints such as cancer. To generalise and predict the outcome of combined exposures, agents will have to be grouped depending on their physical or chemical mode of action on the molecular (genotoxic/non-genotoxic) and cellular level. As a basic requirement, clear concepts and definitions are needed to give the terms interaction, additivity, synergism and antagonism unambiguous meanings.

1. INTRODUCTION

A multitude of natural and man-made agents have the potential to interact with biological materials in ways leading to irreversible changes or reversible deviations from homeostasis. It is well known from epidemiological and toxicological studies that interactions exist between different toxic agents at moderate to high dose levels. Some of these interactions lead to effects which are greater than those which can be predicted from taking into account only single agent exposures and simple addition of effects. The assessment of possible health risks from single agents at the levels of concentration found in environmental and occupational settings is already prone to large uncertainties. For the analysis of combined effects, there is a great scarcity both of appropriate models and of experimental or epidemiological data. Nevertheless, a full assessment of health risks to the public and to the work force has to consider potential interactions between different noxious agents.

Starting from findings from single agent exposures and assuming linear dose effect relationships from high exposure levels down to environmental levels, about 3 to 8 % of all cancers are caused by present day levels of ionizing radiation [1]. However, even in situations where information about effects of several doses at high levels are available and in which a dose effect relationship can be drawn from these data, simple extrapolations from the high dose situation down to low exposure levels cannot be assumed. Therefore, the real dose-effect relationship for most toxicants, including ionising radiation, is unknown for environmental exposure levels.

Spontaneous alterations of and damage to DNA is of prime importance in the assessment of combined effects. Mispairing of bases during DNA synthesis, tautomeric shifts of bases during replication, deamination of bases, incorporation of uracil during DNA synthesis, loss of bases by depurination and depyrimidination show the inherent instability of DNA. Attack by reactive oxygen species is also a major source of spontaneous damage to DNA. From the many products of radical attack on DNA found in human urine, it can be estimated that there are, on average, 10^4 oxidation events per day within the DNA of each of the 6×10^{13} cells in the human body [2]. In comparison, a cell nucleus being affected from the traversal of a single fl-particle - already resulting in a dose to the cell nucleus of 1 to 2 mSv - would only induce about 5 to 10 DNA single-strand breaks and a 5 - 10% chance of one double-strand break.

The universe of chemicals is believed to constitute 9 to 10 million compounds with over 60,000 in commercial use. However, less than 4,000 comprise over 98% of the total production [3]. Trace heavy metals in the environment are another group of potential carcinogens. Although their low level effects are difficult to assess, high binding affinities to receptors in signal pathways and to -SH containing enzymes might show effects at very low concentrations. For some toxic metals, even the global natural fluxes are small as compared to todayís emissions from industrial activities [4]. Most known human chemical carcinogens were first identified in the work-place. However, the burden of occupational cancer in industrial states is generally considered as less than 5% [5-7].

In most instances the importance of potential risks from anthropogenic physical and chemical agents to human health remain unknown, both in absolute terms and in comparison with the myriadís of natural substances and agents interacting with the human body. With the exemption of UV, asbestos, tobacco smoke and may be radon daughters, the projected excess relative risk from environmental exposures for a specific endpoint and even more for the lifetime risk, are generally too low to be directly accessible by epidemiological studies. The same holds for possible deleterious interactions. However, important examples of more than additive interactions have been established with active cigarette smoking.

Both, occupational exposures to asbestos [8] and radon daughters [9, 10] in smokers, respectively, were shown to increase lung cancer mortality well beyond the level expected from an independent action of the two agents. Cigarette smoke also interacts with hard liquor to increase the risk of cancer in the bucal cavity. Smoking alone already confers a very large risk in the range of 30 % lifetime mortality from lung cancer. Tobacco smoke by itself is already a complex mixture and the effects of smoking cannot be traced to a single agent. Benzo[a]pyrene or long-lived radon daughters such as 210polonium or 210lead are present in too minute amounts to be the major carcinogen in this very complex mixture. Even more difficult is the evaluation of interactions between tobacco smoke and additional agents.

2. DEFINITION OF INTERACTIONS AND COMBINED EFFECTS

Many models to describe the combined effects of exposures to different agents have been developed [11-14]. The conceptualisation of combined effects from two or more agents, has to be built from a clear understanding of the terms used and on simple postulates and models.For a discussion of combined effects, it is necessary to provide clear definitions for terms of interaction like antagonism, additivity, and synergism [15]. What may be straightforward for the simple case where two agents each displaying a linear dose effect relationship becomes much more complicated if non-linear dose effect relationships and/or threshold phenomena are involved. If one looks at a quadratic dose effect relationship (Fig.1), additional increments of a single agent can be considered interacting with earlier increments of the same agent because the effect per unit dose of the additional increment is much higher and depends on the earlier exposures. The term synergism has sometimes be used for such situations [16]. Although mathematically correct, this notion would imply that even the same or different agents with the same mode of action produce synergism in any combination of concentrations as soon as their dose effect relationships are non-linear and show an upward curve.

From a mechanistic point of view, synergism should be defined in a more restricted manner. Synergism implies that the combined effect of different agents results from the independent action on different rate-limiting steps of a multistep process, the action on different sites of a molecule or from different molecular mechanisms. It therefore becomes clear that

(a) deviation from additivity is a poor indicator of synergism or antagonism, since this simple definition is only valid in the case of combinations of agents, which all have linear dose effect relationships (this is extremely rare in biological systems),
and
(b) interactions in a simple mathematical sense do not define interactions in a biological or mechanistic sense.

Table 1
Definitions and equations for binary combinations of agents (equations for linear dose effect relationship

Term	Definition	Mathematical equation
Independent effects	No interaction between A and B	$E_{com} = E_A + E_B$
Interaction	Agent B influences effect of agent A or vice versa	$E_{com} \neq E_A + E_B$
Synergism	A and B act cooperatively to increase the total effect above the sum of the single effects	$E_{com} > E_A + E_B$
Antagonism	A and B oppose the action of each other to decrease the total effect below the sum of the single effects	$E_{com} < E_A + E_B$

Efforts to address this ambiguity date back to the first half of this century [17, 18]. To define the range of deviations from additivity which may be caused by non-linear dose effect relationships, the concept of heteroaddition and isoaddition was developed. In the case of non-linear dose effect relationships, heteroadditivity is present when two agents become involved through different pathways, i.e. truly independently, and the single effects can be combined independently of the shape of the dose-effect curves [19] (Fig.1, lower solid line). Isoadditivity, in contrast, is present when agents A and B act through the same or similar mechanisms. In this case, exposure doses from agent B have to be treated like additional dose increments of agent A. For upward inclined dose effect relationships, the additional dose of the "second" agent needed to double the effect, is then much smaller than the earlier dose, because a steeper part of the dose effect curve is involved. The resulting curve for isoaddition (upper solid line in Fig. 1) forms the second part of the envelope of additivity. Any effect resulting from a mixture of agents A and B and adding up to a standardised exposure dose of one, and, falling into this envelope, is then in the realm of additivity. In real life the area delineated by the two curves is enlarged by experimental and statistical uncertainties.

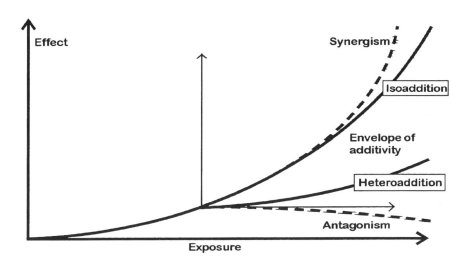

Figure 1.Interaction of two agents having non-linear dose-effect relationships. Isoaddition results from mechanistically similar, heteroaddition from independently acting agents.

In theory, isoaddition in the case of an upwardly inclined dose effect relationship defines an increased risk per additional unit dose, and therefore may call for an adjustment, e.g. of exposure limits. However, in those cases where risk projections and protection from single agents are based on conservative assumptions, using linear dose effect functions derived from high exposure doses, isoaddition will only reduce the safety margin of the linear hypothesis, and never exceed the projected risk.

For many chemicals acting through non-genotoxic/cytotoxic mode of action, important endpoints such as cancer formation, will only result from the interaction with primary events like initiation and subsequent promotional events occurring during regenerative cell growth induced by cytolethality. Under such circumstances, an adverse effect will only occur at exposures above the threshold for cell death. Linear extrapolations to low doses will therefore lead to considerable overestimation of risk.

An increased focus on complex exposures will certainly increase the awareness of the potential risks due to various interactions and combined effects. This awareness will increase in particular due to the growing knowledge about the complexities of biological mechanisms which produce health effects like cancer. A mixture of agents could contain elements affecting different steps in the carcinogenic process (e.g., DNA damage and growth factor secretion) or alternate pathways contributing to the same steps (e.g., different growth factors). For protective purposes, the class of combined effects to be considered in depth arise from real interactive effects. These effects can arise from agents with different action spectra. Predisposition to one agent, e.g. changes in sensitivity or target size for the second agent, might lead to deviations from additivity towards synergism or antagonism. Theoretically, the most critical interactions occur in multistep processes in which two different agents would promote different steps, which normally have a low probability of occurrence. In such a situation, highly synergistic effects from combined exposures could result. Only the understanding of the mechanisms involved in the agents under consideration on the physico-chemical, molecular and cellular level will finally allow a prediction of the effects of combined action, and the rejection of overly conservative assessments.

3. IMPORTANT INTERACTIONS

The potential importance of combined exposures for quantitative risk assessments was first recognised in interactions of noxious chemicals. The strong dependence of asbestos-related lung cancer risk on smoking status in exposed workers [8] is one early example. For the development of a better understanding and of risk models, experimental work on the in vitro and animal level is of outmost importance. The following subsections give examples. Selected examples of interactions between ionizing radiation and other agents leading to non-additive effects are listed in Table 2.

3.1. In vitro
3.1.1. Molecular level

The molecular level is well suited to develop hypothesis and test modes of interactions. The cellular function of a specific molecule can be altered independently by activation or deactivation through binding of stimulators or inhibitors, molecular modification, cellular location and changes in turnover. The reaction between agents and the molecular and cellular structures might involve non-covalent or covalent binding - the latter being irreversible in many cases - or molecular damage. Already at the level of a transmembrane molecule or a 3×10^8 base pair human chromosome, a multitude of functional epitopes (e.g. receptor binding, switching between different conformational states, kinase activities) can be altered or damaged by exposure to an agent. Compounds which bind to the same site and induce the same molecular changes will act in an isoadditive manner. The situation can be best described as the action of different increments of the same agent.

An example of antagonism is the competition of ethanol and methanol for the same active siteof the enzyme alcohol dehydrogenase, which catalyses the degradation of the compounds, leading to the production of the highly toxic methanol metabolite formic acid and the less toxic ethanol oxydation products. This example also shows that when the whole organism is considered, the assessment might change. Only the more benign ethanol would be labelled an antagonist. Although methanol affects ethanol degradation antagonistically, the much higher toxicity of its decay products renders a notion, which is correct on the enzymatic level, senseless.

Table 2
Selected examples of interactions between ionising radiation and other agents leading to clearly non-additive effects.

System	Radiation quality (exposure situation)	Synergism with (risk/effect)	Antagonism with (risk/effect)
in vitro: molecular	gamma, beta, X rays	oxygen (DNA breaks, base damage through the formation of more toxic oxygen radicals)	radical scavengers such as ethanolamine (DNA breaks, base damage)
in vitro: cell culture	alpha rays (hamster cell cultures)	asbestos (transformation rate)	
	X rays (hamster cell cultures)	beryllium (chromosome aberration rate)	
	X rays (hamster cell cultures)	TPA* (transformation rate)	vitamin A, vitamin E, selenium, or benzamide (transformation rate)
	X rays (Mouse thymocytes)		zinc (apoptosis)
Animals	X rays (mice)	TPA* (skin cancer)	kept germ free (lymphoma)
	alpha rays (Radon in beagle dogs)		tobacco smoke (lung cancer)
Epidemiology	alpha rays (Radon in miners)	tobacco smoke (lung cancer)	
	X rays (Breast cancer therapy)	tobacco smoke (lung cancer**)	

* Phorbol acid, a potent tumor promotor;   ** statistical significance questionable

Environmental chemicals with estrogenic effects are an example of synergistic effects on the molecular level. Combinations of two weak environmental estrogens like dieldrin, endosulfan, or toxaphene are about 1000 times as potent in activating human estrogen receptor as any agent alone [20]. It was shown that the agents act in combination with the human estrogen receptor at the hormone binding site.

A screening of potentially critical interactions on the molecular level leads to the detection of bi- and multifunctional molecules, and specific actions of different agents on these different functionalities. If the binding of one agent increases or decreases the affinity or the reactivity of another site of the molecule with a second agent, strongly non-additive effects may result.

3.1.2. Cellular level

On the cellular level, many additional types of interactions are to be considered. Agents may act on different enzymes or DNA segments and still cause functional interaction. If the targets represent different steps of a cellular reaction cascade, e.g. different enzymes needed for the synthesis of a cell component, or different oncogenes / tumor suppressor genes involved in the control of the same class of stem cells, interferences at multiple steps may lead to strongly non-additive interactions.

The emergence of many solid tumors depend on several consecutive genetic changes occurring in the same cell. In the case where two of these changes are rare, and therefore rate and risk limiting, the combined action of two different agents acting specifically on change A or B, respectively, may short-circuit a normally slowly developing process and theoretically lead to a catastrophic increase in risk. Fortunately, few noxious agents meet the specificity needed for such a combined effect. In the real world, heterozygosity for tumor suppressor genes is an example where one rate limiting step has already occurred and is promoted through the germ line. The dramatic increase in cancer risk is for example apparent in the hereditary form of retinoblastoma [21].

Viral DNA or RNA is another example of having potentially the high specificity needed for such events to happen. However, the bulk of genotoxic chemicals and also ionizing radiation are much to unspecific in their attack on DNA to cause selective damage on specific DNA loci.

3.2. Experimental animal systems

Practically all combined exposures of interest can be studied under controlled conditions in animal systems. Therefore they are an important tool in investigating the quality of the interaction and the dependence on the many defence mechanisms available in a complex mammalian organism. Compared to the discovery of interactive effects in human populations through clinical or epidemiological studies, there are a number of advantages in the use of toxicologic approaches. First, they provide a way to identify combined exposure leading to non-additive effects in the absence of human data. For example, this information may be helpful in identifying potential problems in environments, both occupational and environmental, containing mixtures of toxic chemicals and radiation. Second, well-controlled, appropriately designed experiments provide a mean of analysing potential interactions between agents in a precise, stepwise manner, including determination of the relative importance of the constituents toxicologic investigations can be designed to include evaluations of mechanisms acting to produce an adverse response, such as dose to critical sites, toxicokinetics of responses and/or repair of damage; i.e. the full web of mechanisms leading from exposure to adverse health outcomes.

The extrapolation of insights found in such animal systems into quantitative risk projections for low-level exposure of much longer-lived human beings, however, is ridden with many problems. Such extrapolations are strengthened when it can be demonstrated that the same or similar biological mechanisms are rate limiting both in the experimental system and in humans. However, in addition to important differences between laboratory animals and humans that have been recognized for some time, such as differences in physiology, metabolism, and pathologic responses, it is becoming apparent that the spectrum of mutations present in certain genes in human and animal tumors may be different. For example, the lack of concordance between mutations found in lung tumors from humans versus experimental animals argues for caution in extrapolating results across species even in single-agent studies.

3.3. Human experience, epidemiology

Some of the critical combined exposures to the general public also occurred at much higher levels in the workplace. These are, therefore, susceptible to epidemiological studies. Epidemiological studies have already revealed much about carcinogenic interactions, particularly with regard to the interaction of cigarette smoke with either radiation or asbestos.

Use of new tools in molecular biology indicates the path leading to the field of molecular epidemiology, in which investigators will have the ability to use markers of exposure and damage that are much more sensitive than the crude incidence measures of clinical diseases. Such approaches can be expected to yield significant new information regarding interactions between agents at the cellular and molecular levels. Examination of markers of these types can also be included within more classical human epidemiological studies of cancer. In some cases they may well help in probing potential interactions between agents that may have lead to induction of the disease.

The use of combined modalities in cancer therapy covers a large field of mechanistically different interactions of agents where human data is available. However, the high doses and the generally unusual substances used as well as the endpoint considered (tumour cell killing in a limited area of the body) prevent direct inferences to the general public.Epidemiology indicates that indoor radon daughter exposure and cigarette smoking warrants special consideration due to the large fraction of the world population exposed to considerable levels of both toxicants and the clearly over-additive effects found in the meta-analyses of miner studies [9, 10]. Regulatory perspectives, health care policies and mitigation strategies in this extremely important area depend crucially on the judgement of whether tobacco smoke is the main confounding factor in radon progeny induced lung cancer risk.

4. OUTLOOK / CONCLUSIONS

The most important parameters to be considered in the assessment of risk from combined exposures include in depth knowledge of the primary actions of agents involved, quality of interaction, dose, dose rates and time sequences of combined exposures. Important host characteristics include cell types and tissue sites affected, age, sex, genetic traits and predispositions, life style, hormonal conditions and immune status. Time relations between exposures to different agents are critical parameters, too. The collective influence of the above mentioned factors, and possibly others, increases further the uncertainties found, in estimates of risk for populations exposed to low doses of ionising radiation.

There are many examples of interactions between ionising radiation and other toxic agents. Responses observed in humans range from additivity to strong synergism. The available database is not comprehensive and is generally more observational and descriptive rather than mechanistic in orientation. In most cases, findings from existing studies have involved high exposure levels which may occur only in the realm of combined tumor therapies, in accidents, or in exposure situations typical for the time before the implementation of modern protection labor laws. Inferences, therefore, for today's occupational and environmental exposures are difficult to draw.

Only identification and quantitative elucidation of all the mutational or other events which drive cells to malignancy will provide the information needed to make specific predictions on the risk from combined effects. Despite considerable progress in the last decade, risk predictions based on purely mechanistic knowledge remain a remote possibility. Although in some cases, where the minimum number of steps required to lead to specific cancers (for example colon cancer) is known, it has become apparent that as a further complication, there are multiple pathways to the same endpoint to be considered.

Clearly, direct damaging effects to the genome (i.e., genotoxicity) are of primary importance when examining radiation-induced carcinogenesis and when considering how radiation can act in concert with other agents. Over the last few years it has become increasingly apparent that a wide variety of agents can work through non-genotoxic mechanisms. These include toxins that interfere with critical cellular processes such as control of cell proliferation, cell differentiation, cell-cell signalling, repair of damage, and apoptosis. Conceptually, then, one can envision a wide range of ways in which radiation can interact with genotoxic or non-genotoxic agents.

The use of mathematical formalisms to model carcinogenesis using existing human datasets can be supplemented by examining and quantifying the various carcinogenic stages in specific animal tumour models. Carcinogenicity models can and should now be used to study the combined effects of agents acting on different steps within the multistage process leading to cancer. Refined definitions of antagonism and synergism, based more on the mechanisms involved in carcinogenesis rather than on the outcome of high exposures, may arise from the recognition and successful modelling of the way in which combinations of agents may operate at different steps within a chain of events.

Descriptive as well as mechanistic approaches to model combined effects tend to become very complex. At this stage, frustratingly little quantitative inference can be made from these models with regard to exposure situations outside the range where direct experimental or epidemiological information is available. Some general observations, however, can be made. In the case of combined exposures to radiation and non-genotoxic agents having non-linear dose-response relationships or even apparent threshold levels in some cases, a simple linear extrapolation of either antagonistic or synergistic interactions observed at relatively high to low exposure levels is not warranted. For the case of the interaction of radiation with genotoxic chemicals, however, prudent considerations should provide for the potential of risk modifications by interaction down to very low exposure levels. The cautious approach of assuming linear dose-response relationships scaled to results from high exposures for multistep endpoints like cancer takes care of some theoretical notions of synergism. This means that in situations where high dose exposure data indicate a clearly synergistic interaction, the combined effect in the low dose region may well be below the predictions from cautious linear risk models due to pronounced sublinearity of response to one or more single agents in the low dose region for example in the case of sigmoid dose response relationships.

In general, however, it can be said that interactions between radiation and other physical, chemical and biological agents are an important modifier in many biological processes and outcomes. Their implications for the limitation of individual and collective health risks need careful consideration.

Present knowledge on the many qualitatively different interactions already found in biological systems speaks against the emergence of simple unifying concepts to predict modification of risk from combined exposures. Nevertheless, mechanistically based classifications of interactions into groups may be helpful for effect predictions. The development of such comprehensive approaches for the study and quantitative assessment of combined effects is urgently needed to bridge the gap between differing conceptual approaches in the assessment of risk in chemical toxicology and radioprotection.

REFERENCES

1. United Nations Scientific Committee on the Effects of Atomic Radiation, Sources and effects of ionizing radiation , 1993.
2. Ames B. , Environ.Mol.Mutagen. , 14 (1989) 66-77.
3. Blair E, Bowman C. In: Ingle G.W.(eds.), TSCA's Impact on Society and Chemical Industry, ACS Symposium Series, No. 213, 1983.
4. Nriagu J. , Nature, 338 (1989) 47-49.
5. Wynder E, Gori G. , J.Natl.Cancer Inst. , 58 (1977) 825-832.
6. Higginson J, Muir C. , J. Natl. Cancer Inst. , 63 (1979) 1291-1298.
7. Doll R, Peto R., The Causes of Cancer: Quantitative Estimates of Avoidable Risks of Cancer in the United States Today, New York: Oxford Universtiy Press, 1981.

8. Selikoff I, Hammond E, Seidman H. , Ann. N. Y. Acad. Sci., 330 (1979) 91-116.
9. Lubin J, Boice J, Edling C., et al. , NIH Publication No. 94-3644, 1994.
10. Lubin J, Steindorf K. , Radiat.Res. , 141 (1995) 79-85.
11. Mauderly J., Environ. Health Perspect.(suppl.), 101 (1993) 155-165.
12. Scott B. , Radiat. Res., 98 (1984) 182-197.
13. Sipes G, Gandolfi A. , In: Amdur M, Doull J, Klaassen C. (eds.). , Toxicology: The Basic Science of Poisons, 4th edition, New York, Pergamon Press, 88-126, 1992.
14. Zaider M, Rossi H. , Radiat.Res. ,8 3 (1980) 732-739.
15. Streffer C, Müller W. , Adv.Radiat.Biol. , 11 (1984) 173-209.
16. Zaider M, Brenner D. , Radiat.Res. , 99 (1984) 438-441.
17. Frei W. Z., Hyg. , 75 (1913) 433-496.
18. Loewe S. , Ergebn. Physiol. , 27 (1928) 47-187.
19. Steel G, Peckham M. , J. Radiat. Oncol. Biol. Phys. , 5 (1979) 85-91.
20. Arnold SF, Klotz DM, Collins BM, et al., Science , 272 (1996) 1489-1492.
21. Knudson A. , Cancer Res. ,45 (1985) 1437-1443.

Exposure dose from natural radiation and other sources of the population in China

Zhu Changshou[1], Liu Ying[1], Zhang Liang'an[2]

[1]Laboratory of Industrial Hygiene, Ministry of Health, 2 Xinkang Street, Beijing 100088, China

[2]Institute of Radiation Medicine, CAMS, Tianjin, China

1. INTRODUCTION

Throughout the history of the world, mankind has always been exposed to radiation from a variety of natural sources. Within the last fifty years, however, so many additional types of exposure have been developed artificially that it is important to review the amount by which our total exposure has been and will be increased. So the investigation of exposure from varied ionizing radiation sources is a significant background material for assessment dose-effect relationship.

This article systematically summarized the radiation exposure from varied sources of the population living in normal conditions in China.

The contents in this article include a huge and different survey data which is conducted under the guidance and support by national Ministry of Health. The institutions of medical research or the radiation protection agencies of 29 provinces, autonomous regions and metropolitan of the whole-country were cooperated. Laboratory of Industrial Hygiene of Ministry of Health and Institute of Radiation Medicine of Chinese Academic Medicine Science conducted the responsible units for designing the programs and technical guide.

2. EXPOSURE FROM NATURAL SOURCES

Natural radiation and natural occurring radionuclides in the earth's crust provide the major source of human radiation exposure. For this reason natural radiation is frequently used as a basis of comparison for various man-made sources of ionizing radiation.

2.1. External exposure
2.1.1. Cosmic rays

Considering that China is a large country with a vast area of high altitude plateau. So the intensity of ionizing component is varied. The Tibet autonomous region with a high altitude, the average effective dose from cosmic rays is 1.88 mSv/a high. The doses in the regions of east-south seacoast as Guangdong and Shanghai are only 0.23 mSv/a. The average effective dose from cosmic rays in nationwide of China is 0.27 mSv/a.

2.1.2. Terrestrial gamma radiation

The terrestrial gamma radiation in China has been investigated systematically. Measurements of gamma radiation were carried out outdoors, indoors and over roads.

The annual effective dose of the countrywide of China varied from 0.51 mSv/a (Beijing) to 0.88 mSv/a (Fujian) and its average dose is 0.59 mSv/a. The results from different measuring methods are shown in Table 1.

Table 1
External dose from natural radiation using different measurement methods

Methods	No.of site	Annual effective dose (mSv/a)	
		γ	$\gamma + C$ *
Dose rate			
FD-71	92563	0.68	0.95
RSS-111	2000	0.65	0.91
Integrating dose			
TLD in site	5112	0.53	0.82
TLD person wear	1821	0.52	0.82
Soil			
γ-spectrometry	1848	0.59	0.87
Dose rate			
FH-620**	28874	0.55	0.81
Arithmetic mean		0.59	0.86

*γ = terrestrial (radiation, C = Cosmic rays
** referenced from the survey results of NEPA China

The total average annual effective dose from cosmic rays and terrestrial gamma radiation of the countrywide is 0.86 mSv/a and its collective effective dose is 9.7×10^6 man-Sv/a.

2.2. Internal exposure
2.2.1. Inhalation of radon and its daughters

The concentration of radon and its daughters in indoors and outdoors in China have been investigated. Measurements were made in 10654 dwelling indoors and 4269 data of outdoors were obtained. The average concentrations of Rn in indoors and outdoors in countrywide are 20.2 Bq/m^3 and 12.4 Bq/m^3 respectively.

The existence of high Rn concentrations in some dwellings which constructed by using special building materials from stone, brick of coal-ash or mining cinder. The Rn concentration of several hundreds Bq/m^3 was observed, which is expected considering the higher ^{226}Ra contents in these building materials.

The high Rn concentrations in underground buildings and cave dwellings also are obtained in some provinces.

The annual effective dose of Chinese population from inhalation Rn and its daughters ranged from 0.52 mSv/a (Hebei) to 2.3 mSv/a (Fujian) and its average value is 1.2 mSv/a.

2.2.2. Ingestion of radionuclides in foods and waters

In order to provide the internal dose from ingestion of Chinese population, the contents of natural radionuclides in different kinds of foods and waters were investigated.

The ranges and average values of specific activity of natural radionuclides U, Th, ^{226}Ra, ^{228}Ra, ^{40}K, ^{210}Pb and ^{210}Po in 81 catalogs 2840 food samples and 2350 water samples, which represents components of Chinese diets, have been obtained. The annual committed effective dose from intake of these radionuclides is 0.36 mSv/a.

2.3. Total exposure from natural radiation

The exposures resulting from the various sources of natural radiation are summarized in Table 2. The annual effective dose per capita of Chinese population is 2.4 mSv/a and the collective effective dose is 28×10^5 man-Sv. The dose from Rn and its daughters contribute the half of the total dose from all natural sources.

People living in some regions receive the higher exposure dose from natural sources. The annual effective dose to the population in Fujian province is 5.0 mSv/a which is two times of the average values of the countrywide. The dose to Tibet regions is 1.88 mSv/a due to the high dose from cosmic rays, but the contribution of the collective effective dose is lower as the results, of the lower population in Tibet.

3. EXPOSURE FROM MEDICAL RADIATION

The distribution and application of the medical radiation facilities are very non-homogeneous in China. In cities hospitals and other medical services are

Table 2
Annual effective dose from natural radiation to Chinese population

Source	Effective dose (mSv/a)	Distribution (%)
External exposure		
Cosmic rays		
Ionizing	0.24	9.9
Neutron	0.035	1.4
Terrestrial radiation	0.58	24.2
Subtotal	0.86	35.5
Internal exposure		
Ingestion (diet)	0.36	14.9
Rn	1.2	49.6
Subtotal	1.56	64.5
Total	2.42	100

equipped with a good medical facilities, and in the vast countryside the medical facilities is poor relatively.

The frequency levels of medical radiation exposure in China have been surveyed in the century of 1980's. The frequencies of medical exposure diagnostic X-ray, clinical nuclear medicine and radiation therapy in our country were 145, 0.62 and 0.09 case per 1000 persons, respectively.

The effective dose per capita for medical exposure in China is 88 (Sv/a, the GSD is 9.8 x Sv/a. The annual effective dose from clinical nuclear medicine per caput is 22.3 x Sv/a.

Beijing with a good medical facilities, the average effective dose to the Beijing residents from diagnostic X-ray is 0.4 mSv/a.

The annual effective dose per caput from medical radiation is 0.12 mSv/a and the collective effective dose is 1.4×10^5 man-Sv.

4. EXPOSURE FROM MAN-MADE SOURCES

4.1 Atmospheric nuclear tests

Since 1961 China's environmental radiation monitoring actives have been performed. The network is composed of 45 local monitoring stations, distributed all of China, and is headed by Laboratory of Industrial Hygiene, Ministry of Health.

The monitoring results from 1961--1990 show that the peak value of gross β activity and ^{90}Sr appeared in 1963, and then decreased year by year with some variation.

The estimation of the dose from fallout of atmospheric nuclear tests are given in Table 3. The results showed that the effective dose commitment from fallout deposition to Chinese population is similar comparison with the population living in north hemisphere and north temperate zone.

Table 3
Effective dose commitment from fallout to Chinese population (µSv)

Sources	China	North temperate zone	World
External exposure	1000	1000	600
Internal exposure			
^{90}Sr	252	180	102
^{137}Cs	160	280	170
^{131}I	79	79	48
^{3}H	48	48	44
^{14}C	2600	2600	2600
Others	240	240	150
Total	4400	4400	3700

4.2. Exposure from nuclear power generation

The survey of the radiation doses of the inhabitants within 80 km of nuclear facilities have been made in 15 large nuclear facilities. The average annual effective dose of the inhabitants exposure from nuclear power generation is 37 µSv. The highest dose of those inhabitants is 1.8 mSv per year.

The average annual effective dose of the public exposure from nuclear power generation is only 1.7×10^{-3} µSv and the collective effective dose is 1.93 man-Sv which is below 1/10,000 of the dose from nature exposures.

5. OCCUPATIONAL EXPOSURE

The monitoring of the radiation doses of the workers who use ionizing radiation and/or radiation materials in their work have been made. The average annual effective dose of the workers exposure from nuclear power generation is 9.2 mSv and the collective effective dose is 589 man-Sv. The doses of 57% of the monitored person are below 5 mSv/a.

The annual effective doses of the occupational exposure for medical radiation are range from 1.4 to 2.7 mSv and its average dose is 2.1 mSv. The collective effective dose in 1990 is 245 man-Sv.

The average annual effective dose of the other occupational exposures is 1.5 mSv and the collective effective dose is 123 man-Sv.

The total annual effective dose of the all occupational exposures is 1.9 mSv and the collective effective dose is 958 man-Sv.

6. EXPOSURE FROM CONSUMER PRODUCTS

During the last several decades, there has been a wide range of consumer products which are sources of ionizing radiation. Examples include combustible fuels and building materials containing members of U and Th decay series, tobacco products containing ^{210}Pb and ^{210}Po, and radioluminous products containing ^{3}H, ^{147}Pm or ^{226}Ra.

The dominant contributors to the population dose from consumer products are tobacco products and combustible fuels. The average annual effective dose of smokers resulting from the use of tobacco products is 58 µSv and its collective effective dose is about 20,400 man-Sv.

Building materials and combustible fuels, depending on their origin, contribute appreciable quantities of the radon to the indoor air and thus irradiate the lungs. For the non-determinacy of the population, it is not possible at the present time to estimate the collective effective dose from building materials and combustible fuels.

Table 4
The effective dose from all kinds of ionizing radiation of Chinese population

Source	Annual dose		Collective dose
	mSv/a	%	(man-Sv)
Natural	2.42	92.4	27.5×10^5
Medical	0.12	4.6	1.36×10^5
Man-made radiation			
Nuclear tests	0.06	2.3	0.68×10^5
Nuclear industrial	2.0×10^{-5}	<0.01	1.93
Occupational	8.4×10^{-2}	0.03	9.58×10^2
Others			
Air travel	8.7×10^{-5}	<0.01	99
Smoking	0.018	0.7	2.04×10^4
Others	0.002		
Total	2.62	100	29.7×10^5

The average annual effective dose from all kinds of ionizing radiation sources of Chinese population is 2.6 mSv/a and the contribution of dose from natural sources is 92.4% of the total dose. Table 4 gives the results in detail.

Transfer of nuclear fission products to crops and soils and counter-measures

Xu Shiming, Zhao Wenhu, Hou Lanxin and Shang Zhaorong

Institute for Application of Atomic Energy, Chinese Academy of Agricultural Sciences, Beijing 100094, China

Abstract: Many radionuclides, such as fission products ^{90}Sr, ^{137}Cs, ^{59}Fe and ^{60}Co may be released in a nuclear accident and disseminated on soil and vegetation. The accumulation and distribution of the radionuclides in crops and vegetables growing on the soil contaminated by nuclear test are described. The factors, such as chemical forms of nuclides, agrochemical properties of soil, growing stages of the plants etc., may affect the accumulation and the distribution. Possible counter-measures including planting suitable herbage, applying adequate fertilizers, scraping off the regolith, etc. were suggested to eliminate or clear the radionuclides away from contaminated soil.

1. INTRODUCTION

In the past half century nuclear techniques have been applied to industrial, agricultural and medical fields, and have played an important role in the development of science and technology. However, with wide application of nuclear techniques, more radioactive materials may be released and deposited into environment. It is estimated that nuclear power will increase up to 6.5 million kilowatts in China by the year 2,000.

Though nuclear energy is considered as safe and clear, the nuclear accident may occur. If an accident occurs, a large number of radioactive fission products would enter the environment, especially agroenvironment. Such nuclides as ^{90}Sr, ^{137}Cs and ^{131}I are highly toxic. They could directly contaminate water sphere, soils and vegetation, and indirectly contaminate milk, egg, fish etc. The accident in Chernobyl nuclear power plant, which released a great quantity of radioactive materials and resulted in harmful impact far beyond its own country, is an unfortunate event to be ever remembered.

The purposes of this project were to study the accumulation of radionuclides and their biological effects and simulated agroecological environments, to propose the methods of decreasing losses for policy makers' reference, offer useful data for limiting the quantity of radioactive materials in agricultural products and animals, to estimate the absorption coefficients of main nuclides transferred from soils to plants, and to provide suitable parameters for protecting Chinese residents against nuclear contamination.

2. MATERIALS AND METHODS

2.1 On-the-spot tests

On-the-spot pot culture tests were allocated in the leeward area of nuclear explosion test.

2.2 Simulation tests

By using the fallout collected in the area of nuclear explosion test, the simulation tests were made with crop subjected to different radioactivity under protected conditions.

By using fission products, either in solid form or in liquid form, the simulation tests were made with crops, which absorbed ^{90}Sr, ^{137}Cs, ^{59}Fe, ^{60}Co, etc. through roots or leaves.

3. RESULTS AND ANALYSIS

3.1. The fallout (particle diameter < 1mm, solubility 0.5%) dropped in fields in the leeward area of low altitude nuclear explosion may contaminate the growing crop plants and vegetables. The contamination level was found to be: corn (268.03 x 10^4 Bq/kg dry material) > rape (7.96 x 10^4 Bq/kg dry material) > wheat (2.82 x 10^4 Bq/kg dry material) > cotton (0.59 x 10^4 Bq/kg dry material). Since the leaves of corn were broad and rough, and the leaf sheath wrapped tightly around the stems, they could retain more fallout than other plants.

Table 1
Contamination of fallout in crop (x 10^4 Bq/kg dry weight)

Corn	Rape	Wheat	Cotton
268.03	7.96	2.82	0.59

3.2. By using corn as material, an experiment began at the fourth day after fallout contamination, and the experiment was divided into two groups. In one group, the samples were smashed and fixed with acetone solution. The radioactivity was measured at different stages. The result showed that the radioactivity dropped down to 66.23% of the initial radioactivity after 6 months owing to natural nuclear decay.

In the other group, the samples were collected from living plants at different periods. After 6 months, the radioactivity decreased down to 0.15% of the initial value. This was due to mechanical detention and physical adsorption which resulted in external contamination of crop plants by fallout. But such fallout could be removed quickly by shaking, blowing, and shower. This result was proved by washing the contaminated leaves of rape with water. Over 98% of pollutant was eliminated.

3.3. The fallout may give rise to internal contamination by roots of crops. According to the activities of the particles of fallout per square meter an hour after explosion, the radioactive nuclides were applied to pot culture of wheat and rice at four different levels, namely: $3.7 \times 1,010$ Bq, $185 \times 1,010$ Bq, $370 \times 1,010$ Bq and $740 \times 1,010$ Bq. It was showed that the radioactive materials applied at the latter two levels were obviously absorbed by the roots and concentrated in the stems and leaves. The radioactivity in stems and leaves was much higher than that in seeds.

The total Sr content (precipitated with sulphate) of the wheat seed in the group with the largest contamination was 0.47×10^2 Bq/kg, while that in the rice seeds was 0.03×10^2 Bq/kg. The wheat in this test was harvested 92 days after explosion. The radiochemical analysis showed that the ^{90}Sr content at this time was 3% of the total radioactivity.

If we take experimental error and safety factor into consideration, the ^{90}Sr content of the wheat seed was 0.04×10^2 Bq/kg which was lower than that of the national permissible criterion (0.074×10^2 Bq/kg). The plant materials used in these tests were collected from the area where the plants were damaged by nuclear radiation and blast wave, or the external irradiation of the fallout was so serious that men could not enter this area. If the crop plants were polluted again by the fallout, then the food will lose its edibility. If the radioactive fallout in the soils could be fixed or swept off by using film from high polymer solution, then better results could be obtained than rainwash of the dusts.

3.4. By using the radioactive sediments collected near the explosion site, a set of pot cultures was made again. The sediments were smashed and sifted. The particles were within 100 μm in diameter. The radioactivity was 1.07×10^4 Bq/kg. Gamma spectrometry showed that the sediments contained cerium, zirconium, niobium, etc. It was found that the total activity in corn in two contaminated groups had the same grade as C.K. It was proved that few radionuclides were absorbed from fallout by roots of plant. The result was similar to that at the nuclear explosion area.

3.5. Soil polluted by radioactive substances was used continually after plough in the second year. It was shown that if the crops did not have internal contamination in the first year, they were all the same in the next year. This also could be proved by successive dissolution of the fallout. The fallout was dissolved 5 times in distilled water. The quantity of soluble material in the first time was 54% of the total. Only 9% addition was found in the second solution. Other solutions even showed fewer additions.

3.6. The movement of fallout in the soil

The distribution of radioactive materials in different layers of soils in pots planted with corn and wheat in the leeward area of nuclear explosion was studied (see Table 2). 84.5%-99.21% of the content was found in the 0-5 cm layers. In the case of wheat test, 99.21% of the radioactive material was found in 0-3 cm layers. In the case of corn test, 47.19% of radioactive materials existed in the 0 - 3cm layers, while 37.31% was found in the 3 - 5cm layers. As a result of constant loosing the soil during corn growth, the radioactive particles moved downward, whereupon only 15.5% of radioactive content was found under the 5cm layer.

Table 2
Distribution of fallout in different layers of soils(%)

Layer(cm)	Corn	Wheat
0 - 3	47.19	99.21
3 - 5	37.31	0.23
5 - 7	6.58	0.15
7 - 10	5.50	0.09
10 - 20	3.42	0.32

The movement of fallout in soils was studied again by using the soils collected from the explosion test site. It was found that the radioactive materials existed only in the 0-4cm layers. The natural movement of fallout downward was slight. This provides a theoretical basis for eliminating fallout pollution by means of scraping off the regolith, uprooting, piling up the soils and fixing up.

3.7. The soils taken from the 0-2cm layer of grounds polluted by fallout was soaked in distilled water and 0.5 N ammonium carbonate, respectively. The radioactivity was measured after evaporation. The result showed that under normal cultivation, most of the nutrient elements were adsorbed on the surface of soil particles. An exchange took place between crop roots and soil particles in the presence of water, whereby the nutrient was absorbed by the crop plants.

Table 3
The status of fallout in soils

Type of soils	Aquation status		Aquation and exchangeable status		Unexchangeable status	
	dpm	%	dpm	%	dpm	%
Dry land	9.0	0.02	27.3	0.07	37434	99.93
Rice soil	8.9	0.03	19.8	0.067	32825	99.94

The more the exchangeable elements in soil, the more the absorption by crop plants could happen. The radionuclides of fallout in soils were scarcely in exchangeable status (Table 3), while most of them could not contact with roots in the surface layer of soils. This is the reason why very few radionuclides were absorbed by the crop plants near the fallout site.

3.8. The ability of absorbing radionuclides from the solution of the complex fission products by crop plants was much higher than that from the solid fallout. The test result showed that the ability of absorbing radionuclides from soluble fission products by rice plants was much higher than that from solid fallout (see Table 4). The solubility of fallout and the number of radionuclides in exchangeable status in soil were determinant factors for calculating the absorption coefficient.

Table 4
Transfer coefficient of fallout in rice

Status	Contamination level (Bq/kg)	Stems and leaves (Bq/kg)	Coefficient
Solid	6.66×10^5	3.22×10^3	0.0048
Liquid	18.5×10^5	7.93×10^5	0.4286

3.9. Using soils of Qinshan and Daya Bay as an example, the absorption coefficients of ^{90}Sr concerning various crops and vegetables were calculated.
3.9.1. The root of crop plants would not be polluted if soil contamination was 370Bq/kg. The safety criterion of vegetables was 37Bq/kg.

3.9.2. The activity in the contaminated parts of crops was logarithmically proportional to the content of ^{90}Sr in soils.

3.9.3. The soils were different in acidity and basicity. ^{90}Sr could be absorbed much easier by crop plants on acid soils. The absorption coefficient of rice stems and leaves on Qinshan soils was 1.12-7.00 while that of seeds was 0.02-0.062.

The absorption coefficient of ^{90}Sr on Daya Bay soils was 3.6-4.6 times higher for stems and leaves, and 3.4-4.1 times higher for seeds than those on Qinshan soils.

3.10. In an average, after ^{90}Sr was absorbed by roots, its distribution was 60% in leaves, 20% in stems, 16% in shells, and 4% in seeds. As for ^{59}Fe, its distribution was 74.86% in roots, 20.86% in stems and leaves, 3.86% in shells and 0.42% in seeds. With respect to ^{60}Co, its distribution was 61.28% in roots, 31.6% in stems and leaves, 3.9% in shells and 3.22% in seeds.

3.11. Fertilizers had effects on the absorption of ^{90}Sr and ^{137}Cs by wheat. Some plants could decontaminate the soil contaminated by ^{90}Sr.

K_2SO_4 could reduce the accumulation of ^{90}Sr in stems and leaves of wheat by 14.66% of the total and that in seeds by 11.11% of the total. K_2CO_3 could reduce ^{90}Sr by 9.4% in stems and leaves, and by 5.56% in seeds. Fertilizers containing K such as K_2CO_3, K_2SO_4, and KCL could restrain the absorption of ^{137}Cs by wheat. The effect of plant ashes is similar to that of KCl.

3.12. The selection study was made using rice soil around NPP and Beijing drab soil in order to measure the ability of accumulation in plants for ^{90}Sr. The results show that among the tested plants the plants of calabash family have the biggest ability and the plants of grass family have the least one. The absorption coefficients of ^{90}Sr are 16.8-0.6 in rice soil and 6.0-0.5 in drab soil. For decontamination use, it is in good time and economical to harvest for grass family plants during shooting-booting period, and for calabash family and bean cultures before bud-blooming stage.

Dosimetry study of residents near Semipalatinsk nuclear test site

J. Takada, M. Hoshi, S. Endo, M. Yamamoto[a] T. Nagatomo[b], B. I. Gusev[c], R. I. Rozenson[c], K. N. Apsalikov[c], and N. J. Tchaijunusova[c]

International Radiation Information Center, Research Institute for Radiation Biology and Medicine, Hiroshirna University, Kasumi 1-chome, Hiroshima 734, Japan

[a] Kanazawa University, Tatsunokuchi, Nomi-gun, Ishikawa 923-12, Japan

[b] Nara University of Education, Takabatake-cho, Nara 630, Japan

[c] Kazakh Scientiflc Research Institute for Radiation Medicine and Ecology, Semipalatinsk, 490050 Post Box 16, The Kazakhstan Republic

Measurements of environmental radiation doses and thermoluminescence dosimetry in some villages near Semipalatinsk nuclear test site are presented. The radiation levels in several villages are the natural background level in 199_5 six years after the final nuclear explosion. On the other hand thermoluminescence dosimetry exhibits significant radiation exposure to residents in this area due to nuclear test.

1. INTRODUCTION

The effects of radioactive fallout following nuclear weapon tests on the population in the Semipalatinsk region, which is located in the northern part of the Kazakhstan Republic have been studied by a collaboration between Research Institute for Radiation Biology and Medicine, Hiroshima University and Kazakhstan Scientific Research Institute for Radiation Medicine and Ecology since 1995. Our center (Hiroshima) was re-established in 1994 in order to reorganize and expand the Data and Specimens Center of Atomic Bomb Disaster. One of the purpose of the center is research studies concerning worldwide radioactive contamination.

To date, all of the published reports concerning to the effects of radiation on populations near nuclear weapon test sites are based on data provided by the

Defense Department of the former Union of Soviet Socialist Republics (USSR). We are focusing on radiation dosimetry and associated diseases in the population near the Semipalatinsk test site.

In 1995 and 1996 Japanese group visited Semipalatinsk to measure and to sample soil and brick according to the guidance of Dr. Gusev. The sample which weighted about 100 kg each were brought back to Hiroshima and have been measured. We report here concerning the results on environmental dose and on thermoluminescence dosimetry measurements at several settlements from the research in 1995 [1, 2].

2. BRIEF HISTORY OF NUCLEAR TEST AND METHOD OF DOSE EVALUATION

The nuclear test site is located west of Semipalatinsk city, as shown in Fig. 1. A total of 477 nuclear explosions were conducted by the USSR from 1949 to 1989 at the Semipalatinsk nuclear test site, including 87 atmospheric, 26 on the ground, and 364 underground explosions. There are several villages located close to the test site. In Fig. 1 we outline the three types of explosions relating to this research letter briefly [3, 4].

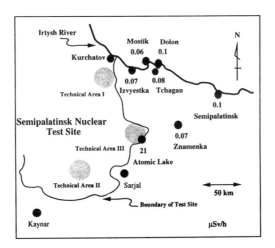

Figure 1. A map of the measued radiation dose rates at the villages and at " Atomic Lake" in October 1995 when is six years after the final test in Kazakhstan Republic. Nuclear weapon test were performed as follows: 26 above ground and 87 atmosphere explosions in Technical Area I from August 29th 1949 to December 30th 1962, 223 underground explosions in Technical Area II from October 11th 1961 to October 10th 1989, 123 underground explosions in Technical Area III from June 19th to October 19th 1989.

On August 29th, 1949, the first atomic bomb test in the USSR was conducted at the Semipalatinsk nuclear test site. The burst height was 38 m above the ground and the energy output was equivalent to 20 kilotons of trinitrotoluene (TNT). Two hours after this atomic bomb test, a huge radioactive cloud attacked Dolon and other villages, which were more than 70 km from the epicenter.

On August 12th, 1953, the first hydrogen bomb was exploded in the USSR. The burst height was 1000 m and the energy output was equivalent to 470 kilotons of TNT.

On January 15th, 1965, a hydrogen bomb equivalent to 240 kilotons of TNT was exploded 100 m under the ground at the east part of the test site. The official reason given for this explosion was the construction of a dam for peaceful use. As a result of this experiment, there exists a lake which is called "Atomic Lake".

The environmental radiation dose measurements in the present study were performed using two detectors. One of the two detectors was the personal pocket monitor (Aloka PDM-101 in which a silicon detector is installed. Measurements were taken with this monitor attached to a shirt pocket, according to the manufacturer's instructions. By this monitor, the environmental dose rate was measured with an error of 10%. The other detector was the spectro survaymeter (Hamamatsu C-3475) which has an NaI scintillator (2.5cmϕ x 5.1cm) and a multi-channel analyzer with 128 channels. The measurements with this device was performed on the ground for 30 min. The geographical points of the measurements were confirmed by the Global Positioning System Satellite Navigator (Magellan Trailblazer) using signals from four satellites.

For TLD analysis we applied quartz inclusion and high temperature techniques. The detail procedure will described elsewhere [2].

3. PRESENT STATUS OF ENVIRONMENTAL RADIATION DOSE AND ACCUMULATED DOSE

The radiation levels in the hypocenter area of the Semipalatinsk nuclear weapons test site are still relatively high. The radiation dose rate is around 10 μSv/h at the edge of Atomic Lake, which is consistent with the findings of Goldman (1994). The value on the ground was 21μSv/h which was obtained using an NaI scintillator. The dose rate at the boundary of the test site of which location is several hundreds meters distant from "Atomic Lake" is less than 1 μSv/h. The total radiation dose for one hour-stay in Technical Area 111 was 5 μSv as shown in Fig.2. However, we clarify that the radiation levels in several villages around test sites are the natural background level which suggests that there are no health effect on the population outside of the boundary of the test site on the present days of 1995.

Radiation dose rates are less than 0.1 mSv/h in some villages and Semipalatinsk city as shown in Fig. I . A radiation level in Znamenka which is located 50 km distant from the hypocenter of "Atomic Lake" is also ordinary level of 0.07 mSv/h. As Gusev reported, there was a heavy fallout in Dolon according to the first atmospheric atomic bomb test. In Do]on and the other villages the radiation levels are less than 0.1 mSv/h 46 years after the hazard. This may be due to environmental decrease of radio activities from the explosion with time dependence of A=AOt-1.2[6].

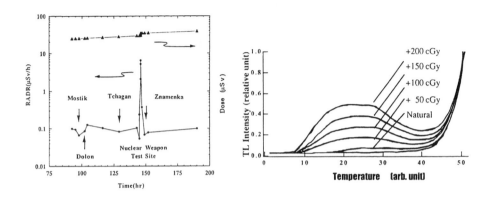

Figure 2. Radiation dose rates measured personal pocket monitor during the period of the research trip in Semipalatinsk area.

Figure 3. Thermoluminescence glow by a curves for a brick-sample from Tchagan. Additional doses are given by Co-60 gamma ray irradiation.

To better understand the radiation levels, comparing them with the natural background levels is meaningful in terms of the radiation protection. These radiation levels in the villages near the test site and in the Semipalatinsk city are the same as in cities like Osaka, Tokyo, Moscow and Almaty. The situation in Hiroshima, where the first nuclear weapon was exploded above the city, is almost the same. Radiation dose rates were less than 0.1 mSv/h around the hypocenter of Hiroshima city in 1995. Radiation levels at high altitudes are around 1.0 µSv/h for international airline flights and around 0.3 µSv/h for domestic flights in the former USSR due to cosmic rays. Higher levels were observed in the Atomic Lake nuclear test site. The total dose was 62 µSv for the whole trip of 11 days.

Dolon is the settlement with the highest dose of 89 cGy among the settlements where we have researched. This is the similar situation to the previous reports[4, 7]. Semipalatinsk city exhibits the secondly highest dose among settlements of the present study. This is completely different from the previous reports order of value to the other reported values. However the value of 41 cGy at Semipalatinsk city . The present TL doses at Dolon and Tchagan are the same is a hundred times higher than previously reported values. This city is the biggest city near

Nuclear test site. Since the population of the city is three hundreds and fifty thousand, this and associated diseases in the population near the Semipalatinsk test site. next time we will sample brick at other several points in Semipalatinsk city. Data on soil will be reported in the near future.

4. SUMMARY

Measurements of environmental radiation doses and thermoluminescence dosimetry in some villages near Semipalatinsk nuclear test site are presented. The radiation levels in several villages are the natural background level in 1995 six years after the final nuclear explosion. However the radiation dose rate is still high level around 10 µSv/h at the edge of Atomic Lake. On the other hand thermoluminescence dosimetry exhibits significant radiation exposure to residents in this area due to nuclear test.

REFERENCES

1. J. Takada, M. Hoshi, R. I. Rozenson, S. Endo, M. Yamamoto, T. Nagatomo, T. Imanaka, B. I. Gusev, K. N. Apsalikov, N. J. Tchaijunusova , Health Pyisics, (in press).
2. J. Takada, M. Hoshi, S. Endo, M. Yamamoto, T. Nagatomo, T. Imanaka, B. I. Gusev, R. I. Rozenson, K. N. Apsalikov, N. J. Tchaijunusova (unpublished data).
3. B. I. Gusev, Radiation Medicine, edited by S. B. Balmukhanov and A. Aitmagambetov, Semipalatinsk Statistic Report Company, Semipalatinsk. (1993). (in Russian)
4. B. l. Gusev, " Medical and demographical consequences of nuclear fallouts in some rural districts in the Semipalatinsk region". Doctor Thesis, Almati, (1993).
5. M. Goldman, Letter from Siberia. HPS Newsletter XXII, 11 (1994) 3.
6. S. Glasstone, and P. J. Dolan, The Effects ofNuclear Weapons, United States Department of Defense and the Energy Research and Development Administration, (1977).
7. A. H. Tsyb, V.F.Stepanenko et al, "Short information about the results of the complex investigations on ecological situation and health status of the inhabitants in Semipalatinsk region of Kazakhstan SSR", Obninsk, (1989). (in Russian)

Radioactive contamination in North Xinjiang resulting from former USSR atmospheric nuclear test in 1962

Hou Jie-li, He Shi-yu, Li Shu-ren, Pi Zheng-zhi, Liu Fu-wen, Xu Ting-gui, Wang Xing-zhi, Liu Xue-fu and Liu Geng-min

Institute of Military Medicine, Xinjiang Military Command, Urumqi 830011, P.R. China

1. INTRODUCTION

From August to December 1962, a series of atmospheric nuclear tests were conducted in the North Pole and Central-Asia by the former Soviet Union (USSR). Their fission yield about 54 % of that given by tests ever conducted in USSR [1]. Owing to the moderate distance from test sites to Xinjiang and to the meteorological characteristics, the north part of Xinjiang was contaminated from radioactive fallout. The level of pollution was the highest among all artificial radioactive pollution ever before. This paper reviewed briefly the result of investigation.

2. MATERIALS AND METHODS

Background radioactivity remaining from 1961 USSR nuclear tests in surroundings and living beings were measured before August 1962. From then, depositions, aerosols, rain and snow water, surface water, soil and living tissues were collected in Urumqi and pollution levels were measured. Some samples were also gathered from other localities of North Xinjiang. In addition, blood cells of normal residents in three areas were examined. For methods and measurement we referred to the literature [2]. Depositions were collected by using gummed paper or acidified water according to the weather. Activities were measured 6 hr. after sampling and 600 °C burned to ash. Aerosols were collected in filters and measured 6 hr later. Acidified water, rain and snow were evaporated to dry and 600 °C burned to ash. Grass, vegetables and corn were 600 °C burned to ash, followed by β activity tests, and K-40 activities were excluded. Thyroids of sheep and cattle (some of humans) and muscle, liver, kidneys and ribs of sheep were collected and burned to ash for measurement (K-40 activities were excluded also), except the thyroids, which were measured directly.

Total β activities were assayed of by means of G-M tubes and counters. The window thickness of G-M tube was <5mg/cm^2. The tubes were protected with chambers made of 4 cm lead and 0.5cm pyrexglass. To calibrate the activities of samples, a known Sr^{90}+Y^{90} source was used to identify in every measurement. It was shown that the efficiency for gross β activity was >20%. The sensibility of measurement was 0.4 Bq/300mg, and the error of counting was less than 5%.

3. RESULTS

3.1. Radioactive background before nuclear tests in 1962

In the period of March to July 1962, radioactive background in northern part of Xinjiang was slightly higher than normal, because of the debris from USSR tests in 1961 and from US tests in Pacific Ocean area. Depositions in Urumqi were 96 MBq/km^2·d on the average. Airborne activities were 0.085-0.70 Bq/m^3. Radioactivity were 0.34-0.85 Bq/L in surface water, 20-106 Bq/L in rain or snow water. The activities in vegetables were similar to K-40 contents. Activities in sheep thyroids were 1.7-2.3 Bq/g, and in sheep ribs were 0.1-0.6 Bq/g.

3.2. Deposition

Following the beginning of nuclear tests, a peak value of radioactive deposition occurred in Urumqi on 8th August, and fluctuated with every new explosions. A higher level up to 30 GBq/km^2·d was observed in late August, and 90 GBq/km^2·d in late September. Simultaneously, the depositions and γ–radiation levels in Tacheng, Karamayi and other areas increased rapidly. The highest level up to 168 GBq/km^2·d appeared in Urumqi. Table I shows the depositions, the dates of occurrence, and the estimated explosion dates according to the decay rates.

According to meteorological data of high altitude, there were three main pathways of USSR radioactive clouds entering Xinjiang: (1) from Ertix valley to southeast, pass through Urumqi, then to east; (2) from going Tacheng County and Altaw pass going across Urumqi, then to east; (3) from Altaw pass and Yili Prefecture going along northern slope of Tianshan Mountains passing across Urumqi to east.

After ceasing of nuclear tests, the deposition levels reduced to 4.6 MBq/km^2·d in the period of November 1963 to October 1965.

3.3. Physical and chemical characteristics of fallout 26 December

Some "hot" particles deposited in the first few days of many peak periods. It could be seen on foliage of sunflower or on the ground surface. For instance, on 18th November, 17 particles were observed on 4 x 4 m^2 snow ground. One of them possessed the size of 170 x 270μm, and the activity was several thousands Bq. All fallout samples in the peak periods were relatively insoluble. 2.5-6.7 % of them were soluble in water at pH 7, and 11-18 % at pH 3 were soluble before August.

Table 1
Peak levels of radioactive deposition in Urumqi in 1962

No.	Date of sampling	Deposition level (GBq/km^{2}d)(2nd day after sampling)	Estimated date of explosion(Beijing time)
1	8 August	6.18	5 August
2	23 August	30.3	Before 23 August morning
3	26 August	29.20	24 August night
4	29 August	6.06	29 August morning
5	18 September	0.27	14 September
6	27 September	89.9	25 September evening
7	24 October	0.94	22 October evening
8	29 October	57.40	28 October evening
9	1 November	12.00	30 October night
10	6 November	3.92	5 November night
11	19 November	168.00	17 November night
12	11 December	0.90	27 November
13	26 December	3.45	24 December night

Besides radioiodine, Sr-89, Ba-140 and rare earth elements in some fallout samples were seen by radiochemical separation. About 25 % to a half were insoluble sediments.

The power indexes for the decay rate of fallout were 1.0-1.8, varying with the time after explosion. Average β energy of 0.4-0.5 MeV was measured, but that in aerosols was about 0.3 MeV.

3.4 Specific activities of aerosols close to ground level

Specific activities of aerosols increased rapidly following every new contamination. Peak level 1.24E^3/m^3 occurred in the contamination period beginning on 27th September, with an average of 750Bq/m^3 in the first 24 hrs, then reduced to 75Bq/m^3 in the second 24 hr.

3.5 Gamma radiation levels

There was no definite relationship observed between γ-radiation levels 1 m above ground outdoors and deposition showed in Table 2. The highest γ-radiation level of 9.7µGy/h had been seen.

Table 2
Maximum γ-radiation levels in three peak periods in Urumqi (subtracted)

1st day of new contamination(date)	Depositions (GBq/km^{2}d)	Maximum γ-radiation level (µGy/h)
27 September	89.9	9.7
29 October	57.4	0.18
19 November	168.0	0.26

3.6 Contamination of water

Gross β activities of river or canal water in Urumqi, Tacheng, and the Ili and Jinghe county between September and October were close to or a little higher than the levels before tests. But the rain or snow water was contaminated obviously; the activities were less than 110 Bq/L pre-tests, 2.03E^3 Bq/L on 30th September and at maximum of 2.01 E^5 Bq/L on 19th November.

3.7 Radioactivity in some species of grass, corn and vegetables

The gross β activities in 9 species of vegetables were slightly higher than β activities of K-40 contents in the 1st half of 1962. Table 3 shows the contamination levels of some species of grass, corn and vegetables sampled between August and October. The bulk of samples were collected after 27 th September. The highest levels were seen in Urumqi and Changji all in northern part of Xinjiang. 2.59E^5 Bq/kg for grass and 4.8E^4 Bq/kg for leafy vegetables. 1-2 orders of magnitude lower activities were seen in root or fruit vegetables, wheat or maize grain, while those in ear of rice were close to the level in grass.

The efficiency of decontamination with water of all kinds of vegetables was 21-83 %. That of leafy vegetables was the highest, about 80 %. Contamination of root or fruit vegetables was slighter, and the efficiency of decontamination was also lower. About 82 % of activities in the wheat threshed by hands were removed.

3.8 Internal contamination of domestic animals and humans

During May and July 1962, the average specific activities in liver, muscles and kidneys of sheep were below 0.12 Bq/g in Changji county, similar to K-40

Table 3
Plant contamination levels in northern part of Xinjinang from August to October 1962

Areas	Vegetables	Sampling date	Date for activ. calibrated	Specific activities (Bq/kg)
Urumqi	Potato carrot beet	11-14 Sept.	30 Aug.	120-366
	Cabbage scallion	11-14 Sep.	30 Aug.	492-1.60E3
	Grass	11-13 Sept.	30 Aug.	6.07E3-1.60E4
Shihezi	Riceear	21 Sep.	30 Aug.	8.04E4
Wulumqi	Grass	5 Oct.	28 Sep.	1.01E5
	Leafy vegetables	28 Sep.-23 Oct.	28 Sep.	4.80E4
	Root and fruit vegetables	5 Oct.	28 Sept.	4.33E2-8.14E3
Changji	Grass	5-11 Oct.	28 Sept.	2.59E5
	Leafy vegetables	7 Oct.	28 Sept.	3.46E4
Kuitun	Grass	29 Sep.	28 Sep.	5.04E4
Tacheng	Grass	26-27 Sept.	28 Sept.	2.23E4
	Leafy vegetables	26-27 Sept.	28 Sept.	2.39E3
Burqin	Grass	25 Oct.	28 Sept.	5.73E4
	Leafy vegetables	25 Oct.	28 Sept.	1.08E4
Jinghe	Grass	27 Oct.	28 Sept.	8.18E4
	Leafy vegetables	27 Oct.	28 Sept.	7.07E3
	Rice (unthreshed)	8 Nov.	28 Sept.	5.18E3
Ili	Grass	17-25 Oct.	28 Sept.	6.32E4
	Leafy vegetables	17-25 Oct.	28 Sept.	6.96E3
	Rice (unthreshed)	5-10 Nov.	28 Sept.	4.88E3

background. Specific activities of sheep thyroids were below 0.2Bq/g. During the period of nuclear tests, specific activities increased rapidly following every new pollution. They usually rose to a peak value 10 days after the beginning of pollution, and then decreased with the half-time of 6 days. The highest average value was $1.04E^4$Bq/g, and the highest individual value was $2.25E^4$Bq/g. All the values from other parts of Xinjiang were less than Changji and close to those in Urumqi. In the same periods, the specific activities in bovine thyroids were about 0.6 fold of those in sheep thyroids. Thyroids of six patients were collected between October and December, showing 4.8 Bq/g in a young woman and 2.1 Bq/g in a 2 month-old baby, about 1/230 of radioactivity of sheep thyroids in the same period. Concentrated samples were measured. According to the decay rate, β energy and the activities of I-131, the specific activities in sheep liver, muscles and kidneys were similar to the levels before tests. β specific activities in sheep ribs increased gradually from May to November, 2 Bq/g in the mid-November. A series new pollution did not affect its increment speed.

3.9 Changes of blood cell counts in population

Following the USSR 1961 nuclear tests, blood cell counts in Xinjiang population changed to a varied extent and did not recover yet in May 1962. As shown in Tables 4 and 5, obvious changes were observed again from September to December in normal young population in Urumqi, Tacheng and Ili county. Leucocytosis ($>10E^9/L$) was manifested mainly by increase in neutrophils and eosinophils.

4. DISCUSSION

4.1. Following a series of USSR nuclear tests in 1962, the northern part of Xinjiang was contaminated once and again. Pollution could be found almost in every surroundings, which occurred 2-3 days following every new tests, entered living bodies, and followed by changes in blood cell counts in population.

4.2. Because of topography and high altitude meteorological characteristics, the radioactive clouds having entered northern part of Xinjiang from northwest, regardless of the pathway taken, all passed through Urumqi area and caused the highest level of contamination. According to data of fallout depositions, γ-radiation levels, pollution in living bodies and the wind stream in high altitude, taking the pollution level in Urumqi and the vicinity(Changji, Miquan, Fukann) as 1.0, the corresponding levels were as follows: Shihezi 0.70, Karamay 0.50, Kuytun 0.45, Jinhe 0.30, Bole 0.25, Burqin 0.25 and Ili 0.15.

4.3. The obvious features were as follows. Depositions appeared as less soluble and fused particles. Biological absorption of radionuclides were rare. Short-lived radionulides vanished in a few months, but long-lived radionuclides remained in living surroundings for several years. The important radionuclides affecting human health were iodine, strontium, barium and others and the mixed fission products irradiating the digestive tract and respiratory tract.

Table 4
Percentages of leucocytosis (>10E9/L) cases in some areas of northern Xinjiang

Area	Period of examination	Monophils (>0.9E9/L)	Eosinophils (>0.6E9/L)	Neutrophil (>8E9/L)	Monocytes (40%)
Urumqi	May 1962	1.6	30	1.6	6.1
	Nov. 1962	14(P<0.05)	58(P<0.01)	14(P<0.05)	18(P<0.05)
Tacheng	Sep. 1962	26(P<0.01)	52(P<0.05)	20(P<0.05)	16(P<0.05)
Ili	Oct. 1962	6.7(P>0.05)	32(P>0.05)	9.9(P<0.05)	10(P>0.05)

Table 5
Percentages of some types of leucocytosis cases in some areas of north part of 50 Xinjiang

Area	Period of examination	No. of person examined	Percentage of leucocytosis cases
Urumqi	Late May 1962(pre-tests)	60	4.9
	Nov. 1962	50	38 (P<0.01)
Tacheng	Late Sept. 1962	50	48 (P<0.01)
Ili	Oct. 1962	50	30 (P<0.01)

4.4. Dose estimates for population in Urumqi and the vicinity

4.4.1. Absorbed dose from external exposure. The following factors were taken from the reference(1) supplement A in calculation: γ-exposure rate, and ratio of absorbed dose from γ-rays indoors to outdoors, ratio of stay time indoors to outdoors, conversion factors from exposure rate into tissue effective dose. The estimated effective doses of Y-radiation from USSR 1962 tests were as follows: whole body $1.2E^{-1}$mSv, gonads $2.3E^{-2}$mSv, thyroids $2.8E^{-3}$mSv and skin $9.3E^{-3}$ mSv.

4.4.2. Internal dose equivalents due to inhalation of polluted air and ingestion of polluted vegetables and corn. They were estimated according to the references [3,4]. Models of transfer and deposition in respiratory tract were considered as Y type, AMAD=1μm. Only vegetables and corn consumed before October 1962 were considered, because of no crops growing in the fields after that period. 500 g vegetables were consumed per day, in which 30010 were leafy, and 80 % of contamination might be washed off. The contamination levels in root and fruit vegetables and corn were 1/50 of that in leafy vegetable, and 500 g corn were considered to be consumed per day. Also considered were the 18 days of half live in fallout particles from foliage (without physical decay rate), 10 % solubility of fission products(FP) in gastrointestinal tract, 80 % absorption of soluble part, by iodine 10 % of FP contributed, contributions of I-131, I-132, I-133 varying with time after detonation. 7.8 % of FP came from Ba-140 and 2.9 % from Sr-89. A few herdsmen consuming 2L snow water in two events of highest pollution were contaminated. The results are shown in Tables 6-9.

Table 6
Internal dose equivalent and sources to the thyroids of population in Urumqi (mSv)

Nuclides	Inhalation	Ingestion	Inhal.+inges.	Snowwater	Inh.+ing.+snow water
I-131	0.561	3.74	4.30	0.345	4.65
I-132	5.71E-3	3.07E-5	5.74E-3	2.83E-2	3.40E-2
I-133	4.50E-2	9.10E-3	5.50E-2	0.383	0.438
Total	0.613	3.75	4.36	0.756	5.12

Table 7
Internal dose equivalents and sources to gastrointestinal tract of population in Urumqi(mSv)

	Inhalation	Ingestion	Inhal.+inges.	Snow water	Inh.+ing.+snow
Stomach	1.84E-2	8.31E-3	2.67E-2	0.160	0.188
Intestines	3.33E-2	1.52E-2	4.85E-2	0.290	0.339
ULI	0.168	7.89E-2	0.250	1.45	1.70
LLI	0.483	0.239	0.722	3.97	4.69

Table 8
Internal dose equivalents and sources to bone tissue of population in Urumqi(mSv)

Tissue	Inhalation	Ingestion	Inhal.+inges	Snow water	Inh.+ing.+snow
Bone surface	1.64E-3	0.430	0.432	0.115	0.547
Red bone marrow	1.64E-3	0.410	0.412	0.114	0.526

Table 9
Internal dose equivalents and sources to Urumqi residents due to USSR nuclear tests in 1962 (mSv)

Organs	External dose equi.	Internal		Internal(who drunk snow water)	
		Dose equi.	Weighted	Dose equi	Weighted
Whole body	0.12				
Gonads	2.3E-2				
Skin	9.3E-3				
Thyroids	2.8E-3	4.36	0.22	5.12	0.256
Lungs		0.74	8.9E-2	0.74	8.9E-2
Nasopharynx		0.76		0.76	
Stomach		2.67E-2	3.2E-3	0.188	2.3E-2
Intestine		4.85E-2	2.4E-3	0.339	1.7E-2
ULI		0.25	1.5E-2	1.70	0.102
LLI		0.72	4.3E-2	4.69	0.281
Bone surface		0.43	4.3E-3	0.547	5.5E-3
Red bone marrow		0.41	4.9E-2	0.526	6.3E-2
Effective dose	0.12		0.42		0.84

4.5. Due to USSR nuclear tests in 1962, 0.5 mSv average effective dose received by Urumqi residents was estimated, 2/3 of which were received in a few months. Investigation showed restorable m shorter term blood changes. As shown in Tables 4 and 5, the percentage of leucocytosis cases in Urumqi residents was less than that in Tacheng, because the examinations were conducted earlier in Urumqi, which was contaminated by the tests in August only, while the examinations in Tacheng followed the heavier contamination in the late September.

4.6. For the critical group of population, two factors should be considered. One is the herdsmen consuming snow water, which led to higher internal dosage, but less consumption of vegetable, lead to less internal dosage. The other is the age-dependent factor according to the reference (5), the ratio between babies 3 months to 1 year old and adults was not greater than 10, but the babies consume milk mainly less that in vegetables two orders of magnitudes at least. The highest weighted dose or no vegetables, and the specific activities in milk were lower than equivalent of thyroids contributed only a smaller fraction to effective dose; so the difference between ages might be small. The effective dose to critical group was estimated to be not greater than 3-fold average population dose.

5. CONCLUSION

Northern part of Xinjiang was contaminated owing to USSR nuclear tests in 1962. The highest pollution level occurred in Urumqi and adjacent areas. Deposition, activities of aerosols, foliage of plant, sheep thyroids and γ-exposure rate rose rapidly 2-3 days following every new explosions. Radioiodine was also found in human thyroids. The average effective dose to Urumqi residents was estimated to be 0.5mSv, and several folds above the average to a few people was possible. Soon after the radioactive pollution, blood changes in the population restorable a shorter term were observed.

REFERENCES

1. United Nations, United Nations Scientific Committee on the Effects of Atomic Radiation, Ionizing radiation: sources and biological effects, United Nations, New York, 1982.
2. ICRP Publication 30, Lists for Intake of Radionuclides by Workers, Pergamon Press, 1976.
3. ICRP Publication 60, Recommendations of the International Commission on Radiological Protection. Pergamon Press, 1991.
4. ICRP Publication 56, Age-Dependent Doses to Member of the Public from Intake of Radionuclides, Pergamon Press, 1990.

Observations in Kyoto, Japan of Radioactive Fallout from Chinese Atmospheric Nuclear Explosion Tests

Naoto Fujinami

Kyoto Prefectural Institute of Hygienic and Environmental Sciences 395-Murakami-cho, Fushimi-ku, Kyoto, 612 JAPAN

1. INTRODUCTION

Neither making nor having any nuclear weapons is the national policy of Japan, and there have been no nuclear tests. Hence, there is no direct object of the research on consequences of nuclear explosions in Japan, except for the experience of Hiroshima and Nagasaki. The present paper, therefore, deals with the monitoring system for radioactive fallout in Japan and observations in Kyoto of the radioactive nuclides released from Chinese atmospheric nuclear tests.

2. ENVIRONMENTAL RADIOACTIVITY MONITORING SYSTEM OF KYOTO PREFECTURE IN JAPAN

Kyoto Prefectural Institute of Hygienic and Environmental Sciences (KPIHES) is carrying out the analysis of radioactivity in various environmental samples such as airborne particulates, deposited materials, soil, water, vegetables, milk and many kinds of foodstuff, in order to examine the influence of radioactive fallout. Each prefectural institute has a similar environmental radioactivity monitoring system, and is carrying out monitoring in its jurisdiction. These monitoring systems constitute a nationwide survey network, whereby results of these surveys are collected and compared in the Japan Chemical Analysis Center, providing a nationwide data base of environmental radioactivity in Japan [1-2].

In addition to the aforementioned monitoring against global fallout, KPIHES is also conducting the monitoring in the vicinity of a nuclear power station. This is to verify that individuals living around the station are not exposed to more than 1 mSv per year, of radiation attributed to that facility.

3. MEASUREMENT METHODS

Deposited material is collected continuously in a basin 5000 cm^2 in area. Each month, this sample is concentrated by evaporation and then counted with a gamma spectrometer. After determining the deposited amounts of individual gamma emitting nuclides, these samples are sent to the Japan Chemical Analysis Center, where the contents of Sr-90 and Cs-137 are obtained by radiochemical analysis.

Airborne particulates are collected continuously with a filter air sampler. The mass of sampled air is about 10,000 m^3/month. Concentrations of individual gamma emitting nuclides, in surface air, is determined from gamma spectrometry of the sample for each month.

4. RESULTS

Japan is situated about 4000 km east of the Chinese nuclear test site, as shown in Figure 1. When an atmospheric nuclear test is conducted there, the released radioactive nuclides are usually detected in Japan [3-7]. For instance, Figure 2 shows various radionuclides observed in the fallout collected in Kyoto after the test conducted by China on October 16, 1980. Some radionuclides were also observed in surface air, as shown in Figure 3. The presence of U-237 and Y-88 indicates that a hydrogen bomb was exploded. It is also seen from Figs. 2 and 3 that deposited amounts and air concentrations of relatively long-lived nuclides

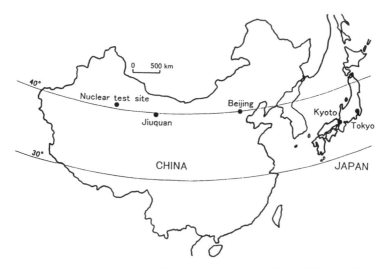

Figure 1 Locations of the Chinese nuclear test site and Kyoto, Japan.

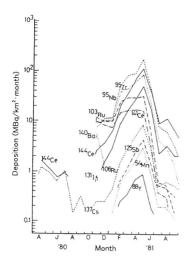 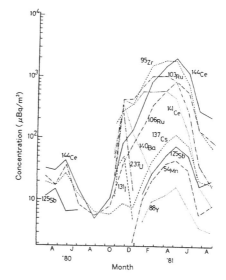

Figure 2 Monthly deposited amounts of radionuclides in Kyoto, Japan from April, 1980 to September, 1981.

Figure 3 Variations of concentrations of radionuclides in surface air in Kyoto, Japan from March, 1980 to September, 1981.

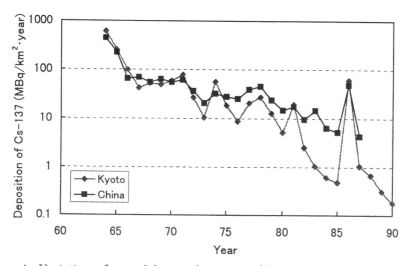

Figure 4 Variations of annual deposited amounts of Cs-137 in Kyoto and China.

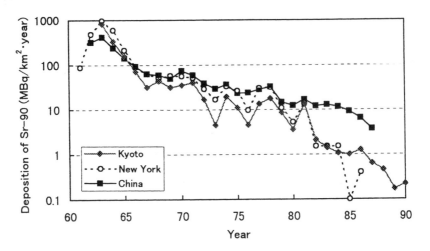

Figure 5 Variations of annual deposited amounts of Sr-90 in Kyoto, New York and China.

have maximum values in the spring of 1981. This fact means that a large fraction of debris, injected into the stratosphere by the nuclear detonation, was deposited in this season [8]. Figures 4 and 5 illustrate variations of deposited amounts of Cs-137 and Sr-90 respectively. Kyoto is compared with China [9-10] in both figures, and with New York [11-13] in Fig. 5. In these figures, China's deposited amounts are nationwide mean values. While the deposited amounts of Cs-137

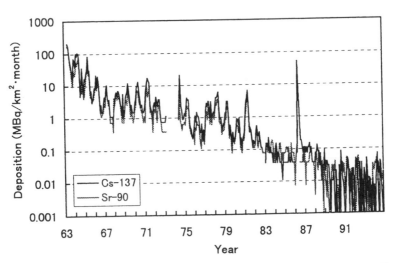

Figure 6 Variations of monthly deposited amounts of Cs-137 and Sr-90 in Kyoto.

rose transitorily in 1986 because of the Chernobyl accident, both the amounts of Cs-137 and Sr-90 have been generally decreasing in each place since 1963. The decrease in China was slower than those in Kyoto and New York in the 1980's. This difference may be ascribed to resuspension of the nuclides within China and effects of Chinese underground tests in this period. The causes, however, are currently being investigated. Except for this point, the variations in China are similar to those in Kyoto and New York. The similarity suggests that the deposition averaged all over China is governed by global fallout rather than by local.

Figure 6 shows variations of monthly deposited amounts of Cs-137 and Sr-90 in Kyoto. These monthly variations indicate several sharp peaks corresponding to Chinese nuclear tests from 1964 to 1980 besides the Chernobyl accident in 1986 more obviously than the annual in Figs. 4 and 5.

5. DISCUSSION

No atmospheric nuclear explosion test has been conducted in the world since 1981, and 10 years have passed since the Chernobyl accident. Because of decay of the radionuclides originated from those sources, fission products continuously detected in environmental samples in Japan are only Cs-137 and Sr-90. Furthermore, a content of Cs-137 can be determined by gamma spectrometry without complicated analytical procedures. Cesium-137 is therefore an appropriate nuclide for examination of the trend in accumulation of artificial radionuclides in the environment.

Actually, no gamma emitting nuclide other than Cs-137 was detected in environmental samples collected in the vicinity of our target nuclear power station in recent years. The observed concentrations of Cs-137 indicated similar levels and variations to those of other regions without any nuclear facility in Japan. Hence, there is no evidence that the nuclear station has influenced the environment. Moreover, the public exposure evaluated from the data of detected Cs-137, was consistently far less than 1 mSv per year, even after including the contribution from the other typical fallout nuclide, Sr-90.

Although the environmental radioactivity monitoring was begun in order to obtain data for countermeasures against environmental contamination by radioactive fallout from nuclear tests, its role has changed. As mentioned above, its main purpose is to discriminate between the influence of target nuclear facilities and others, which include past nuclear tests and severe nuclear accidents at a great distance.

6. CONCLUSIONS

When deposited amounts of Cs-137 and Sr-90 were compared between China, Kyoto and New York, it was found that there was no significant difference between them. Therefore, it appears China was hardly affected by local fallout. This is because data available regarding China was only that on long-term nationwide effect [9-10].

In order for more extensive international collaborative studies to be conducted, concerning the transport, deposition, and human health effects of the radioactive fallout from nuclear tests, more data on local fallout needs to be published. The off-site radiation exposure review project of the U.S.DOE performed the first phase of dose reconstruction at locations within approximately 300 km of the Nevada test site [14]. In the case of Chinese nuclear tests, there exists a report which briefly mentions influence of fallout on Jiuquan Region, which is located 420 to 750 km away from the test site [9], much farther away than the survey area of the DOE project. It is desirable that detailed data from regions closer to the test site in China will be made available, if further research is to be of value.

REFERENCES

1. Aochi, T., Nishiyama, M., Hashimoto, T., The nationwide surveillance for environmental radioactivity in Japan, JCAC M-9302, 1993.
2. Aochi, T., Nishiyama, M., Hashimoto, T., An overview of surveillance for environmental radioactivity in Japan, The proceedings of JCAC 20th anniversary symposium on environmental radiation monitoring technology, JCAC M-9401, pp.36-67, 1994.
3. Kojima, S., Furukawa, M., The measurement of neutron-induced radionuclides from Chinese nuclear weapons tests, J. Radioanal. Nucl. Chem., 100(2), 231 - 240, 1986.
4. Matsuoka, N., Hirai, E., Momoshima, N, Takashima, Y., Radioactivity of pine needles of Kyushu Island, Japan, after the 26th Chinese nuclear explosion test, Radiochem. Radioanal. Lett., 57(3), 161-168, 1983.
5. Takagi, S., Ohtou, Y., Nakaoka, A., Inoue, T., Kanbe, H., Fukushima, M., Koyama, H., Determination of fallout nuclides by the 19th Chinese nuclear explosion test, J. Nucl. Sci. Technol., 15(12), 926-934, 1978.
6. Yamato, A., Fallout Pu-239, Pu-240 and Sr-90 in surface-air in Ibaraki-prefecture, Japan, Radioisotopes, 30(2), 104-108, 1981.
7. Yamato, A., Recent radioactive fallout at Tokai, Ibaraki Prefecture, Radioisotopes, 31(11), 587-590, 1982.
8. Miyake, Y., Saruhashi, K,. Katsuragi, Y., Kanazawa, T., Seasonal variation of radioactive fallout, J. Geophys., Res., 67(1), 189-193, 1962.
9. Wei, K., Ren, T., An overview of surveillance of environmental radiation in China, The proceedings of JCAC 20th anniversary symposium on

environmental radiation monitoring technology, JCAC M-9401, pp.28-35, 1994.
10. Zhu, C., Zhu, G., Gu, Z., Liu, Y., Environmental contamination by radioactive fallout and evaluation of population dose in China, Proceedings of the International Symposium on Radiological Protection, Beijing, China, Nov. 1989.
11. Larsen, R.J., Worldwide deposition of strontium-90 through 1982, US DOE Rep., EML-430, 1984.
12. Juzdan, Z.R., Worldwide deposition of strontium-90 through 1985, US DOE Rep., EML-515, 1988.
13. Monetti, M.A., Larsen, R.J., Worldwide deposition of strontium-90 through 1986, US DOE Rep., EML-533, 1991.
14. Church, B.W., Wheeler, D.L., Campbell, C.M., Nutley, R.V., Overview of the Department of Energy's off-site radiation exposure review project, Health Physics, 59(5), 503-510, 1990.

V Epidemiological study

Methodology of epidemiology in low dose radiation exposure

H. Kato

Consultant, Department of Eidemiology, Radiation Effects Research Foundation, Hiroshima 732, Japan

SUMMARY

In order to estimate the risk of low dose radiation exposure, the following two epidemiological methods are commonly used: A) extrapolation from data of high dose radiation; B) direct observation on data of low dose (dose- rate) radiation exposure. Both methods have weakness as well as strength. For direct investigation of low dose or low dose- rate exposure, both controlling the confounding factors other than radiation and availability of accurate individual dose estimate in addition to a large sample size of the cohorts are the key to succeed in the epidemiological studies on low dose radiation exposure.

It should be emphasized that radiation effects on man, especially the risk estimate, can only be achieved by epidemiological studies. However, great care must be taken to reduce the uncertainty involved in this epidemiological approach in the estimation of risk.

1. INTRODUCTION

In order to estimate the risk from low dose radiation exposure in man, we should rely solely on epidemiological studies, since it is difficult to estimate by other methods (e.g., animal experiments). There are two methods in this epidemiological approach as shown in Table I : A) Extrapolation from estimated risk of high dose radiation exposure(e.g., A- bomb survivors study). B) Direct observation on the population exposed to low dose radiation, Representative examples of this approach are : 1) The high background radiation area study (HBRA study) in the Province of Guangdong, China [1]. 2) The combined analysis on radiation worker studies in Canada, UK and USA [2]. These two studies are classified as cohort studies in which a large, fixed, exposed population is followed for a long period of time. 3) An epidemiological study on the effect of residential

Table 1. Epidemiological methods in estimation of risk from low dose radiation exposure

---

A) **Extrapolation** from Estimated Risk for **High Dose** Radiation Exposure

Example: A-bomb Survivors Study

B) **Direct Observation** on the Population Exposed to **Low Dose** Radiation

Example: 1. High Background Radiation Area Study
2. Radiation Workers Study (Combined Analysis)
3. Residential Radon Study (A Case-Control Study)

---

radon in USA [3]. Another epidemiological method, a case - control study was used in the study. Methods A) or B) have weaknesses as well as strengths.

2. METHOD

2.1. Method A. Extrapolation from high dose radiation exposure

ICRP has estimated the lifetime risk of low dose or low dose-rate radiation exposure for radiation protection purposes. It is assumed that the risk from a high dose or high dose-rate would be around 10×10^{-2} Sv^{-1} for the general population, as indicated in the report of UNSCEAR 1988, BEIR V(1990) and ICRP (1991). They also assume the risk for a working population as 8×10^{-2} Sv^{-1}. They estimated the risk for low dose or low dose-rate exposure from the high dose risk by dividing the risk by 2, assuming DDREF (i.e. Dose and Dose-Rate Effectiveness Factor) is 2. As a result, the risk for low dose exposure is estimated to be 5 and 4×10^{-2} Sv^{-1} for general population and working population respectively [4]. The dose limit used by ICRP is based on this risk estimates.

The low dose extrapolation factor (LDEF) which is equivalent to DDREF and its 90 % confidence interval in the A-bomb survivors cohort was estimated [5], Because A-bomb radiation consists of a high dose- rate even in the low dose range, they preferred to call it LDEF to DDREF , even though both have equivalent meaning. For leukemia, the mean of LDEF is 2.2, but its 90 % confidence interval is wide, 1.1 - 6. For all cancer except leukemia, the mean is 1.3 and its confidence interval is 0.8-3.6. Thus, in both cases, the means of LDEF include value of 2 ,but the confidence interval are fairly large .

The A-bomb survivors cohort study is a representative example of high dose exposure, and it becomes a major source of estimating the risk from radiation exposure in man. Since most methodological problems involved in risk estimation of this cohort are similar to these of other cohort studies of low dose radiation exposure, some of the uncertainty involved in risk estimation in the A - bomb survivors study are shown in Table 2.

Table 2. Uncertainty involved in risk estimation in the A-bomb survivors study

Uncertainty

1. Extrapolation from high to low dose/dose rate
2. Epidemiological uncertainty
3. Dosimetrical uncertainty
4. Projection to life time risk

Epidemiological uncertainty

1. Statistical uncertainty (Sample size)
2. Accuracy of cause of death (Under reporting of cancer)
3. Causative factors other than radiation (e.g. smoking, socio-economic)

The first uncertainty is derived from extrapolation from high dose exposure to low dose using LDEF. The second item which is the epidemiological uncertainty is classified into three : 1) statistical uncertainty including sample size ; 2) accuracy of cause of death ; and 3) confounding factors (e.g, smoking). These are particularly important for low dose risk estimation . Other sources of uncertainty include dosimetrical uncertainty and uncertainty related to projection to life time

Table 3. Accuracy of cause of death

Autopsy Diagnosis	Death Certificate Diagnosis		Total
	Disease X	Other Disease	
Disease X	a	b	a + b
Other Disease	c	d	c + d
Total	a + c	b + d	a+b+c+d

1. Confirmation Rate: a/(a + c) .... 91% (all cancer)
2. Detection Rate: a/(a + b) 76%
3. False Positive Rate: c/(c + d) 3%
4. False Negative Rate: b/(a + b) 24%
5. Correction Factor: [(Detection R./Confirmation R.) -1]
            [(a+c)/(a+b) -1] .... -17%

risk estimation [4] . In the A -bomb survivors survey autopsies were performed whenever possible for pathological study. During the period 1950-1987, there were 6,613 autopsies performed among 46,331 deaths (14.8 %). Accordingly cause of death in death certificate was compared with the autopsy diagnosis to determine the accuracy of the cause of death among the autopsied cases [6]. As shown in Table 3, the confirmation rate (a/a+c) was 91 % for all cancer, and the detection rate (a/a+b) was 76%. Accordingly, the false positive rate (c/c+d) and false negative rate (b/a+b) were 3 and 24 % respectively. Correction factor for number of deaths with disease x (all cancer) should be -17% . Thus the total number of all cancer deaths in death certificates was underreported by 17 % in this example .

In order to determine the effect of misclassification of cause of death on risk estimation, change(s) in parameter estimates in regression models for cancer mortality were compared with and without correction for misclassification of diagnosis . When correction for 20 % of misclassification was made, the ERR becomes 0.55, whereas when no correction was made, ERR was 0.49. Thus ERR increased by 12%, 20 % of misclassification was corrected. Similarly, when 36% of misclassification was corrected, ERR increased by 13 %.

2.2. Method B. Methodological problems in direct observation of the cohort exposed to low dose radiation

As shown in table 4, the two representative studies in this approach, HBRA study [1] and Nuclear workers study [2] were compared in terms of the strength

Table 4. Epidemiological studies on low dose, dose rate radiation exposure

	HBRA Study[1]	Nuclear Workers Study[2]
1. Exposure	Natural Radiation Exposure	Occupational Exposure
2. Type of Studies	Cohort Studies on wide range of age, sex pop.	Cohort Studies on working age and male pop.
3. Sample Size	106,517	95,673 (10 Facilities in Canada, UK, USA)
(Cum. average dose)	(52.6 mGy)	(36.6 mSv)
4. Confounding Factors: Occupation Socio-economic, Life style Migration rate	Farmers (Majority) Homogeneous Small (less than 1%)	Nuclear Workers Varied Relatively large
5. Dosimetry Individual dose Measurement	LImited subjects. (Ave. Dose of hamlet. Indirect estimation of individual dose)	All subjects

Source 1. Workshop: China-Japan Cooperative Studies 1995, Kyoto
        2. IARC Technical Report No. 25, 1995

and weakness of the method. Both are cohort studies. In the HBRA study, the majority of the subjects for the study are farmers, with a wide range of age and sex distribution. The nuclear workers study, on the other hand, involved occupational exposure ; subjects had a limited age range and were mostly male.

Sample size, confounding factors and dosimetry are particularly important for the evaluation of epidemiological studies on low dose or low dose-rate radiation exposure. Sample size; subjects numbered around 100,000 in both studies, the largest cohort among ongoing epidemiological studies. Mean cumulative dose in both studies was roughly the same, when the excess dose from background dose in control area is considered in the HBRA study [1,2].

Confounding factors; where the radiation dose is small, the importance of controlling the effects of factors other than radiation, such as smoking and occupation, comparatively increases in low dose exposure. In this respect the majority of the HBRA cohort consisted of farmers, and their life style and nutrition were more similar than in the other cohorts (i.e. nuclear worker study). The HBRA cohort was very stable; migration rate was much lower (less than 1 %) than that of the nuclear worker cohort. These are the strengths of the HBRA study.

Individual dosimetry; it is essential to measure the individual exposure doses. In this regard, individual dose is available for all subjects used in the analysis for the nuclear worker study, whereas, in the HBRA study, it is available only for limited subjects. Although, indirect estimation of the individual dose is under way using the environmental dosimetry and occupancy factor, it remains debatable whether accurate dose estimate will be possible.

Firstly , the problem regarding sample size is discussed, taking the HBRA study as an example. A total of 100,000 subjects were divided into 4 dose groups (each group consisting of 25,000 subjects). A cancer mortality in each high dose group was compared with that of the control group which is 55 per 100,000 per year. False positive (Type I) error (α) and false negative (Type II) error (β) are assumed to be 0.05 and 0,20 , respectively.

The sample size should also depend on the difference in mortality between the two comparison groups. The sample size to detect the difference should be larger, when the difference in mortality is small. In this case, three examples are assumed in which the relative risk (ratio of mortality rate to that of control group) is 2, 1.5 and 1.2, respectively. The sample size to detect the statistically significant difference in cancer mortality between comparison groups is then calculated under these conditions (Table 5) . In example 1 in which the relative risk is assumed to be 2, the required sample size is 42,727. In example 2, where relative risk is 1.5, the required sample size will be 142,446 or 3.3 fold larger than that of example 1 . In the example 3, where relative risk is 1.2, the required sample size is around 780,000, which is 18-fold larger than that of example 1. As we assume that number of subjects in one group is 25,000, the required follow up period will be 1.7 years in example 1 (number of person years reach to the requir-

Table 5. Estimation of sample size

Assumption:

$P_1 = 0.00055$ per year
$\alpha = 0.05, \beta = 0.20$

Estimated Sample Size

Example	Relative Risk	No. Subjects (person years)	Ratio	Average* follow up (year)
1	2	42,727	1	1.7
2	1.5	142,446	3.3	5.7
3	1.2	783,524	18.3	31.3

* With the assumption of 25,000 cohorts size

ed sample size). The required numbers of years of follow-up will be 5.7 and 31.3 years for example 2 and 3, respectively. This gives an idea of the sample size necessary to detect the small difference in mortality between the comparison groups.

Though limitation of sample size of the cohort may preclude to detect the small difference in mortality between comparison groups, these studies should be useful in setting the upper bounds of estimated risk.

In low dose radiation exposure, since radiation effects should be small, the importance of confounding factors comparatively increases.

In the nuclear worker study [2], ERR (excess relative risk) per Sv were compared when SES (socio-economic status) was or was not adjusted. For all cancer excluding leukemia, when SES was not adjusted, ERR is 0.20, though this is not statistically significant. When SES was adjusted, ERR becomes -0.07, much smaller, and the coefficient was negative. For leukemia, ERR does not change whether SES was adjusted or not. Hence SES is not a strong confounding factor for leukemia mortality. In either case increase in ERR 2.2 was suggested. Thus far, effect of confounding factors was shown with SES as an example. It is important to examine the role of confounding factors (e.g., life style and nutrition as well as medical source of radiation).

In order to estimate the risk of low dose or low dose-rate radiation exposure, a large cohort study (e.g. HBRA study in China [1] or nuclear workers study [2]) is best. However, the weakness of the cohort studies under discussion may preclude the success of the cohort study. As an alternative epidemiological method, a case control study in which subjects (i.e., both cases and controls) are selected from the cohort, could be made to supplement the cohort study. This is called as a case control study nested in the cohort. As an example, a case-control study on effects of residential radon conducted in Missouri USA in 1993 is introduced [3]. The purpose of the study was to clarify the dose-response relationship between residential radon dose and lung cancer incidence. The significance of the method in this study could be summarized in the following three points: A) Non-smoking women were selected because they offered the best opportunity to detect radon-related risk with high indoor occupancy, while minimizing the potential confounding influences of cigarette smoking and occupation. A total of 538 lung cancer cases were ascertained from the Missouri tumor registry and 1,183 age-sex matched control were selected randomly from the same residential area. B) For all subjects, case and control, indoor radon (kitchen and bed room) was measured at nearest time when diagnosis of lung cancer was made. In addition, average cumulative dose of indoor radon for past 30 years was estimated. C) Detailed information on confounding factors other than smoking (passive smoking, intake of saturated fat, educational history etc.) was obtained by questionnaire. No significant dose response was observed at the level of indoor radon in this study.

In general, as the number of cases was limited the number of subjects including control in the case- control study was much smaller than that of the cohort study. Accordingly, the case-control study has the advantage to conduct very detailed survey with less expense including individual, dose measurement and controlling the confounding factors as illustrated in this example. A similar case-control study nested in the cohort is now being planned in the ongoing HBRA study in Guangdong Province, China.

REFERENCES

1 . High Background Radiation Research Group, China (L.Wei), Science, 209, 22 Aug. 1980, 877-880.
2. E.Cardis, E.S.Gilbert et al., 1ARC Technical Report No.25(1995).
3. C.R.Michel, R.C.Alvanja et al., JNCI. 86 (1994), 1829-1887.
4. W.K.Sinclair, NCRP. Proceedings of the twenty-ninth annual meeting 7-8 April, 1993, 209-244.
5. R.Sposto, D.L.Preston et al., RERF TR 4-91 (1991).
6. E.Ron, R.L.Carter et al., RERF CR 6-92 (1992).

Recent advances in dosimetry investigation in the high background radiation area in Yangjiang, China

Y.-L Yuan[a], H. Morishima[b], H. Shen[a], T. Koga[b], Q.-F. Sun[c], K. Tatsumi[d], Y.-R. Zha[e], S. Nakai[f], L.-X. Wei[c] and T. Sugahara[f]

[a]Labor Hygiene Institute of Hunan Province, Changsha 410007, China

[b]Kinki University Atomic Energy Research Institute, Osaka 577, Japan

[c]Laboratory of Industrial Hygiene, Ministry of Health, Beijing 100088, China

[d]Kinki University Life Science Research Institute, Osaka 577, Japan.

[e]Guangdong Institute of Prevention and Treatment of Occupational Diseases, Guangzhou 510300, China

[f]Health Research Foundation, Kyoto 606, Japan

1. INTRODUCTION

Since 1991, the High Background Radiation Research Group of China (HBRRG) has cooperated with the Health Research Foundation of Japan (HRF) for the joint research project Epidemiological Study in High Background Radiation Area (HBRA). For the purpose of getting a complete knowledge of the radiation levels in every hamlet in the investigated areas, and for a reasonable classification of dose groups for the cohort members, which will be very useful to the analysis of dose-effect relationships, the following work has been done in this period of investigation: (1) additional measurements of radionuclides in soil and construction materials; (2) in addition to the measurements of environmental ionizing radiation (indoors, outdoors, on the field etc.) conducted in the previous study (before the Joint Research Project), all the rest of hamlets in the investigated areas (about 2/3 of the total hamlets) were measured with the same items as before; (3) in addition to the measurements of individual cumulative dose conducted in the previous study, additional thermoluminescent dosimeters (TLD) were used for measure-ments of cumulative dose for the individuals and for the environment; (4) factors which might influence the results of dose assessments were surveyed, among which the occupancy factor was used for estimation of individual annual doses with the results obtained from environmental measurements; (5) the results obtained from both TLD measurements and dose-rate meters in the same hamlets were compared and analyzed with the coefficients of correlation; (6) four dose groups were classified based on the above-mentioned data; (7) factors which might influence environmental radiation levels were investigated. These factors include change in different seasons, kinds of structures, construction materials and distributed locations of houses and range of radiation levels within the hamlet etc.; (8)quality control of the measurements were conducted.

2. CONTENTS OF NATURAL RADIONUCLIDES IN SOIL AND CONSTRUCTION MATERIALS

For a better understanding of the causes of fluctuation in dose-rates of environmental radiation in the investigated areas, a complementary survey on the sources of higher background was conducted, which included (1) measurements of the contents of natural radionnuclides in soil samples, and (2) measurements of the contents of natural radionuclides in the construction materials. The sampling of local soil was somewhat different from that in the previous study: in the previous study the samples were taken from the uncultivated soil far from the residential area; for comparison, in this study the samples were taken from the cultivated land in the residential area. 26 samples from 13 hamlets in HBRA, 8 samples from 4 hamlets in control area (CA) were taken; every sample contained 5 sub-samples from a land of 10 square meters in area. Samples were analyzed by chemical processes for uranium and thorium and by scintillation emission method (after chemical preparation) for radium-

Table 1
Contents of natural radionuclides in soil of the investigated areas

Area	Year	No. of samples	Contents in Bq kg^{-1}		
			U-238	Th-232	Ra-226
HBRA					
	1979	30	93.5±20.9	248.2±115.1	144.3±55.5
	1993	26	127.1±50.8	237.4± 91.7	187.1±66.9
CA					
	1979	30	20.9± 8.6	32.5±12.7	29.6±11.0
	1993	8	37.4±13.0	33.9±18.3	36.8± 8.9
HBRA:CA					
	1979		4.5	7.6	4.9
	1993		3.4	7.0	5.1

226. Table 1 shows the results of measurements. For comparison, the results obtained from previous study (1979) are also shown in the table.

The results listed in Table 1 indicate that the contents of natural radionuclides in soil of HBRA were significantly higher than those in CA, especially those of thorium were about seven times higher than in CA.

The samples of local construction materials were taken from the same hamlets where the samplings for soil measurements were conducted. The samples included bricks and tiles. 57 samples from 27 households of 13 hamlets in HBRA and 10 samples from 5 households of 4 hamlets in CA were taken, respectively. The methodology for laboratory analysis was the same as that for the contents of natural radionuclides in soil. The results of the analyses are listed in Table 2.

Table 2
Contents of natural radionuclides of construction materials in HBRA and CA

Area	No. of samples	Contents in Bq kg^{-1}		
		U-238	Th-232	Ra-226
HBRA	57	208.2±57.4	327.9±83.5	251.1±52.5
CA	10	80.6±35.9	78.3±39.4	87.8±56.3
HBRA:CA		2.6	4.2	2.9

The results in Table 2 show that the contents of natural radionuclides in construction materials in HBRA were higher than those in CA. If we compare the results in Table 1 with those in Tables 2, it is obvious that the contents of natural radionuclides in construction materials were higher than those in local soil. Very probably the processes of fabrication from local soil to bricks and tiles concentrated the contents of natural radionuclides.

In order to study whether the activity of radionuclides changed in different depth of soil samples, 6 samples were collected from uncultivated soil in HBRA and CA, respectively. Every sample was collected from five layers. The samples were analyzed with chemical procedures and -spectrometry to measure concentrations of U-238, Th-232 and Ra-226. The results are shown in Figure 1.

The results show that the distribution of the concentrations for the above three natural radionuclides did not change with the change of soil depths; their distributions were even. It proves that the formation of higher background in HBRA took a long time, and increase of environmental dose-rate in this area is not caused by pollution of human activities.

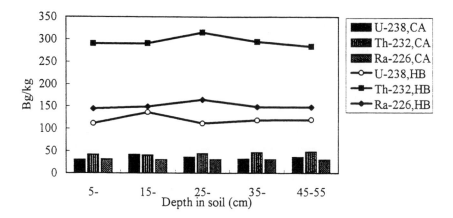

Figure 1 Contents of natural radionuclides by depths in soil

3. MEASUREMENT OF ENVIRONMENTAL RADIATION LEVELS

We have measured the environmental radiation levels with dose-rate-meters in all hamlets in the investigated areas. There are 384 hamlets in HBRA and 142 in CA. One third of the households undergoing stratified random sampling in every investigated hamlets were measured for indoors (bedroom, sitting room and kitchen), outdoors (alley, yard, pond, well and lane) and farmland (rice field and dry field). The absorbed dose rate in air was measured 1m above ground. In order to ensure the reliability of the measurement results, we used different instruments and methods to repeat independent measurements. The results are listed in Table 3.

In order to compare the difference of absorbed dose rate in air from terrestrial gamma radiation between different dosimeters in Table 3, the contributions of cosmic rays were subtracted. It is known from the measured results in Table 3 that although different investigators used different instruments and methods, the results obtained were distributed in a very close range. This proves that the results were comparable. It is reasonable to use the arithmetic mean of the results. The values for indoors and outdoors in HBRA were 33.42×10^{-8} Gy/h and 16.51×10^{-8} Gy/h, respectively. The values for indoors and outdoors in CA were 9.47×10^{-8} Gy/h and 4.50×10^{-8} Gy/h, respectively. The ratio of the dose rate between HBRA and CA was 3.5 for indoors, and 3.7 for outdoors.

For estimating the annual effective doses, several parameters are needed. UNSCEAR recommended that the occupancy factor is 0.8 for indoor stay, and is 0.2 for outdoor stay, on an average, around the world. However, in our case, the

Table 3
Comparison of results from environmental measurements with different dosimeters in HBRA and CA

Measurement instrument	Absorbed dose rate in air (10^{-8}Gy·h^{-1}*)			
	HBRA		CA	
	Indoor	Outdoor	Indoor	Outdoor
FD-71	34.53±4.89	16.13±3.59	9.49±1.63	4.12±0.96
RSS-111	34.77±6.01	15.63±3.69	9.97±1.87	4.49±0.69
LiF(Mg Cu P)	34.63±5.38	16.65±3.98	9.26±2.18	4.40±1.22
CaSO$_4$(Tm)	-	17.00±4.25	-	4.49±1.08
CaSO$_4$(Dy)	31.69±3.98	14.46±2.49	8.67±2.03	4.18±0.85
RPL	31.49±4.31	20.44±2.52	9.96±2.09	6.09±1.46
Spectrometer	-	15.28±4.19	-	3.75±0.69
Average	33.42±4.91	16.51±3.53	9.47±1.96	4.50±0.99

* Figures in Table 3 are average absorbed dose rates from terrestrial radiation. It does not include contribution from cosmic rays.
Indoors : HBRA:CA=3.5:1.0
Outdoors: HBRA:CA=3.7:1.0

investigated areas are located in the rural area of South China and most of the pepole we observed are farmers. Thus we analyzed the data of residential time for sleep, indoor stay, outdoor stay of 5,291 cohort members in the investigated areas. The data were collected during our dosimetry survey there. Table 4 shows the data of residential time we collected.

The results in Table 4 indicate that the difference between both sexes was small (for males, indoors: outdoors = 0.69 : 0.31; for females, indoors : outdoors = 0.71 : 0.29), but the difference between the age groups was prominent. Figure 2 gives the distribution of time in one day for sleep, indoor and outdoor stays.

From Fig. 2 we see that no great difference exists in time of sleep, indoor and outdoor stays in the age from 20 to 65 years for males, and from 25 to 50 years for females. But other age groups have great difference.

Table 5 shows the estimated annual effective doses received by the cohort members based on the data of above-mentioned environmental measurements and considering the occupancy factors for males and females of different age groups. For comparison, the data obtained before the cooperation are also listed.

Table 4
Residential time of the cohort members of both sexes in different age groups

Age range	Residential time (hours per day)					
	Sleep		Indoors		Outdoors	
	Females	Males	Females	Males	Females	Males
0-	11.00	11.11	8.29	8.04	4.71	4.85
5-	9.61	9.61	8.51	8.11	5.88	6.28
10-	8.94	9.10	8.86	9.06	6.20	5.84
15-	8.82	8.86	9.03	8.90	6.15	6.24
20-	8.29	8.46	8.96	7.21	6.75	8.33
25-	8.20	8.30	8.20	7.19	7.60	8.51
30-	8.07	8.03	7.76	7.06	8.17	8.91
35-	8.09	8.06	7.48	7.15	8.43	8.79
40-	8.13	8.13	7.54	6.91	8.33	8.96
45-	7.95	8.07	7.92	6.95	8.13	8.98
50-	8.08	7.92	7.75	7.50	8.17	8.58
55-	8.02	7.90	8.56	7.15	7.42	8.95
60-	8.29	7.96	9.09	7.57	6.62	8.47
65-	8.69	8.24	9.05	8.04	6.26	7.72
70-	8.84	8.67	9.16	8.95	6.00	6.38
75+	9.08	9.11	10.33	8.67	4.59	6.22

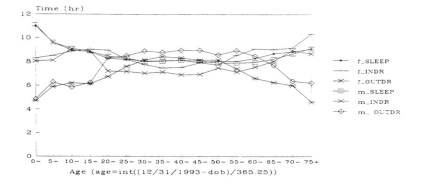

Figure 2 Time spent on bed, indoors and outdoors of cohort members by sex and age

Table 5
Estimated annual effective doses in HBRA and CA based on the data of environmental measurements of radiation level

Area	No. of hamlets measured	No. of people involved	Annual effective dose (10^{-5} Sv.a^{-1})			
			Weighted mean	Mean	Max.	Min.
HBRA						
1*	161	34446	210.49±25.86	212.27±26.56	308.04	160.22
2*	223	44168	212.92±30.67	213.57±30.31	292.58	125.29
1 and 2 combined	384	78614	211.86±28.72	213.03±28.77	308.04	125.29
CA						
1*	48	9797	69.61±9.28	67.96±8.56	85.23	50.43
2*	94	18106	68.06±9.41	67.91±9.30	95.67	50.54
1 and 2 combined	142	27903	68.60±9.35	67.92±9.02	95.67	50.43

1 * Measurements before the Joint Research Project
2 * Measurements after the beginning of the Joint Research Project

4. RESULTS OBTAINED FROM MEASUREMENTS OF INDIVIDUAL CUMULATIVE DOSES

Before the Joint Research Project the HBRRG in China has measured the cumulative exposures with thermoluminescent dosimeters (TLD) for the inhabitants in the investigated areas. In the Joint Research Project, additional TLD were used for measurements of cumulative doses for the individuals. It should be noted that part of the people wore two or more than two dosimeters for the purpose of intercomparison or for quality control; thus, the numbers of persons available for dose assessments were 3,900 in HBRA and 1,304 in CA. Table 6 lists the estimates of annual effective doses received by people in HBRA and CA based on the data of TLD measurements.

Table 7 shows the estimates of annual effective doses received by males and females, based on the data of TLD measurements.

Table 8 shows the estimates of annual effective doses received by people of various age groups, based on the data of TLD measurements.

Tables 7 and 8 indicate that the females in HBRA had slightly higher dose than the males, but the difference was not significant statistically (P=0.18). However, the individual cumulative dose received by the 0-7 years age group in HBRA was highest among all age groups (P<0.05), while the individual cumulative dose for age group of 41-60 years was lower than those for other four age groups. In CA, there was no statistically significant difference between both sexes, nor between various age groups.

The influence of construction structures on the dose estimates in the investigated areas was also studied. The inhabitants' houses were divided into four kinds of structure: houses built of adobes, houses built of bricks and tiles,

Table 6
Estimates of annual effective doses based on the data of TLD measurements

Period of study and area	No. of people wearing TLD	No. of hamlets measured*	Annual effective dose (10^{-5} Sv·a^{-1})
Before the collaboration			
1 HBRA	585	37	205.23±38.19
2 CA	161	10	71.88±12.23
After the starting of collaboration			
3 HBRA	3315	35	207.02±33.18
4 CA	1143	14	66.44±12.53
1 and 3 combined	3900	72	206.75±36.85
2 and 4 combined	1304	24	67.11±12.14

*7 hamlets in HBRA and 1 hamlet in CA had been measured both before and after the collaboration.

Table 7
Estimates of annual effective doses received by males and females, based on the data of TLD measurements

| Area | Annual effective dose (10^{-5} Sv·a^{-1}) | | | | P-value |
| | Males | | Females | | |
	No. of people	Mean±S.D	No. of people	Mean±S.D	
HBRA	2010	205.98±38.29	1890	207.56±35.24	0.18
CA	688	67.34±12.17	616	66.85±12.12	0.47

Table 8
Estimates of annual effective doses received by people of various age groups based on the data of TLD measurements

| Age group (in years) | HBRA | | CA | |
	No. of people	Annual dose (Mean±S.D) in 10^{-5} Sv·a^{-1}	No. of people	Annual dose (Mean±S.D) in 10^{-5} Sv·a^{-1}
0-	311	212.49±38.14	77	67.14±12.92
8-	953	208.15±34.67	326	67.34±12.00
17-	999	207.65±37.36	406	67.00±12.29
41-	752	202.94±36.82	276	67.49±12.47
60-	885	205.88±37.92	219	66.47±11.47

houses built of both unbaked bricks and baked bricks, and others (including concrete structure and houses built of stones). 3900 people in HBRA and 1304 in CA who wore TLD for 2 months, were inquired about their house structures. The annual cumulative doses estimated by TLD measurements of the above-mentioned subjects and the structures of their houses are listed in Table 9.

The results in Table 9 demonstrate that people living in houses of brick structure in HBRA received the highest doses among all the subjects. However, this difference was not revealed in CA.

Table 9
Estimated annual doses of the inhabitants living in various structures of houses

House structure	Annual effective dose(10^{-5}Sv.a^{-1})			
	HBRA		CA	
	No. of subjects	Mean±S.D	No. of subjects	Mean±S.D
Adobes	84	198.13±26.22	386	68.36±14.10
Bricks and tiles	1727	212.13±38.24	326	68.49± 9.60
Bricks and unbakerd bricks	1006	201.01±36.54	294	62.31± 9.54
Others*	496	202.92±36.29	124	65.20±13.79
Unknown	587	205.22±32.57	174	71.21±11.85

* Others included concrete structure and stone structure.

5. CLASSIFICATION OF DOSE GROUPS OF THE COHORT MEMBERS

For the purpose of studying the dose-effect relationships among the cohort members, it would be ideal that every subject have their own records of individual accumulated doses received before the evaluation. However, it is very difficult to realize it in practice(each of 106,517 persons should wear TLD). Thus we have planned two sets of measurements. Firstly, we measured the environmental dose rate (indoors, outdoors and over the bed) in every hamlet of the investigated areas, considering the occupancy factors for males and females of different age groups to convert the data of dose rates into the annual effective doses. Secondly, we measured the individual cumulative dose with TLD for part of the subjects in the investigated areas. All the 526 hamlets in the investigated areas were measured with scintillation dose-rate meter; among them 88 hamlets were measured and estimated for external dose with both the scintillation dose-rate-meter and individual cumulative dosimeter (TLD). To compare the results,

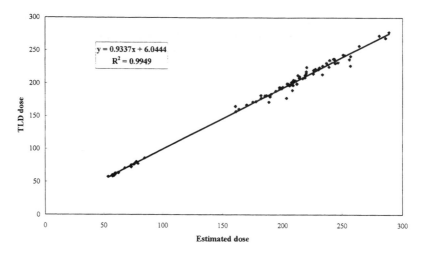

Figure 3 Scatter of hamlet-specific means of TLD and estimated dose ($10e^{-5}Sv/a$)

hamlet by hamlet, obtained by scintillation dose-rate-meter and by TLDs in the same hamlets, a curve was plotted as shown in Figure 3.

The plot demonstrates that the results of the two sets of measurements were in good correlation(y=0.9337x+6.044, coefficient of correlation is 0.9974).As described above, we are able to use the data from environmental dose-rate measurements of all the 526 hamlets in the investigated areas to classify the dose groups for the cohort members. Table 10 shows these 4 groups divided by the

Table 10
Classification of dose groups for the cohort members of investigation

Dose group	Annual effective dose ($10^{-5}Sv/a$)			No. of hamlets	No. of people
	Range	Weighted average*	Mean		
High	224.10-308.04	245.48	246.07	124	23718
Medial	198.07-224.09	209.85	210.19	135	28803
Low	125.29-198.06	183.50	183.31	125	26093
Control	50.43- 95.67	68.60	67.92	142	27903

*Weighted by the population size of hamlet

ranges of annual doses received by the people. The number of hamlets, population sizes and the averages of dose are also noted. The Table indicates that most inhabitants (66.81%) in HBRA were in "high" and "medial" groups.

6. CONCLUSIONS

6.1. The results of analyses of soil samples showed that the contents of radionuclides in soils of HBRA were significantly higher than those in CA, especially those of thorium, which were about seven times higher than those in CA. Construction materials of houses in HBRA had higher contents than those in CA. Construction materials of houses in HBRA were also contributors to the higher background of gamma radiation because they mostly came from the local earth.

6.2. The values of absorbed dose rate in air from terrestrial gamma radiation for indoors and outdoors in HBRA were 33.42×10^{-8} Gy/h and 16.51×10^{-8} Gy/h, respectively. The values for indoors and outdoors in CA were 9.47×10^{-8} Gy/h and 4.50×10^{-8} Gy/h, respectively. The ratio of the dose rate between HBRA and CA was 3.5 for indoors, and 3.7 for outdoors. Estimated annual effective doses in investigated areas based on the data of environments of radiation levels, considering the occupancy factors for males and female of different age groups in HBRA and CA were 211.86×10^{-5} Sv/a and 68.60×10^{-5} Sv/a, respectively.

6.3. According to the results from TLD measurements, the estimates of annual average effective doses received by cohort members in HBRA and CA were 207.56×10^{-5} Sv/a and 66.85×10^{-5} Sv/a, respectively. It should be noted that they are only external doses from cosmic rays and terrestrial gamma radiation, and do not include the contribution of the neutron components of cosmic rays.

Study of the indirect method of personal dose assessment for the inhabitants in HBRA of China

H. Morishima[a], T. Koga[a], K. Tatsumi[a], S. Nakai[b], T. Sugahara[b], Y. Yuan[c], Q. Sun[d] and L. Wei[d]

[a]Kinki University, 3-4-1, Kowakae, Higashi-Osaka, Osaka 577, JAPAN

[b]Health Research Foundation, 103-5 Tanaka-Monzen-cho, Sakyo-ku, Kyoto 606, JAPAN

[c]Hunnan Institute of Labor Hygiene and Prevention, Changsha, Hunnan 410007, CHINA

[d]Laboratory of Industry Hygiene, Ministry of Health, Beijing 100088, CHINA

1. INTRODUCTION

As a part of the China/Japan cooperative research on radiation epidemiology, we carried out a dose-assessment study to evaluate the exposure to natural radiation in the HBRA of Yangjiang on the Guangdong province and Enping prefecture as the control area from 1991 to 1995 [1,2]. Representative international survey studies have been conducted to evaluate the exposure due to natural radiation, the Chernobyl accident and radioactive fallout [3-6]. As it is very difficult to directly measure using a dosemeter the personal doses for all inhabitants of the areas exposed to high natural radiation, we must estimate them by an indirect method using the data of environmental dose rates measured up to now by survey meters as the most feasible method. There are several factors inherent in the life styles of the inhabitant and environmental radiation dose. Accurate personal dose assessment for inhabitants in HBRA is an essential requirement of the present investigation. For this purpose, we have quantitatively examined the distribution characteristics of the numerical measurements and analyzed the variables concerned, directly using several dosemeters in order to obtain the detailed information on certain hamlets with regard to better accuracy of the environmental doses and personal exposure doses.

2. MATERIALS AND DOSE MEASUREMENT METHODS

Environmental indoor and outdoor dose rates had been measured using a NaI (Tl) scintillation survey meter (Aloka TCS-166), and for personal doses with electronic pocket dosimeters (Aloka PDM-101) and thermoluminescence dosimeters (TLD, National Co. UD-200S) which were exposed for 24 hours and 2 months, respectively. We selected the Madi hamlet with about 200 houses from the many hamlets on the HBRA for the environmental dose measurement and further selected a large family containing three generations of parents and children for personal dose assessment.

3. RESULTS AND DISCUSSION

3.1. Nonhomogeneous spatial distribution of radiation exposure

Average indoor and outdoor environmental radiation dose rates in the different hamlets of the HBRA were measured using a NaI (Tl) scintillation survey meter (TCS-166). It was remarkable that the indoor dose rates are about 2 times higher in the Madi hamlet, while the indoor dose were almost similar to the outdoor one in the control area (CA). The indoor dose rate has a good correlation with the dose rate for the house wall. These results strongly suggest that radionuclide concentrations in the house walls are deeply connected with the radiation dose exposure. We have performed a radionuclide analysis of the building materials, and nuclides of Th-232 and U-238 decay products in the HBRA were several times higher than those of the CA. Personal doses largely depend upon exposure originating from radionuclides in the building materials. Variations in the indoor radiation doses in the Madi hamlet were observed such that the relative standard deviation was several percent in the bed, the living room and kitchen, while the other rooms were not more than 20%. The spatial distribution of the environmental dose showed that the center of the hamlet was slightly higher than the periphery. There was a little inverse correlation between indoor radiation doses and ages of the building as shown in Figure 1. The indoor doses for the new homes were lower.

We observed there was a fairly large variation of indoor doses with different homes in the same hamlet, presumably due to the different housing conditions such as levels of radionuclides in the building materials, age of the building and spatial location in the hamlet.

3.2. Occupancy factors and variation in the estimated personal doses

The personal doses will be affected by the indoor dose levels and occupancy factors which are reflected from personal behavior in the radiation environment. This information was obtained from the inhabitants of 10 selected homes with large family containing three generations, of parents and their children in the Madi hamlet.

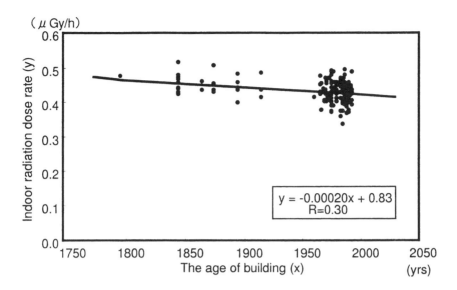

Figure 1. Relation between the mean indoor radiation dose rates and the ages of the building in the Madi hamlet

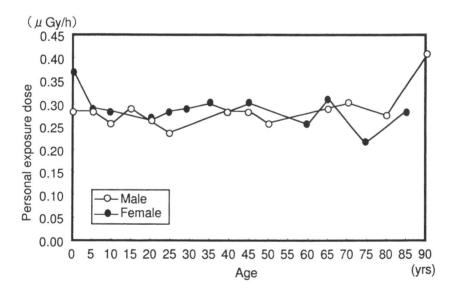

Figure 2. Age distributions of the personal exposed doses by the direct method using PDM

In order to obtain occupancy factors of the inhabitants, we conducted an interview questionnaire after the personal dose measurements. The mean indoor occupancy factors for the baby and the older members were about 0.8, a little higher than the other age group (about 0.6) while the difference in sex did not show any correlation. The distribution of the occupancy factors obtained is more like the pattern for the normal curve. According to the UNSCEAR [7], the mean occupancy factor used for dose assessment was 0.8.

3.3. Estimation of personal dose; direct method and indirect method

We have examined the variations in the personal doses measured by TLDs as well as PDM in the Madi hamlet. As shown in Figure 2, the personal dose rates were nearly uniform with age except for the baby and the older inhabitants. To estimate the personal doses of the hamlet inhabitants, two different approaches were used the direct method measured using a dosimeter such as a TLD and PDM and the indirect method using the occupancy factor and environmental radiation doses. If the difference between the indoor and outdoor dose rates is large, personal doses will be significantly affected by the occupancy factor. This relation is shown by the following equation;

$$P = \sum_{i=1}^{n} (q_i \times A_i)$$

where P is the personal exposure dose, q_i are occupancy factors in the bedroom, living room, kitchen, school, shop in hamlet and outdoors; and A_i are their environmental dose rates in air, respectively. To estimate the personal dose, variations in the environmental radiation dose on almost all houses and the occupancy factors of the inhabitants in the Madi hamlet were observed. A variation of the mean estimated personal dose by age groups for the bedroom, gross indoor and outdoor in the hamlet are demonstrated in Figure 3.

Furthermore, we have measured their exposure dose directly using a dosimeter. A comparison has been made of the personal dose assessment by the direct method (PDM) and indirect method (fine calculation and rough calculation). There is a good correlation of the estimated values. As shown in Figure 4, the mean variation coefficient of the ratio of the measured values to the estimated was about 10 % and ranged less than 27 %. The variation coefficients of the measured exposure dose by PDM repeated three times was 8 %. It is sufficient to estimate using the mean occupancy factors which are specific for the sex and age groups, and the mean indoor and outdoor environmental doses.

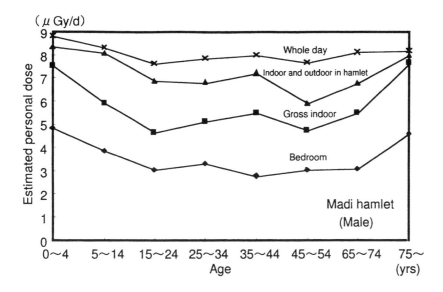

Figure 3. Age distributions of the estimated doses using the indirect method

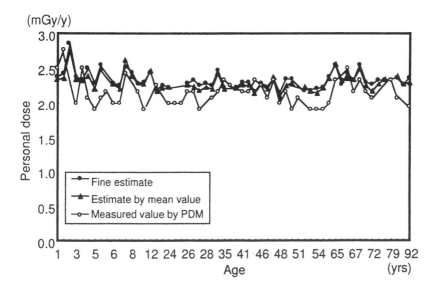

Figure 4. Comprison of the measured personal doses and the estimated doses

4.CONCLUSIONS

We have selected the Madi hamlet for the large scale and high level radiation dose investigation in the HBRA to determine the distribution characteristics of environmental radiation on the periphery of the hamlet and the personal exposure dose assessment. It is sufficient for the personal dose assessment to estimate it indirectly using the mean occupancy factors specific for the sex and age groups, and the mean environmental doses. Further studies to elucidate seasonal effects on their occupancy factors and distribution characteristics for different levels of environmental radiation doses require a more accurate personal dose estimation.

REFERENCES

1. H. Morishima, T. Koga, K. Tatsumi et al., Proc. Asia and Pacific Basin Regional Congress on Radiation Protection, 18 Oct. 253-256 (1993)
2. H. Morishima, T. Koga, K. Tatsumi et al., Proc. of Int. Congress Radiation Protection IRPA-9, 2, 213-215 (1996)
3. Y.E. Dubrova, V.N. Nesterov, N.G. Krouchinsky et al., Nature, 380, 25 April (1996)
4. S.L. Simon, R.D. Lloyd, J.E. Till et al., Health Physics, 59, 5, 669-691(1990)
5. W.J.Schull et. al., Radiation dose reconstruction for epidemiologic uses,
6. F. Bochicchio, G.C. Venuti, F. Monteventi et al., Proc. of IRPA-9, 2-190 (1996)
7. UNSCEAR, UNSCEAR 1993 Report - Exposures from natural sources of radiation

Databases and statistical methods of cohort studies (1979-90) in Yangjiang

Quanfu Sun[1], Suminori Akiba[2], Jianming Zou[3], Yusheng Liu[1], Zufan Tao[1], and Hiroo Kato[4]

[1] Laboratory of Industrial Hygiene, Ministry of Health, Beijing 100088, China

[2] Faculty of Medicine, Kagoshima University, Kagoshima 890, Japan

[3] Guangdong Institute of Prevention and Treatment of Occupational Diseases, Guangzhou 510310, China

[4] Radiation Effects Research Foundation, Hiroshima 732, Japan

1. INTRODUCTION

In 1972, epidemiological cancer study of high background radiation (HBR) was started in Tongyou and Donganling districts in Yangjiang county[1]. Most of the residents in the HBR area are farmers who grow grain and have lived there for more than five generations. A control area, consisting of several towns selected from Taishan and Enping counties, neighboring Yangjiang, is similar to the HBR area in terms of social and economic status, including lifestyles and living standards.

Between 1970 and 1974, preliminary survey was conducted in the HBR and control areas in order to examine the feasibility for mortality follow-up survey in both areas[1]. In 1978, retrospective follow-up study was conducted to gather mortality data of the period 1975-78[2]. In 1979, our study group established Health Household Registry, a village-based mortality survey system, in the HBR and control areas and started prospective mortality follow-up[2]. In 1990, a fixed cohort was established to collect mortality data during the period 1987-95. The subjects who were living in the HBR and control areas covered by the Health Household Registry and alive as of January 1, 1987 were recruited to the fixed cohort. One town (Chengnan) in Enping and all the towns in Taishan were excluded from the control area, since they underwent more rapid economic developments in recent years when compared to the rest of the study areas.

This paper, supplement to materials and methods in ZF Tao's paper[3], will describe our mortality follow-up, computer databases and methods of statistical analysis.

2. MORTALITY FOLLOW-UP SURVEY

2.1 Morality follow-up system

The study areas, with 130,000 subjects in 20,000 families, cover 526 hamlets of 54 villages in 8 towns. The administrative hierarchy which implemented in the study areas was shown in Figure 1.

Fig. 1 Administrative hierarchy in study areas

Mortality follow-up system, consisting of demographic survey and ascertainment of cause of death, was set up in each village in 1979[2,4]. In demographic survey, Fundamental Registrars (FRs), who were village doctors selected by our research group, collected subjects' demographic information in their own villages. FRs filled in the form of Health Household Registry for each subject. Task Group (TG), consisting of three physicians from Guangdong Institute of Prevention and Treatment of Occupational Diseases and trained in the Guangdong Province Tumor Hospital, went to each village once a year to check up the records of Health Household Registry.

In ascertainment of cause of death, TG obtained the name list of the deceased from FRs, and interviewed the family members and neighbors of the deceased as well as village doctors to collect information on the hospitals the patients visited, the examinations they underwent and their possible causes of death. Then they filled in Death Registration Cards for all the deceased cases. TG also reviewed the name lists of biopsy, X-ray and ultrasonic examinations as well as casebooks in all the county hospitals and other major hospitals to identify the patients from the HBR and control areas.

A similar follow-up method was used for the cohort study of the period 1987-95 with some modifications. There were three major changes. First, a unique 10-digit individual identification (ID) code (see Table 1) was assigned to each study subject. The ID code tells the county (first 1 digit), town (first 2 digits), village (first 4 digits), hamlet (first 6 digits), family (first 8digits), and an individual (full 10 digits). Second, research group, rather than FRs, visited each village once a year to collect demographic information and filled in Death Registration Cards with local doctors' help. Third, in addition to Death Registration Cards, Questionnaires on Cause of Death were used to record all information related to causes of death, such as a brief description of the medical history, signs and symptoms, diagnosis, pathological findings and so on.

Table 1
Individual ID code used in cohort study of Yangjiang

Digit of ID code	Explanation
1st	County
2nd	Town
3rd-4th	Village
5th-6th	Hamlet
7th-8th	Family or household
9th-10th	Family member

The underlying cause of death was determined based on the WHO rules and coded according to the 9th revision of the International Classification of Diseases and Injuries (ICD)[5].

2.2 Migration and completeness of cancer-case finding

The migration rate was about 2 per thousand person-years. For the details of migration refer to the paper by Zha YR et al[6]

Generally, patients saw their village physicians first and then were, if necessary, introduced to higher level hospitals including county hospitals. Unfortunately, however, a large part of cancer patients went home after short periods of hospitalization, because they could not afford surgical operations, or advanced medical examinations and treatments. Therefore, most of cancers in the study subjects are fatal. Once patients returned home, information on their health conditions, and medical treatments and examinations spread all over the village by various means of local communications. We are sure that no deceased case left unreported to Health Household Registry. Due to poverty suffered by local people and a lack of effective medical service, however, some cases of cancer might have been misclassified as non-cancer cases, particularly, among the old people. That kind of misclassification may explain the decreases of cancer mortalities in the subjects of about 70 years or older, which were shown in Figure 2.

The proportions of cancer with pathology diagnosis varied with cancer sites (Table 2). As can be judged from the data presented in this table, the accuracy of diagnosis in some cancer sites may not be good enough for site-specific cancer risk analysis.

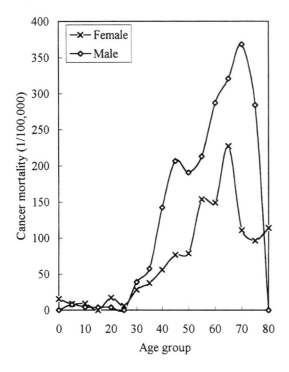

Fig. 2 Cancer mortality in HBR and control areas combined

Table 2
Methods of diagnosis by cancer sites in cohort study (1987-90)

Site of cancer (ICD-9 codes)	No. of cases diagnosed by corresponding methods		
	Pathology	X-ray or ultrasonic exam.	Others
Leukemia (204-208)	9	0	0
Nasopharynx (147)	22	14	2
Liver (155)	1	61	7
Lung (162)	6	13	0
Stomach (151)	11	8	3
Esophagus (150)	1	5	3
Breast (174)	2	1	2
Others	20	17	23
Total	72	119	40

3. COMPUTERIZED DATABASES

3.1 Cohort data sets

There are three sets of mortality data covering different periods, i.e., 1970-78, 1979-86 and 1987-95. There was no computerized database available for the period 1970-78. Data for the period 1991-95 were still in the process of follow-up.

3.1.1 Fixed cohort data set of the period 1987-90

The are two databases for this period, one is for demographic survey, and the other for cause of death. The information stored in the two databases are listed in Tables 3-1 and 3-2. Both databases were kept in the format of Foxbase+ Rev 2.00 (Fox Software Inc., 1987).

Table 3-1
Demography database for the period 1987-90

Individual ID code (see Table 1)
Name (family name and given name)
Date of birth
Address
Place of birth
Date of moving out for the migrant
Place to moving out for the migrant
Dead or alive
Date of death for the deceased

Table 3-2
Database of cause of death for the period 1987-90

Individual ID code (see Table 1)
Name of the deceased
Hospital name
Date when the patient saw doctors in the hospital
Diagnosis
Cause of death
Underlying cause of death
ICD-9 code

3.1.2 Data set of the period 1979-86 and its record linkage with the data of 1987-90

The data set of the period 1979-86 was stored in two databases in the format of Rbase 5000 PC-DOS version (Microrim Inc., 1985)[7]. First one was for demographic survey, and the second for cause of death. Both databases were exported into Foxbase+ in order to combine them with the data covering the period 1987-90.

Since the 1979-86 database had neither individual ID codes common to the 1987-90 database nor names in Chinese, we tried to link the records of the 1979-86 and 1987-90 databases using family-ID (first 8 digits of individual ID, see Table 1), sex and date of birth; if a record from the 1987-90 database shared the same family-ID, sex and date of birth with a record in the 1979-86 database, it was judged to be an identical subject. The subject in the 1979-86 database was assigned the same individual ID code as the one in the 1987-90 database.

3.2 Dosimetry data sets

Since 1972, outdoor doses in the study areas have been measured by our dosimetry group, as described by Yuan et al[8]. Outdoor environmental radiation exposure levels were measured in all studied hamlets, which numbered 526. The information stored in the database are given in Table 4-1.

Table 4-1
Outdoor dosimetry database

Hamlet-ID (first 6 digits of individual ID)
Name of hamlet
Exposures in pubic places
Alley
Pond
Water-well
Road
Rice paddy
Dry land

Indoor environmental radiation doses were measured in 5,990 houses, which accounted for about 30% of all the houses in the study area. The indoor dosimetry database had the information listed in Table 4-2.

Table 4-2
Indoor dosimetry database

Family-ID (first 8 digits individual ID)
Name of head of a household
House-type
Year of construction of house
Size of house
Radiation exposures in house
Bed
Siting-room
Kitchen

Personal cumulative doses were measured in 5,273 subjects using thermoluminescent dosimeters (TLD). The database for this individual dosimetry stored the information listed in Table 4-3.

Table 4-3
Individual dosimetry database

Individual ID
Job
Hours for sleep
Hours for indoor activities except sleep
Hours outdoor activities
Date of starting to wear TLD
Date of stopping wearing TLD
TLD dose measured

4. RISK COMPARISON AMONG DIFFERENT DOSE-RATE GROUPS

To estimate cancer risks related to different levels of exposure to high background radiation, we divided the whole HBR area into three groups based on kerma dose in each hamlet. Person-years at risk was accumulated over time for each subject from the date of entry into cohort to date of exit (defined as the earliest one among date of death, date of emigration and December 31, 1990). Person-years and number of deaths were stratified by sex, age at risk (0-, 5-, 0, ... 70-, and 75+), calendar years (1979-86 and 1987-90), and dose-rate group of the hamlet (control, low, middle or high). Risk comparisons among different dose-rate groups were conducted by means of relative risk. Point estimate of relative risk and its 90% confidence interval based on maximum likelihood techniques were calculated using AMFIT program in Epicure[9].

REFERENCES

1. High background radiation research group. Health survey in high-background radiation areas in China. Science 1980; 209: 877

2. Wei LX, Zha YR, Tao ZF, He WH, Chen DQ, and Yuan YL. Epidemiological investigation of radiological effects in high-background radiation areas of Yangjiang, China. J Radiat Res 1990; 31(2): 119

3. Tao ZF, Kato H, Zha YR, Akiba S, Sun QF, He WH, Lin ZX, Zou JM, Zhang SZ, Liu YS, Sugahara T, and Wei LX. Study on cancer mortality among the residents in high background radiation area of Yangjiang, China. Proceedings of 4th International Conference on High Levels of Natural Radiation. Beijing, China, 1996

4. Zhai SJ, Lin XJ, and Pan TM. Preliminary report on cancer mortality 1970-80 in high background radiation areas. Chin Radiol Med Prot 1982; 2(2): 48

5. World Health Organization. Manual of the International Statistical Classification of Diseases, Injuries, and Cause of Death. Ninth revision. Geneva, World Health Organization, 1975

6. Zha YR, Zou JM, Lin ZX, He WH, Lin JM, Yang YH, and Kato H. Confounding factors in radiation epidemiology and their comparability between the high background radiation areas and control areas in Yangjiang, China. Proceedings of 4th International Conference on High Levels of Natural Radiation. Beijing, China, 1996

7. Zhou SY, Guo FR, and Liu YS. Setup of database for epidemiological investigation in high background radiation research, China. Chin Radiol Med Prot 1992; 12(5): 316

8. Yuan YL, Morishima H, Shen H, Koga T, Sun QF, Tatsumi K, Zha YR, Nakai S, Wei LX, and Sugahara T. Recent advances in dosimetry investigation in the high background radiation area in Yangjiang, China. Proceedings of 4th International Conference on High Levels of Natural Radiation. Beijing, China, 1996

9. Preston DL, Lubin JH, and Pierce DA. EPICURE user's guide. HiroSoft International Corp, Seattle, USA, 1992

Study on cancer mortality among the residents in high background radiation area of Yangjiang, China

Z-F. Tao[1], H.Kato[2], Y-R. Zha[3], S. Akiba[4], Q-F. Sun[1], W-H.He[3], Z-X.Lin[3], J-M. Zou[3], S-Z. Zhang[1], Y-S.Liu[1], T. Sugahara[5] and L-X. Wei[1]

[1]Laboratory of Industrial Hygiene, Ministry of Health, Beijing 100088, China

[2]Radiation Effects Research Foundation, Hiroshima 732, Japan

[3]Guangdong Institute of Prevention and Treatment of Occupational Diseases, Guangzhou 5103300, China

[4]Faculty of Medicine, Kagoshima University, Kagoshima 890, Japan

[5]Health Research Foundation, Kyoto 606, Japan

1. INTRODUCTION

The cancer mortality study conducted in the high background radiation area (HBRA) of Yangjiang and the neighboring control area (CA) of Taishan and Enping in Guangdong Province of China was started in 1972. The major objective of the study is to estimate cancer risk associated with the low level radiation exposure in HBRA. Early cancer mortality data (1970-1978) were collected by a retrospective survey and the data from 1979 were obtained by a prospective survey through a death registry system established in the investigated areas. Up to the end of 1986, 467 cancer deaths were found among 100,8769 person-years (PYr) at risk in HBRA. The corresponding figures in CA were 533 cancer deaths and 995,070 PYr. The general conclusion was that the cancer mortality in HBRA was lower than that in CA, though the difference was not statistically significant[1].

The study has been carried out collaboratively between Chinese and Japanese scientists since 1991. The major purpose of the study in current stage is to accumulate further person-years of observation for improving the statistical power of test, and to see the reproducibility of the previous results. Compared with previous stages of the study, the main features of the study in current stage is to follow-up a fixed cohort and to have a internal comparison by classifying the cohort into four groups according to the intensity of dose rate. The results of analysis of the new data during the period of 1987-1990 obtained in the current cooperative stage and of the combined data with previous stages (1979-1990) are presented in this paper.

2. METHODOLOGY

2.1. Mortality follow-up

As mentioned above, the cancer mortality data of the current stage were collected from a fixed cohort instead of a dynamic cohort as in previous stages. There were 106,517 members in the fixed cohort at the beginning of 1987. Of them, 78,614 were in HBRA and 27,903 in CA. They are distributed over 384 hamlets in HBRA and 142 in CA.

The follow-up study in current stage consists of two phases, i.e. the survey of demography and the ascertainment of death causes. The demography survey was based on the name list of inhabitants who were living in the investigated areas on 1st January 1987. The Demography Task Group visited the hamlets and identified all the members in the cohort and collected relevant information, especially on deaths and migrants once a year at least. A Registry Card of Death was set up for each of the deceased. According to the Registry Card of Death, the Mortality Task Group visited the hamlets and interviewed the local doctors and reviewed the medical records to check the cause of deaths. The main purpose of this step was to pick up all the cancer cases and the cases suspected to be cancer. Therefore, if the cause of death was judged to be a non-cancer disease, no further investigation was made. However, if the cause of death was recorded to be cancer or suspected cancer or could not be denied to be cancer, the hospitals concerned were visited by the task group to identify the death causes of them by means of reviewing the medical records. If neither medical records nor other relevant information was obtained at the hospitals, the task group revisited the hamlet doctors and/or family members to interview them for obtaining the information on the signs and symptoms of the deceased patients. All the information collected in this phase of the survey was filled in a Questionnaires of Death Causes and reviewed by an expert group to determine the underlying cause of death for each deceased subject.

2.2. Dose estimate

Two approaches were used for estimating the individual gamma ray accumulated dose in the cohort. One was to measure the environmental gamma ray dose rate and to convert it into the annual absorbed doses taking into account the occupancy factors. The other was to measure the individual cumulative dose with TLD for part of members in the cohort. Finally, a hamlet-specific environmental exposure individual dosimetry system served to the data analyses of cancer mortality. Based on the hamlet- specific average annual gamma ray absorbed doses, the cohort members were classified into four groups and the individual gamma ray accumulated doses were estimated for each member in the cohort [2,3]. The average dose rates and their range (10^{-5} Gy/a) for each group are 246.04 (224.10-308.04, high), 210.19 (198.07-224.09, medial), 183.31 (125.29-198.06, low) in HBRA and 67.92 (50.43-95.67, control) in CA.

2.3. Statistical methods

The methods of the data linkage of the databases of demography and of the causes of death and of dosimetry, the methods of combining the data set of 1987-1990 with that of 1979-1986 are presented in a separate paper [3]. The estimates of relative risk (RR) were calculated by AMFIT in Epicure (Hirosoft International Corp., 1988-1992) [4].

3. RESULTS

3.1. Analysis of mortality data for the period of 1987-1990

There were 231 cancer deaths (160 in HBRA and 71 in CA) and 1990 non-cancer deaths (1,441 in HBRA and 549 in CA) among 421,640 PYr at risk in the cohort (311,237 in HBRA and 110,403 in CA) during the period of 1987 to 1990. The total death rate (per 100,000 PYr) during this period is 514.40 in HBRA, and 561.58 in CA. The first six causes of deaths are the same in each group. They are diseases of the circulatory system, diseases of the respiratory system, neoplasms, infection and parasitic diseases, injury and poisoning, and diseases of the digestive system. This is also the same as previous results obtained from the data of 1984-1986.

The crude mortality rate of overall cancers (per 100,000PYr) is 51.41 in HBRA and 64.31 in CA. The order of first four site-specific cancer deaths in HBRA is cancers of liver, nasopharynx, stomach and lungs. It is almost the same in CA, only the sequence of nasopharynx cancer and stomach cancer changed each other.

The RRs adjusted for age group and sex for each dose group in HBRA compared with CA for overall cancers and for all cancers except leukemia are less than 1 except for low dose group, and there seems to be a trend of RR decrease with the increase of exposure dose from natural radiation, though there is no statistically significant difference. As for the site-specific cancer studied, the cancers of lungs, liver, stomach, nasopharnyx and leukemia, the RRs are also less than 1 except for nasopharnyx cancer and for leukemia in high dose group (Table1). Although the sample size in each group was not large enough to make definite conclusion, the observation of current follow-up period reproduced the previous results obtained from the data up to 1986 that the cancer mortality in HBRA is generally lower than that in CA.

3.2 Analysis of cancer mortality for combined data from different periods of 1979-1986 and 1987-1990

For the purpose of extending the sample size and improving the statistical power of test, the cancer mortality analysis was conducted by combining the data

Table 1

RRs(90% CI) adjusted for age group and sex site-specific cancers by dose group (1987-1990)

Site of cancer	CA				HBRA						
	No.	RR	No.	Low RR	No.	Medial RR	No.	High RR	No.	Subtotal RR
All cancers	71	1.00	61	1.0850(0.8133-1.447)	56	0.8764(0.6529-1.176)	43	0.7766(0.5649-1.0670)	160	0.9113(0.7203-1.1530)
All Cancers except leukemia.	68	1.00	60	1.1150(0.8326-1.494)	54	0.8825(0.6536-1.192)	40	0.7527(0.5422-1.0450)	154	0.9155(0.7201-1.1640)
Leukemia	3	1.00	1	0.4239(0.06322-2.843)	2	0.7583(0.1683-3.417)	3	1.3820(0.3591-5.3170)	6	0.8370(0.2603-2.6930)
Nasopharynx	8	1.00	11	1.7360(0.8074-3.732)	11	1.5140(0.7044-3.252)	8	1.2630(0.5546-2.8760)	30	1.5040(0.7811-2.8960)
Lungs	7	1.00	4	0.7741(0.2752-2.177)	4	0.6687(0.2382-1.878)	4	0.7591(0.2705-2.1300)	12	0.7308(0.3335-1.6010)
Liver	26	1.00	16	0.7313(0.4329-1.235)	19	0.7745(0.4709-1.274)	8	0.3768(0.1937-0.7322)	43	0.6354(0.4217-0.9574)
Stomach	10	1.00	7	0.8566(0.3802-1.930)	4	0.4387(0.1657-1.162)	1	0.1291(0.02298-0.7247)	12	0.4789(0.2365-0.9697)

Table 2

RRs(90%CI) adjusted for age group and sex for site-specific cancers by dose group(1979-1990)

Site of cancer	CA		Low		Medial		High		Subtotal	
	No.	RR	No.	RR	No.	RR	No.	RR	No.	RR
All cancers	193	1.00	160	1.121(0.9398-1.337)	157	0.9551(0.8002-1.140)	130	0.9177(0.7613-1.106)	447	0.9959(0.8641-1.148)
All cancers except leukemia.	186	1.00	157	1.150(0.9615-1.374)	147	0.9336(0.7785-1.120)	124	0.9126(0.7540-1.104)	428	0.9954(0.8613-1.150)
Leukemia	7	1.00	3	0.4884(0.1566-1.523)	10	1.447(0.6415-3.262)	6	1.041(0.4159-2.605)	19	1.010(0.4867-2.097)
Nasopharynx	32	1.00	31	1.314(0.8674-1.990)	32	1.173(0.7769-1.770)	27	1.137(0.7395-1.748)	90	1.206(0.8590-1.693)
Lungs	19	1.00	9	0.6683(0.3432-1.301)	10	0.6396(0.3363-1.217)	12	0.8972(0.4889-1.646)	31	0.7298(0.4517-1.179)
Liver	56	1.00	39	0.9351(0.6631-1.319)	37	0.7655(0.5401-1.085)	25	0.5979(0.4024-0.8884)	101	0.7658(0.5819-1.008)
Stomach	21	1.00	12	0.7873(0.4338-1.429)	21	1.195(0.7189-1.986)	10	0.6617(0.3515-1.245)	43	0.8972(0.5787-1.391)

obtained from current stage (1987-1990) with those from previous stages (1979-1986). There were 88946 subjects in the combined cohort (64,070 in HBRA and 24,876 in CA) at the beginning of 1979. Of them 640 cancer deaths (447 in HBRA and 193 in CA) occurred among 949,018 Pyr at risk (696,181 in HBRA and 252,837 in CA) during the period of 1979-1990. The RRs adjusted for age group and sex for overall cancers and for all cancers except leukemia for each dose group in HBRA are also less than 1 except for low dose group and also there seems to be a trend of RR decrease with the increase of dose, but the difference is not significant statistically. The RRs for site-specific cancers of lungs, liver and stomach are also less than 1, but the RRs for nasopharynx cancer and for leukemia are larger than 1 in general (Table2). Even for the combined data, the sample size in each group was not large enough to make definite conclusion. However, the results from the combined data confirmed previous results with stronger power of test that the cancer mortality in HBRA is in general lower than that in CA.

4. PROPOSALS FOR FURTHER STUDY

To improve the further estimates of cancer risk in this study, the following points are proposed.

A. To continue the follow-up for extending the person-years and improving the statistical power of test.

B. To improve the individual gamma ray accumulated dose estimation for reducing the uncertainty of dose estimate.

C. To consider the internal exposure doses especially for some site-specific cancers (e.g. lung cancer) risk estimate.

REFERENCES

1. Wei, L. X., Zha, Y. R., Tao, Z. F. et al., Epidemiological investigation in high background radiation areas of Yangjiang, China. In: High levels of Natural Radiation -Proceedings of an International Conference, Ramsar, 3-7 November 1990, organized by AEO-IRAN, IAEA, WHO, UNEP and INTS pp.523-547.
2. Yuan, Y. L. , Recent advances of dosimetry investigation in the high background radiation area in Yangjiang, China. (in preparation)
3. Sun , Q. F., Akiba, S., Zou, J. M. et al. , Data linkage and statistical methods for cancer risk of cohort study in high background radiation area of Yangjiang, China. (in preparation)
4. Preston, D. L., Lubin, J. H. and Pierce, D. A., Epicure user's guide. Hitosoft International Corp., 1992.

Infant leukemia mortality among the residents in high-background-radiation areas in Guangdong, China

Suminori Akiba[a], Sun Quanfu[b], Tao Zufan[b], Zha Yongru[c], Lin Xiujian[c], He Weihui[c], Hiroo Kato[d]

[a]Department of Public Health, Faculty of Medicine, Kagoshima University

[b]Laboratory of Industrial Hygiene, China

[c]Guangdong Occupational Diseases Prevention and Treatment Center

[d]Consultant of Radiation Effect Research Foundation

1. INTRODUCTION

Recently, a study in Greek showed that infants prenatally exposed to ionizing radiation from Chernobyl accident had 2.6 times risk of infant leukemia when compared with unexposed children[1]. Since the estimated levels of radiation exposure were lower than those observed in high-natural-radiation areas, we evaluated infant leukemia mortality in the high-background-radiation (HBR) areas in Yangxi and Yangdong Counties, Guangdong Province, China.

2. MORTALITY FOLLOW-UP

The HBR area, i.e., Yangxi and Yangdong Counties, have 158 and 226 hamlets, respectively. As a control area, we selected Enping County with 142 hamlets located east to Yangdong County. In total, there were 526 hamlets in our study areas[2].

Detail descriptions of the subjects and mortality follow-up of this study are given elsewhere [3]. The follow-up of the subjects in this study consisted of two phases. The first phase, to be referred to as the demographic survey hereinafter, was to identify the residents in the study areas and those deceased in the past years since the previous survey. In a typical survey, a group of physicians visited hamlets and interviewed hamlet leaders and senior residents to collect information on the deaths and migration of the residents during the period since their last visit. Then, they visited local physicians and/or family members of the deceased to ascertain the date of death. For the migrants, the date of migration was determined mainly by interviewing their family members.

The second phase, to be referred to as the mortality survey hereinafter, was a survey to ascertain the cause of death, and was usually conducted a few months after the demographic survey. In the first step, our mortality task group visited the hamlets and interviewed the local physicians and reviewed their medical records. If the cause of death was suspected to be cancer, our task group visited district hospitals and higher level hospitals, including Yangjiang City Hospital, Enping County Hospital, and Guangzhou Provincial Hospital, to review medical records. All the information collected in the mortality survey was filled in a four-page record form and reviewed by a committee consisting of four physicians. The underlying cause of death determined by the committee was coded using the 1975 edition of International Classification of Disease (ICD 9th).

3.STATISTICAL METHOD

Person-years and number of deaths were aggregated and stratified by sex and age (0,1-4,5-9,10-14). In the calculation of the expected numbers of childhood leukemia and cancer, we used the mortality statistics during the period between 1975 and 1978, specific for sex and age (0,1-4,5-9,10-14), obtained by a nation-wide survey conducted in 29 provinces [4]. Poisson distribution was assumed in the calculation of p values for the tests comparing the observed and expected numbers. Using Epicure statistical package [5], Poisson regression analysis was conducted to compare the cancer mortalities in the HBR and control areas.

4.RESULTS

We analyzed the mortality of childhood cancer and leukemia in the HBR and control areas during the period between 1979 and 1986. As shown in Table 1, there were three infant leukemia deaths in the HBR area, where the expected number was 0.4. The risk increase, compared to the expected number, was statistically significant at 1 % level. There was no clustering of leukemia cases; the three leukemia deaths occurred in different regions in the HBR area. The field dose of their hamlets ranged between 1.7 and 2.0 mSv/year. In the control area, where the expected number was 0.1, no infant leukemia death was observed. In the children aged 1-14, there were slight increases of leukemia deaths in the HBR and control areas when compared with the corresponding expected numbers. Neither of differences were statistically significant.

Table 2 compared the risk of cancer except leukemia in the HBR and control areas. There was a 2.2 fold increase of childhood cancer other than leukemia in the HBR area when compared to the expected number (p value < 0.01). In the control area, there was only one childhood cancer death, while 1.5 deaths were expected. When the mortalities of childhood cancer except leukemia in the HBR and control areas were compared using Poisson regression analysis, we obtained the relative risk of 3.2 (95% confidence interval=0.6-59), which was not statistically significant.

Table 1.
Leukemia mortality during the period 1979-86.

	AGE 0		AGE 1-14	
	HBR	CONTROL	HBR	CONTROL
Rate (10^{-5})	19.7	0	3.2	4.8
Person-years	15198	5859	221630	63049
Observed deaths	3	0	7	3
Expected	0.4	0.1	5.1	1.5

Table 2.
Mortality of all cancer except leukemia in children aged under 15 years

	HBR	CONTROL
Rate (10^{-5})	4.6	1.5
Observed	11	1
Expected	4.9	1.5
O/E	2.2	0.7

Age-specific mortalities of childhood cancer except leukemia are shown in Table 3. After the first birthday, the excess risk of the childhood cancer was observed in all age groups. In sex specific analysis, there were 6 and 5 deaths of childhood cancer except leukemia in boys and girls, respectively, while two deaths were expected for each sexes.

Table 3.
Age specific mortalities of all cancer except leukemia in the children aged under 15 years in the HBR area

Age	Person-years	Observed	Expected
0	15198	0	0.3
1-4	60379	4	1.5
5-9	78871	2	1.4
10-14	82382	5	1.6

4. DISCUSSION

There were increased mortality rates of infant leukemia in the HBR area when compared to the expected numbers. It should be emphasized here that the results of infant leukemia depend on only three deaths. It is difficult to draw any conclusion unless further epidemiological studies are conducted. Cytogenetic examination is also necessary since the majority of infant leukemia is reported to have some molecular rearrangements in the MLL gene[6].

The significance of the elevated mortality rate of childhood cancer except leukemia depends on the validity of comparison between the observed and expected numbers of deaths. There are two major problems in the comparison: differences in place and time. Since childhood cancer mortality in Guangdong Province was not available, we used the national mortality statistics during the period between 1975 and 1978 obtained by a nation-wide survey conducted in 29 provinces [4]. It is difficult to evaluate the magnitude of potential biases introduced into our expected numbers by the use of national mortality. Table 4 shows cumulative mortalities for leukemia and cancer in the children aged less than 15 years, which were obtained from a nation-wide survey for the period 1973-75 [5]. There were 2-3 fold differences in childhood cancer mortalities among different counties in Guangdong Province. If the expected numbers had been calculated using the data obtained from the county showing the highest childhood cancer mortality in Table 4, the expected numbers would have been about 1.5 fold larger than those numbers used in Table 2 and, consequently, the difference between the observed and expected numbers in the HBR area would not have been statistically significant.

Table 4.
Cumulative rate of childhood leukemia mortality in 65 counties across the nation (aged under 15, 1973-5)

	Cancer		Leukemia	
	Male	Female	Male	Female
Median	0.77	0.60	0.39	0.29
counties in Guangdong Province				
Sihui	0.59	0.97	0.46	0.61
Panyu	0.90	0.70	0.61	0.32
Shunde	1.13	0.96	0.60	0.55
Zhongshan	1.02	0.94	0.49	0.46
Wuchuan	0.48	0.33	0.35	0.19

The second problem is that our observation period is different from the period covered by the referent statistics used in the expected-number calculation. Since detail information on the time trend of childhood cancer mortality in China are not available, it is difficult to address this problem. The only statistics in China that enabled us to evaluate the time trend of childhood cancer was that obtained from tumor registry in Shanghai. The incidence of cancer of all sites and leukemia in the children in Shanghai decreased in the 1970s and 1980s[7]. If that was also the case in our study areas, the expected numbers of childhood cancer and leukemia were probably overestimated in our calculation. In addition to the two problems discussed above, there is another problem, i.e., the quality of national statistics. We can't deny the possibility that relatively low expected numbers of cancer deaths can be explained by the incomplete ascertainment of childhood cancer in the 1975-78 nation-wide survey.

By the comparison between the mortalites of childhood cancer except leukemia in the HBR and control areas, we obtained a relative risk of 3.2. Here, one may raise a question that the two areas are different in terms the completeness in case ascertainment and accuracy of cancer diagnosis. Indeed, there is evidence indicating that the two areas are somewhat different in terms of socioeconomic status, and therefore the level of medical care. For example, the mortality of children aged less than a year was 24.5 / 1000 and 18.1/1000 in the HBR and control areas, respectively, in our study period (data not shown). Together with the observation that the HBR area had a slightly low cancer mortality in the adults when compared to the control area [9], we can't deny the possibility that the conditions of welfare and health care in the HBR area are not as good as those in the control area.

Even if the elevated risk of childhood cancer except leukemia in the HBR area is confirmed, that does not necessarily mean that natural-radiation exposure causes the cancer. There are at least three problems that should be pointed out here. First, there is no information on the exposure level of each mother. It is necessary to estimate individual radiation doses during pregnancy. Second, confounding biases might have been at work. Although there is no established risk factor of infant leukemia and childhood cancer, prenatal X-ray radiation exposure, pesticide use, and viral infection during pregnancy and/or after birth might have different distributions in the HBR and control areas and therefore confounded our results. Several surveys including the one reported at this conference compared medical exposure, life styles and other factors in the HBR and control areas and concluded that there were small differences, if any, in the two areas[10]. Although the above discussed biases due to confounding and incomplete case ascertainment are unlikely to be large enough to create a relative risk of 3.2, its wide confidence precludes us from drawing any conclusion.

If radiation needs a latent period of 10 years to increase solid tumor risk, the excess risk of cancer other than leukemia observed in the children younger than 10 years old in the HBR area is not attributable to radiation exposure. The levels of radiation doses in the HBR area also make it unlikely that natural radiation causes the increase of childhood cancer risk. In the study of children prenatally exposed to the atomic bomb, Yoshimoto et al reported that no leukemia cases was found in the children aged under 15 years[11]. There were two cases of cancer other than leukemia only in the children with uterine doses of 0.3 Gy or over. On the other hand, there is evidence suggesting that prenatal X-ray exposure is related to increased risk of childhood cancer. According to the study by Monson and MacMahon, an elevated risk of leukemia was observed in the children delivered by the mothers who underwent X-ray examination during pregnancy in the third trimester [12]. On the other hand, solid tumor risk was increased by X-ray exposure in the 1st and 2nd trimester. The latter findings was confirmed by the OSCC (Oxford Survey of Childhood Cancer) study, which reported that childhood cancer risk was increased by the exposure in the 1st trimester [13]. It is of interest that the magnitude of risk increase in our study was similar to those obtained in the studies of prenatal X-ray examination.

In conclusion, our results presented in this paper do not support or deny the possibility of an elevated risk of infant leukemia in the HBR area. There was weak evidence indicating the elevation of childhood cancer other than leukemia in the HBR area when compared to the expected number based on nation-wide statistics. Further follow-up of the population is necessary to evaluate the possible association of natural radiation with infant leukemia and childhood cancer risk. It is also desirable to establish the system that enable us to conduct cytogenetic and molecular examinations of specimens obtained from the cancer patients in the study.

REFERENCES

1. Petridou E, Trichopoulos D, Dessypris N, Flytzani V, Haidas S, Kalmanti M, Koliouskas D, Kosmidis H, Piperopoulou F, Tzortzatou F: Infant leukemia after in utero exposure to radiation from Chernobyl. Nature, 382, 352-353, 1996
2. High background radiation research group, China: Health survey in high background radiation areas in China. Science, 209, 877-880, 1980
3. Sun QF, Akiba S, Zou JM, Liu YS: Data linkage and statistical methods for cancer risk of cohort study in high background radiation area of Yangjiang, China. Proceedings of 4th International Conference on High levels of Natural Radiation. Beijing China, 1996
4. Office of Tumor Prevention Studies, Department of Health: China malignancy mortality survey report. People's Medical Publishing House, 1979
5. Preston DL, Lubin JH, Pierce DA. EPICURE: risk regression and data analysis software. Seattle : Hirosoft International Corporation, 1990
6. Pui CH, Kane JR, Crist WM: Biology and treatment of infant leukemias. Leukemia, 9, 762-769, 1995
7. Chen J, Campbell TC, Li J, Peto R: Diet, Life-style, and mortality in China. a study of the characteristics of 65 Chinese counties. Oxford University Press, Beijing, 1991
8. Coleman MP, Esteve J, Damiecki P, Arslan A, Renard H: Trends in cancer incidence and mortality. IARC Scientific Publications No 121, International Agency for Research on Cancer, Lyon, 1993
9. Tao ZF Kato H, Zha YR, Akiba S, Sun QF, He WH, Lin ZX, Zou JM, Zhang SZ, Liu YS, Sugahara T, Wei LX: Study on cancer mortality among the residents I high background radiation area of Yangjiang, China. Proceedings of 4th International Conference on High levels of Natural Radiation. Beijing China, 1996
10. Zha YR, Zou JM et al: Confounding factors in radiation epidemiology and their comparability between the high background radiation areas and control areas in Guangdong, China. Proceedings of 4th International Conference on High levels of Natural Radiation. Beijing China, 1996
11. Yoshimoto: Mortality rates and cancer incidence in prenatally exposed atomic bomb survivors. In "Effects of A-bobmb Radiation on the Human Body." Ed by I Shigematsu, C. Ito, N. Kamada, M. Akiyama, H. Sasaki. Bunkodo Co., Ltd Tokyo, Japan, 1995
12. Monson RR, MacMahon B: Prenatal X-ray exposure and cancer in children. in Radiation Carcinogenesis: Epistemology and Biological significance, ed by JD Boice and J Fraumeni jr. Raven Press, New York, 1984

13. Mole RH: Childhood cancer after prenatal exposure to diagnostic X-ray examinations in Britain. Brit J Cancer, 62, 152-168, 1990

Confounding factors in radiation epidemiology and their comparability between the high background radiation areas and control areas in Guangdong, China

Y. -R. Zha[a], J. -M. Zou[a], Z. -X. Lin[a], W. -H. He[a], J. -M. Lin[a], Y. -H. Yang[a], H. Kato[b]

[a]Guangdong Institute of Prevention and Treatment of Occupational Disease, 165 Xingangxi Road, Guangzhou 510310, China

[b]Radiation Effects Research Foundation, Hiroshima 732, Japan

Selection of a suitable control area (CA) and investigation on its comparability of confounding factors with those in the high background radiation area (HBRA) are most important for evaluation of the results obtained from radiation epidemiological study on cancer mortality. Previous results of the investigation have been reported by the High Background Radiation Research Group (HBRRG) elsewhere. In this paper we present the progress since 1990.

1. POPULATION STRUCTURE AND CONSTITUENTS

Demographic data is very important for analyzing risk of cancer mortality induced by radiation, and is also an important criterion to evaluate the comparability between HBRA and CA, or between groups of different dose-rate levels. The subjects we observed in this period were members of a fixed cohort, who lived in the investigated areas on January 1st, 1987. Based on the official permanent residence booklet, the general data including name, sex, date and place of birth, occupation and address were registered. The total number of fixed cohort was 106,517, including 78,614 in HBRA and 27,903 in CA. A census taker (they were local inhabitants) was invited by the Task Group for about every 2,000 persons, they were responsible for recording the moving-out, death of the cohort members and any change of the residence. The information of cohort members were established as a set of computer database.

Based on the annual effective dose ($\times 10^{-5}$ Sv$\cdot$a^{-1}), four dose groups (224.10-308.04 for high-dose group, 198.07 - 224.09 for middle-dose group, 125.29 - 198.06 for low-dose group and control-dose group for 50.43 - 95.67, respectively) were classified. 23,718, 28,803, 26,093 and 2790,3 people were in "high", "middle", "low" dose and control groups respectively, of whom the members of high, middle, low dose groups were from HBRA, and the members for control dose group were from CA. The population structures and constituents (Jan. 1, 1987) are shown in Table 1.

Table 1
Age-sex distribution in different dose rate groups (Jan. 1, 1987)

Age groups	High		Middle		Low		Control	
	Male	Female	Male	Female	Male	Female	Male	Female
0-	169	146	213	169	161	163	236	234
1-	1041	762	1238	1043	1088	910	1564	1420
5-	1436	1330	1826	1621	1654	1465	1433	1348
10-	1465	1303	1834	1613	1614	1612	1035	1066
15-	1667	1297	1876	1700	1708	1546	1248	1153
20-	1485	1174	1717	1500	1637	1345	1476	1827
25-	823	715	1048	869	950	716	1133	1165
30-	955	712	1126	933	1016	872	1331	950
35-	724	425	849	594	753	578	1073	624
40-	594	364	707	452	605	466	709	388
45-	543	415	588	470	585	450	613	385
50-	500	441	619	547	529	482	515	531
55-	510	490	537	543	471	488	601	653
60-	368	401	403	463	321	402	540	587
65-	286	299	296	343	244	271	388	448
70-	217	208	246	271	203	264	279	328
75-	79	151	131	197	113	187	137	251
80+	82	141	55	166	69	155	78	156
Total	12944	10774	15309	13494	13721	12372	14389	13514

Table 1 shows that the age distributions for males and females in different dose categories are similar to each other except for the age groups of 5-10 years, 10-15 years and 15-20 years. In these age groups, the numbers of persons were larger in three dose groups from HBRA. It indicated that the populations in these dose-groups of HBRA were slightly younger than the population in CA group. Therefore, the difference among them need to be considered in estimating cancer risk.

People of 35 years old and above are in the age of frequently developing cancer. Figure 1 shows the age distribution for age groups of 35 years old and above in accumulated person-years for males and females in different dose-rate categories during the period 1987-1995. The results showed that they were similar to each other.

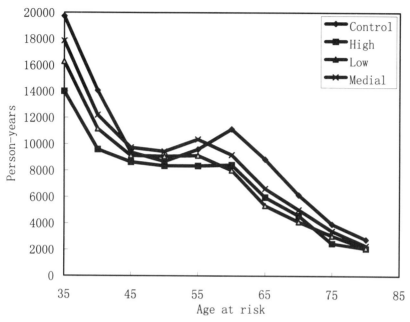

Figure 1 The age distribution for age groups of 35 years old and above in accumulated person-years during the period 1987-1995

In the fixed cohort followed up during the period 1987-1995, there were 5,164 death cases, and the death rate was 551.5 x 10^{-5}. The number of lost follow-up was 2,088 cases, and the rate of lost follow-up was 1.96% (1.50-2.80%). The main cause for loss was the moving-out of subjects to places other than investigated areas (Table 2 and Table 3).

Table 2
The numbers of lost follow-up (1987-1995)

Dose group	Subjects	No. of death	No. of loss	Rate of loss (%)
HBRA				
High	23718	1136	357	1.49
Medial	28803	1332	463	1.61
Low	26093	1222	486	1.86
CA	27903	1474	782	2.80
Total	106517	5164	2088	1.96

Table 3
The constituents of lost follow-up during 1987-1995

Causes	No. of loss	Rate (%)
Moving out from investigated areas	2050	98.18
Whereabouts unknown	28	1.34
To be further confirmed	10	0.48
Total loss	2088	100.00

2. LIFE EXPECTANCY

The aim of calculating life expectancy was to evaluate the health status of the members of this cohort (35 years old and above) in investigated areas and to analysis the comparability in different dose groups. Using Chiang Long Chiang method, the life expectancies in four dose groups were calculated. The results are shown in Table 4.

Table 4
Life expectancy of 35 years old and above persons in different dose groups by age-sex in accumulated person-years (1987-1995)

Age group	High-dose		Medial-dose		Low-dose		Control	
	Male	Female	Male	Female	Male	Female	Male	Female
35-	39.53	45.81	40.11	44.95	39.17	44.80	39.62	46.89
40-	34.99	41.11	35.62	40.50	34.57	39.97	35.08	42.19
45-	30.66	36.60	31.21	35.76	30.30	35.39	30.70	37.35
50-	26.28	31.96	27.01	31.25	26.45	30.94	26.63	33.03
55-	22.13	27.52	23.29	26.81	22.28	26.47	22.50	28.96
60-	18.40	23.39	19.31	22.53	18.46	22.14	18.91	24.74
65-	14.79	19.46	15.63	18.68	14.91	18.18	15.49	20.89
70-	11.95	15.96	13.00	14.99	11.97	14.85	12.31	17.35
75-	9.39	13.21	10.65	12.33	9.82	12.21	10.25	14.85
80+	7.71	11.62	9.64	10.47	9.11	10.01	8.75	13.67

The results obtained from three dose groups of HBRA and from CA showed that their life expectancies were similar to each other in general, but their life expectancy for males in HBRA was slightly higher than that in CA, and their life expectancy for females in HBRA was slightly lower than that in CA.

3. COMPONENTS OF DIET

It was reported by WHO that the foodstuff can influence the incidence of malignancies all over the world, and etiologically 30-40% of male cancer cases and 60% of female cancer cases are related to diet components [5]. On purpose to ascertain whether the foodstuff might influence the incidence of malignancies in the investigated areas, a survey of daily diet was conducted from 120 households in 24 hamlets (30 in each dose group) by multistage random sampling method based on the methodology in the " National nutrition survey program 1992, China".

We worked out three inquiry forms: condition of household, persons and times per day to have meals, and the dietary components in three days. The survey was carried out from October to December, 1993. All data were classified based on " Diet, life-style, and mortality in China" (Table 5).

Table 5
Dietary components and quantities of daily intake in HBRA and CA

Components*	Dietary intake (g/person · day)**			
	Group 1	Group 2	Group3	Group4
Rice wheat flour	434	450	447	456
Sweet potato and other grains	93	124	165	---
Bean and bean products	37.8	35.7	51.4	9.4
Vegetables	423	427	447	480
Fruits	39.7	37.1	48.7	49.4
Sugar	7.5	11.5	8.5	6.1
Milk and milk products	Trace	Trace	Trace	Trace
Eggs	5.1	13.3	8.4	7.3
Meats	84.3	89.9	105.5	109.7
Aquatic products	71.7	69.6	84.4	86.7
Edible vegetables Oils	6.8	7.0	7.5	13.2
Soy sauce	4.7	3.4	2.9	5.7
Salt	9.2	9.5	8.8	9.3

The results showed that the diet constituents in the four dose groups were similar: sufficient grains and vegetables, less meat and milk, which are typical dietary components of peasants.

4. LIFE STYLE

Life style, especially tobacco smoking and alcohol drinking, has close relation with the incidence of some cancer. During the dietary survey, an inquiry on cigarette smoking and alcohol drinking for cohort members of 16 years old and above was also made. The total number was 385 persons, of whom 101 and 101 were in high dose and in medial dose groups respectively, and 100 persons and 83 persons were in low dose and control group respectively.

The results of tobacco smoking showed that the smokers were mostly males, the percentages being 54.9-59.6%. The percentages in different does groups were similar (Table 6).

Table 6
Percentage of cigarette smokers in different dose groups

Dose group	No. of persons inquired			No. of smokers			Percentage		
	M	F	Subtotal	M	F	Subtotal	M	F	Total
High	52	49	101	31	0	31	59.6	0	30.7
Medial	51	50	101	28	0	28	54.9	0	27.7
Low	50	50	100	29	1	30	58.0	2.0	30.0
Control	45	38	83	26	0	26	57.8	0	31.3

Table 7 shows the percentages of alcohol drinkers. It indicates that the percentages of alcohol drinkers in HBRA and in CA (4.0-6.9%) are low. The numbers of alcohol drinkers in four dose groups are also similar.

Table 7
Percentage of alcohol drinkers in different dose groups

Dose group	No. of persons inquired	No. of drinkers	Percentage
High	101	7	6.9
Medial	101	4	4.0
Low	100	4	4.0
Control	83	4	4.8

5. CONCLUSION

5.1 The population structures and constituents, the life expectancy, the dietary components and life style between HBRA and CA, or in four dose groups were comparable to each other.

5.2 The fixed cohort of 106,517 members (males and females) is large and stable, containing all age groups. The rate of lost follow-up was less than 3.0%.

5.3 The diet pattern in the investigated areas provided sufficient plant foodstuffs, less meat and milk. It is suggested that it may be one of the reasons for low cancer incidence in those areas.

REFERENCE

1. Zha Yongru, et al., Studies on carcinogenic and mutagenic factors in high background radiation area. China J. Radiol. Med. Prot., 2 (1982) 45.
2. Zha Yongru and Lin Zuanxuan, Investigation on carcinogenic and mutagenic factors in high background radiation area. China J. Radiol. Med. Prot., 5 (1985) 127.
3. Tao Zufan, Li Hong, Zha Yongru et al., Comparative study of mutational factors in high background radiation and control areas. China J. Radiol. Med. Prot., 5 (1985) 127.
4. Wei Luxin, Zha Yongru, Tao Zufan et al., Recent advances of health survey in high background radiation area in Yangjiang, China. Chin. J. Radiol. Med. Prot., 7 (1987) 145.
5. WHO, Diet Nutrition and the Prevention of Chronic Diseases. Geneva, 1991.
6. Institute of Nutrition and Food Hygiene, Chinese Academy of Preventive Medicine. Manual on National Nutrition Survey. Beijing, 1992.
7. Richard Doll, Peter Greenwald, Chen Chun-ming. Diet, life-style, and mortality in China. People's Medical Publishing House, Beijing, 1991.

Epidemiologic study of cancer in the high background radiation area in Kerala

M. Krishnan Nair[a], N. Sreedevi Amma[a], P. Gangadharan[a], V. Padmanabhan[a], P. Jayalekshmi[b], S. Jayadevan[b] and K. S. Mani[b]

[a]Regional Cancer Centre, Trivandrum-6950 11, Kerala, India

[b]Natural Background Radiation Registry, Karunrgappally-690528, Kerala, India

1. INTRODUCTION

It is now almost four decades that a W.H.O. Expert Committee observed that the Chavara-Neendakara belt in coastal Karunagappally taluk in Kerala is a potential area for meaningful epidemiologic studies of High Natural Background Radiation [1]. Under the aegis of the Department of Atomic Energy, Government of India and some other organization several basic research studies, and radiation level measurements have been conducted in the area as also some epidemiologic studies [2-6].

Due to the persistent apprehension expressed by the society and the scientific inquisitor to study the health effects of such high natural radiation, the Regional Cancer Center initiated an epidenliologic study of cancer in the Karunagappally thank (NBRR Project) with financial and technical support from the Department of Atomic Energy, Government of India in 1989. The field activities were started in 1990.

2. THE STUDY OBJECTIVES OF THE NBRR PROJECT

1. What is the cancer incidence in the area.
2. Is it different form other population groups .
3. What is the pattern of cancer in the area.
4. Is it different from other areas.
5. Are the incidence and pattern of cancer related to background radiation.

2.1 Karunagappally, the land and the people

Karunagappally taluk (map) spread over an area of almost 192 Sq·km has a population of 385,103 (1991 census), of which around 100,000 people lived in

MAP: NOTIONAL MAP OF KARUNAGAPPALLY- SHOWING PANCHAYATHS

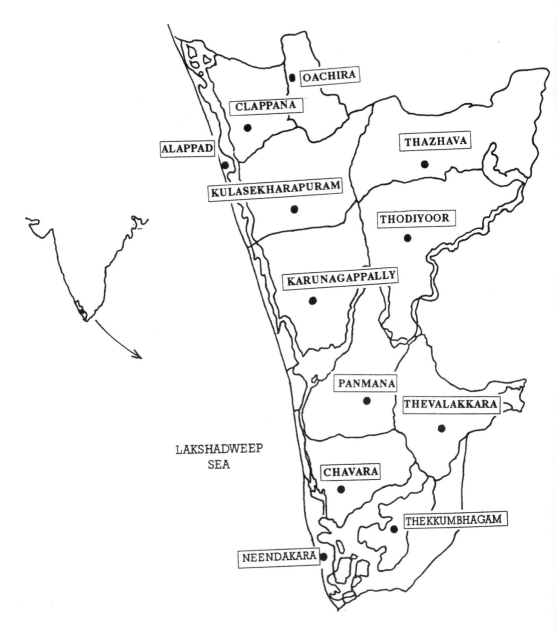

High Natural Background Radiation Area (HBAR). The density of population in the Talky was around 2000/sq·km. Along the coastal belt of this thank stretching over 25 km. in length and never more than half a km in width interrupted deposits of limonite sands containing radioactive material occur [1] The radioactivity is due to the monazite component (1 % of coastal sand) which contains thorium and to a limited extent uranium, along with several other rare earth minerals, such as rustle, sillimanite, zircon and titanium-bearing ores . In this monazite sand-the thorium content ranges from 8% to 10.5% as against 5-6% in the Brazilian monazite deposit. Most of the radioactivity (95%) arises from thorium and its decay products [1]. This observation was based on studies in the coastal areas. The interior had large areas with normal background also and this helped to meet the study objectives better.

As per the Governmental census classification this is a rural area and major occupations of the people are agriculture, fishing, coir making and in cashew nut industry. A majority of the population (65 to 70%) are Hindus . Apart from Hindus in certain areas there were 25 to 30,010 Muslims and 5 to 10,010 Christians, whereas in some areas there were 25 to 30% Christians and 5 to 10% Muslims. Almost 10% had college education. About 11 to 16% of the population was aged 55 years and more.

In Kerala, there is a female preponderance; in Karunagappally the sex ratio was 985 males to 1,000 females; and the expectation of life at birth among males was 69 years and in females it was 70 years . The annual growth rate of the population in Khnlnagappally was 1.09%. The population of the Thank is stable and people have been residing in the area since several generations .

For local administration, the Kerala state is divided into 14 districts and each district is divided into talks. Taluk are further divided into Panchayats, each Panchayat will have 30,000 to 40,000 population. Every Panchayat is further divided into wards, for census purposes also division of the Panchayat into wards is done. We have used the census wards for the study. Population in each ward will be between 2,000 to 3,000 people. Karunagappally taluk has 12 Panchayats, and is in Kollam (Quilon) District. The Regional Cancer Center at Trivandrum (RCC) is the comprehensive cancer Gentle nearest to the study area and is at a distance of about 100 kms.

2.2 Study methods

Three types of activity ware undertaken to meet the study objectives, They were-

(a) Complete enumeration of the population of the talky and base line data collection.
(b) Radiation level measurements as guided by the Bhabha Atomic Research Center, Bombay .

(c) Cancer registration according to standard registry system methodology adopted to the area with guidance from RCC and assistance from the Hospital Cancer Registry (ICMR) Trivandrum .

2.2.1 Population Enumeration and base hone data collection

Information was obtained from every resident of the taluk by house-to-house visit by trained enumerators and interviewing all the members in the house-hold using a pre-coded questionnaire. Information was thus obtained on several socio-economic-demographic variables, occupation, personal habits of tobacco usage, alcohol consumption, dietary characteristics, marital & pregnancy factors, health status etc. By a unique numbering system all individuals enumerated are numbered and this helps to analyze the effects of radiation exposure against various competing risk factors. The cohorts thus identified will be an portent study material for several years to come and with a good surveillance system data accrual will enhance the testing power of hypothesis generated. This is primarily because the population is stable and very few migration to & from the area takes place. The entire work is conducted through a field office established in the taluk.

2.2.2 Radiation level measurements

Radiation level measurements are conducted with the advice and technical assistance from the Bhabha Atomic Research Center, Bombay. . These include spot reading of radiation levels, (Gamma Radiation) both inside and outside of houses by micro R Scintillometres. These are used to estimate annual dose received afire obtaining cumulated annual gamma dose measurements inside the houses by TLD (Thermo Luminescent Dosimeter) placed in chosen houses in the area. CaF_2 filled brass capsules were used for TLD. The correlation 'r' between the estimated annual dose from scintillometer readings and the observed TLD cumulated dose has been high, $r > 0.98$. While recording the radiation levels the house number generated during population enumeration is used . This procedure helped to link all information. Radon, Thoron and their daughter products are measured by SSNTD (Solid State Nuclear Track Detector). Thoron-in-breath studies. and soil analysis (by taking soil samples from each 1 sq·km. grid in the taluk) for Uranium, Thorium and Potassium are also conducted. These values will be used to estimate the total ingested dose and body burden in a given individual.

2.2.3 Cancer registration

Recognizing the limitations of the health care facilities in the taluk a cancer registration system was initiated from 1.1.1990. This was an active registration system and according to the standard cancer registry system methodology. For the study of cancer occurrence vis-a-vis radiation levels emphasis is placed on 'cancer incidence', as, cancer mortality data of any area or population was influence by treatment, cancer site/type distribution and the quality of death reporting and registration system. An ongoing cancer registry is now in vogue in the taluk covering the entire taluk population. Mortality from cancer as recorded in the official dearth records in the population is also simultaneously obtained. Information on cancer cases are collected from several sources within the taluk and processed by computer for duplicate elimination, information merging and tabulation. Almost 60% of cancer cases were obtained from the Hospital Cancer Registry, Medical College Hospital, Trivandrum. In order to enhance cancer diagnostic capabilities, a cytology laboratory with facilities for Pap Smear and other cytological examination is functioning in the field office.

The observations from 6 (out of the total 12) Panchayat areas & 3NBRA are provided in Table 1.

Table 1
Some observations on high background and normal background radiation exposed populations

Parameters studied	Panchayats	
	Neendakara, Chavara and Alappad(HBRA)	Clappana, K.S.Puram and Thazhava(NBRA)
Population-(92) July	79,450	100,213
Number of houses	14,368	19,174
External -range (mSv/yrs)	0.70-110.9	0.30-22.2
Soil Bq/Kg-range	96-9,070	31-922
Uranium Bq/Kg	27-1,453	5-174
Potassium Bq/Kg	5-206	5-175
Tobacco Smoking, Men 14+ yrs	55%	50%
Alcohol Habituees, Men 14+yrs	45%	36%
Average age at death: M	62 yrs	62yrs
all causes:F	67yrs	67yrs
Average age at death:M	59yrs	59yrs
cancer:F	64yrs	61yrs
Age Standardised (Wp) Incidence Rete / 100,000		
All cancer Male	105.1	105.7
All cancer Female	86.8	71.8
Ca Lung Male	17.3	16.9
Ca Breast Female	8.5	12.9
Ca Cervix	21.2	14.0
Ca Thyroid Female	3.6	4.5
Leukemia Male	4.2	4.5
Leukemia Female	2.3	1.8

3. DISCUSSION

This is the first ever large scale population based study of High Natural Background Radiation exposed population in India. This is the only study which has similarity with studies in High the Karunagappally area. This heterogeneity now prohibits any firm conclusion regarding Background areas of China. The observations indicate the heterogeneity in radiation levels in Cancer occurrence. The effort is now to classify individuals with similar exposure levels and build up specific cohorts. Observations on health effects will reach sound conclusion only when such cohorts are compared afire recognizing the role of competing risk factors. Estimations of total body burden will require assessment of contribution from all pathways of radiation. This will be attempted with universally acceptable standards. Cancer registry operations in the area win be continued and over the years stabilizer rates of acceptable quality will be obtained.

ACKNOWLEDGMENT

We wish to express our thanks to Dr. K.S.V. Nambi, Head Environmental Assessment Division BARC, Bombay, (Late) Dr. T.P. Ramachandran, Directors of Hospitals in Karunagappally Taluk and Quilon Town and Medical Doctors of Regional Cancer Center, Trivandrum for their cooperation in the study.

REFERENCES

1. World Health Organization ,Technical Report series No. 166, 4-9, 1959.
2. Gopal-Ayengar , A. R., Effect of Radiation on Human Heredity: WHO, 115-124, 1957.
3. Gopal-Ayengar, A.R., Sundaram, K., Misery, K.B., Sunta, C. M., Nambi ,K. S. V., Kathuria, S. P., Basu, A. S. and David, M., UN Conference on Peaceful uses of Atomic Energy, 11 (1971) 31.
4. Gruneberg, H., Baine B.H., Riles, L, Smith C. A. B. and Weise, R. A., Medical Research Council Report No.SRS-307, Her Majestys Stationary Office London, 1966.
5. KochuPillai, N., Versa, I.C., Grewal, M.S. and Ramalingaswamy ,V., Nature , 262 (1976) 60-61.
6. Kochu Pillar, N., Thangavelu, M. and Ramalingaswamy, V. , Indian J. Med. Res., 64 (1976) 537.
7. Wei, L.,-X., Zha, Y.-G., Tao, Z.-F., He W.-H. , Chen D.-I. and Yuan Y.-G., J. Rad. Res., 31 (1990) 119-136.

Radiation associated lung cancer in northern Iran*

R.M.Raie and Z. T. Boroujeny

Dept. of Biology, Faculty of Science, Tehran University, Iran

To investigate radiation associated lung Cancer in northern Iran, medical records and tissue slides from 291 lung cancer patients being treated at four hospitals in Tehran were reviewed. In addition 34 well distributed surface water samples were analyzed for radium - 226, using co-precipitation and radon emanation technique with 0.1 p Ci/Lit MDL.

In this study the frequency distribution of various histological subtypes of lung cancer was determined for three provinces with high level of natural radioactivity in the north of Iran (Mazanadaran, Azarbayejan, Gilan). Then the results were compared with the frequency distribution of Tehran in central Iran. In summary, it was concluded that:

1) The level of natural radioactivity declines from west to the east.
2) Eighty five per cent of the patients of both sexes were over 45 years old, and the frequency of lung cancer was highest between the ages of 55 to 65 years.
3) The frequency of afflicted men was three times higher than afflicted women.
4) Among all provinces in Iran, Azarbayejan has the highest frequency of lung cancer at 15.9%. Tehran and Mazandaran shared second ranking, each at 13.8%, and 11.3% of the patients were from Gilan.
5) The most frequent subtype of lung cancer in the three northern provinces of Iran, is the small-cell carcinoma subtype. Thirty eight percent of afflicted individuals in northern Iran shows this kind of neoplasisia.

1. INTRODUCTION

Lung cancer is one of the relatively few cancers for which specific carcinogens have been linked to specific histological subtypes. Two such associations, those

---

*This investigation was carried out with financial support of the Research Deputy of Tehran University. Thanks are also due to Mr., K., and Mr. Banijamali, J. from Nuclear Waste Department of AEOI for their valuable help in the analysis of water samples.

pertaining to smoking history [1,2] and those pertaining to ionizing radiation [3,4] have been reported by many investigators.

The Mazandaran province of Iran has the world's highest reported level of natural radioactivity in hot springs and surface waters [5]. The radiation level from radium-226 in Ramsar from the soil, mineral water etc. is 0.08-9 mR/h [6]. In addition it was reported that drinking water of Villadarreh north of Ardebil in Azarbayejan province has the highest value of Ra-226 (182.1 mBq/lit)[7]. The nobel gas radon-222 is itself a decay product of Ra-226. The alpha particles produced in the radon decay chain have high energy(6.0 and 7.7 MeV) and low penetrance (40-70 mm) [8]. These alpha particles cause a dense region of tissue damage within a few cell diameters of their source. Radon associated lung cancer is therefore believed to result from the local action of radioactivity decay of radon and radon daughters(Po-214 and Po-218), introduced into the bronchioles by inhalation. To study the possible effect of ionizing radiation, especially alpha particles, in producing specific type of lung cancer in northern Iran, medical records and tissue slides from 291 lung cancer patients attending four hospitals in Tehran were reviewed. In addition 34 well distributed surface water samples, only from Mazadaran province were analyzed for Ra-226 and the results are reported. The frequency distribution of various histological subtypes of lung cancer was determined for three provinces in the north of Iran. Then the results were compared with the frequency distribution of Tehran.

2. MATERIALS AND METHODS

Medical records and tissue slides from 291 lung cancer patients attending four hospitals of Tehran (Imam Khomeini, Masih Daneshvary, Moddaress, and Loghman Hakim) were reviewed. The diagnosis of the type of lung cancer was made by specialists and Oncologist at the above mentioned hospitals. The tissue slides were classified according to the most recent WHO classification of pulmonary neoplasms [4,9]. The principal subtypes were squamous-cell, small-cell, adenocarcinoma, anaplastic carcinoma, large-cell carcinoma and unclassified carcinoma. Less frequent subtypes were also reported. This classification was applied without the knowledge of sex, age, smoking habit and region of residence. After typing, the results were correlated with smoking history, region of residence, age and sex. The major constraints were the frequent inadequacy or unavailability of smoking histories. Therefore results on smokers and nonsmokers are presented on 132 tissue slides. It was found that smoking history was highly correlated with sex. In addition 34 well distributed surface water samples from all over the province of Mazandaran were analyzed for Ra-226, using a radiochemical assay technique (co-precipitation) and a radon emanation method with minimum detection limit of 0.1 pCi/Lit [7], with the help of Nuclear Waste Department of AEOI.

3. RESULTS

The frequency distribution of lung cancer according to age is shown in the form of histogram in Fig. 1. The effect of sex on the specific histological subtypes are presented in Table 1. Fig. 2 shows the frequency distribution of various histological subtypes of lung cancer for Gilan, Azarbayejan, and Mazandaran, three provinces in the north of Iran in comparison with Tehran. Correlation between smoking history and various types of lung cancer is also given in Fig. 3.

Table 1
Distribution of lung cancer according to sex and specific histological subtypes.

	Male		Female		Total	
	No.	%	No.	%	No.	%
Squamous cell	41	18.3	5	7.5	46	15.8
Small cell	65	29.0	22	32.8	87	30.0
Anaplastic Carcinoma	31	13.8	8	11.9	39	13.4
Large cell	19	8.5	3	4.3	22	7.6
Adeno carcinoma	7	3.0	4	6.0	11	3.8
Unclassified	23	10.3	10	14.9	33	11.3
Other types	38	17.0	15	22.4	53	18.2
Total	224	100.0	67	100.0	291	100.0

The level of Ra-226 for 34 water samples from all over the province of Mazandaran is summarized in Table 2.

4. DISCUSSION

As Table 2 indicates the level of natural radioactivity declines from west to the east, which is in agreement with the results reported by Sohrabi et al [7]. Eighty five percent of the patients of both sexes were over 45 years old, and the frequency of lung cancer was highest between the ages of 55-65 years (Fig 1). This might be the result of a long latent period (30 years) for this kind of cancer [10].

Table 2
The levels of Ra-226 in water samples from Mazandaran Province

	Activity pCi/lit	Reported levels [7]
Ramsar		
Old Ramsar Hotel	94.0 ± 4	
Hot mineral water	22.1 ± 0.62	
River	14.6 ± 2	<0.2-123.63
Hot water for Sona	12.4 ± 0.5	
Hot mineral water of Sadatmahaleh	9.7 ± 0.4	
Drinking water	0.8 ± 0.2	
Tonekabon + Khorramabad		
River	1.5 ± 0.2	
Drinking water	<0.25	
Noor		
Drinking water	0.4 ± 0.1	
Seliakesh River	<0.25	<0.2
Hashtrood		
Spring	<0.25	
Well	<0.25	
River	1.5 ± 0.2	
Amol		
Drinking water	0.4 ± 0.1	
River	1.3 ± 0.1	2.4 ± 0.2
well	<0.25	
Aleshrood		
Well	<0.25	
spring	0.6 ± 0.2	
Drinking water	1.1 ± 0.3	
Gonbad		
River	<0.3	
Drinking water	<0.3	6.1 ± 0.4
Makhtoom spring	0.4 ± 0.2	
Azadshahr		
Drinking water	0.9 ± 0.2	
Well	1.4 ± 0.3	
Minoodasht		
Well	0.4 ± 0.2	
Drinking water	<0.3	
Gogol spring	<0.3	
Galikesh		
Well	0.4 ± 0.2	
Drinking water	0.5 ± 0.3	
River	<0.3	
Colaleh		
Seyed Abad spring	<0.3	
River	<0.3	

The frequency of afflicted men was three times higher than afflicted women (Table1).

Among all provinces in Iran, Azarbayejan had the highest frequency of lung cancer patients at %15.9. This is consistent with the level of natural radioactivity from Ra-226 in this province. Tehran and Mazandaran shared second ranking, each at %13.8, and %11.3 of the patients were from Gilan.

The most frequent subtype of lung cancer in the three northern provinces of Iran, where a high level of natural radioactivity exist, is the small-cell carcinoma subtype. More than %30 of afflicted individuals in northern Iran show this kind of neoplasia. This is in agreement with previous proposal [4] that alpha radiation induced lung cancer is of the small-cell subtype.

In the northern provinces, the most frequent subtype of lung cancer among men after the small-cell carcinoma is the squamous-cell carcinoma, whereas in women, it is the anaplastic carcinoma. There is strong evidence that smoking related lung cancer is of squamous-cell origin [4]. Therefore the high frequency of squamous cell carcinoma in men may be related to the higher smoking rate in this sex. As Fig 3 indicates the frequency of squamous-cell carcinoma in smokers is significantly higher than non-smokers.

REFERENCES

1- Parkin, D.M.et.al.,Int.J. Cancer, 59 (1994) 494-504.
2- Agudo,A.et.al.,Int.J.Cancer, 59 (1994) 165-169.
3- Neuberger, J.S., Health Physics, 63 (1992) 503-509.
4- Land,C.E. et.al. Radiation Research, 134 (1993) 234-243.
5- Mishra, U.C., Proceeding of ICHLNR, Ramsar, 29-35, 1990.
6- Sohrabi, M. et. al., Proceeding of ICHLNR, Ramsar, 9-45, 1990.
7- Sohrabi, M. et. al., Proceeding of ICHLNR, Ramsar, 425-432, 1990.
8- Harley, N.H., Health Physics, vol.55 (1988) 665-669.
9- Horwich, A., " Oncology ", Chapman and Hall, Chapters 40 and 41 (1995) 585-598 and 599-620.
10- Frank, L. M. and Teich, N. M. " Cellular and Molecular Biology of Cancer. " 2nd.ed., Oxford University Press., 1991.

A survey of senile dementia in the high background radiation areas in Yangjiang, Guangdong Province, China

J. Li[1], H. Sugasaki[2], Y.-H. Yang[3], M. Mine[2], Y.-R. Zha[3], M. Kishikawa[2], Q.-F. Sun[1], Y. Nakane[2], J.-M. Zou[3], M. Tomonaga[2], Z.-F. Tao[1] and L.-X. Wei[1]

[1]Laboratory of Industrial Hygiene, Ministry of Health, Beijing 100088, China.
[2]Nagasaki University School of Medicine, Nagasaki 852, Japan.

The 1993 report of United Nations Scientific Committee on the effects of Atomic Radiation (UNSCEAR) noted that the human brain is relatively sensitive to ionizing radiation at certain stages in its prenatal development. Data from Hiroshima and Nagasaki, as well as from a few other studies, indicate that there can be consequences to the central nervous system from exposure to radiation at these stages. Both severe mental retardation and lower intelligence test scores were observed [1]. Senile dementia is a degenerative disease of central nervous system. Its etiology is still unidentified. The main manifestation of this disease is overall intelligence decline, and pathological studies showed that critical pathological changes were set in cerebral cortex. It is very significant to study whether the prevalence rate of senile dementia will increase in the persons who have been exposed to low dose and low dose-rate ionizing radiation all their lives. A survey of senile dementia was carried out in high background radiation areas (HBRA) in Yangjiang, Guangdong Province, China. The annual effective dose from natural radiation sources in HBRA is about three times than that in nearby control areas (CA), being 6.4 vs 2.4 mSv a^{-1} [2].

The purposes of this survey were to obtain the prevalence rate of senile dementia in the HBRA and CA, study the effects on senile dementia of exposure to this low-level radiation, go into the other risk factors of senile dementia, and thereby provide the direct observational data of human beings for the purpose of further assessing the effects of low-dose radiation exposure on central nervous system and studying the etiology of senile dementia.

1. SUBJECTS AND METHODS

1.1 Subjects and method of sampling

Subjects of the survey were the inhabitants aged 65 or above in HBRA (high-dose group) and CA. All the male subjects of HBRA are native, about half of the female subjects of HBRA are born in the locality and the other half moved in from other places for marriage during the age of 15~16 years. The average annual effective doses from external gamma radiation are 240.68 ± 16.09 (x 10^{-5}) Sv in investigated HBRA, and 72.70 ± 9.17 (x 10^{-5}) Sv in CA [3].

Method of sampling: First, the demographic database of the inhabitants aged 65 or above in both areas was set up (up to January 1, 1994, for age counting) from the population database of the high background radiation research which had been established earlier (cut-off date was August 1, 1994). It consisted of 1599 inhabitants in HBRA and 2209 in CA. Next, the method of two-stage sampling was applied to the study. In the first stage system sampling was conducted and the sampled unit was Guanqu (village). Through this stage, the primary samples were obtained. In the second stage, the simple sampling was conducted and 700 inhabitants were selected in each of the areas, consisting of the final samples of 1400 subjects.

1.2 Procedure of survey

The survey was conducted in two stages: screening and diagnosis stages. For minimizing the rate of lost follow-up, the two stages were merged. Once the subject whose score of screening questionnaire was lower than the criteria, the second stage of examination and diagnosis was entered. A pilot survey was carried on before the formal investigation. By way of the pilot survey, the formal questionnaires and criteria were defined.

1.3 Questionnaires and criteria

Hasegawa Dementia Scale (HDS) was used for all subjects in screening [4]. In accordance with the analysis of results of the pilot survey and the special conditions of the investigated areas, HDS used in this survey was revised. According to the education levels, if a illiterate subject whose score of HDS was lower than 16, primary school education level lower than 20 or middle school education level lower than 24, he or she is assumed to be a suspected case, needing interview again and examination in the second stage. In the second or diagnosis stage, the special questionnaires of health status of old people provided by Nagasaki University School of Medicine (Japan) was used, involving the fundamental status, nervous system and psychiatric examinations of the subjects. The test scales include HDS, activities in daily life (ADL), Huchinski ischemic index and GBS scale. The final diagnoses were made according to the American Psychiatric Association's third revised edition of the Diagnostic and Statistical Manual of Mental Disorders (DSM III-R). For comparing with other studies easily, the diagnoses were also given on the basis of DSM IV,

International Classification of Diseases (ICD)-10 and Chinese Criteria of Mental Diseases (CCMD)-II-R. This paper reports the results in accordance with DSM III-R.

1.4 Quality control

The survey was organized and coordinated by two experts of radiation epidemiology. The technical personnel participating in the investigation are specialists in epidemiology or psychiatrists. Before the formal investigation, all investigators took part in a special training in the use of HDS. A test of using HDS was conducted and the results showed that the reliability of the administration of HDS was high among the investigators. The diagnosis stage was carried out by two experienced psychiatrists and guided by a psychiatric as technical adviser. To examine the HDS screening efficiency, 10% of samples were selected from the final samples in the two areas, HDS was conducted for every subject of the two subsets, and no matter how large the score of HDS was, the special questionnaires were used and the diagnoses made directly.

1.5 Establishment of the database and method of statistics

All the data were put into computer, the database of this survey was established by using dBase V for Windows (1994, Borland International, Inc.), the statistical tools provided by dBase V and Excel 5.0 (1985-1994, Microsoft, Inc.) were used for statistical analysis.

2. RESULTS AND ANALYSIS

2.1 Rate of successful follow-up

Because there was an interval of about half year between the cut-off date of database (August, 1994) and the date of investigation, the number of the samples changed owing death and migration (out of town) . The number of inhabitants that should be observed in HBRA was 594, and 513 of them were successfully followed up (86.36 percent), while there were 570 to be observed and 505 followed up (88.60 percent) in CA. To sum up the numbers of two areas, the corresponding numbers were 1164 and 1018 (87.46 percent). The numbers of subjects lost to follow up were 81 in HBRA and 65 in CA. A total of 146 subjects could not be successfully followed up. The major reason was that the subjects had occasionally gone out (Table 1).

2.2 Distribution of the subjects by age and sex

The average age of the subjects was 73.3 years. It was 73.4 in HBRA and 74.0 in CA. The oldest subject was 94 years old, and the youngest was 66. The distributions of the subjects by age and sex in both areas are basically consistent (Table 2).

Table 1
The reasons of lost follow-up in HBRA and CA

Reasons	HBRA	CA	Total
Occasionally going out	64	54	118
Refusal of being interviewed	3	1	4
Being critically ill	3	4	7
Blindness and deafness	6	5	11
Others*	5	1	6
Total	81	65	146

* Including errors of age counting (the true age was not in the range of investigation), mental diseases and unavailability of any data.

Table 2
Distribution of subjects by age and sex in HBRA and CA

Age groups (Years)	HBRA			CA		
	Males	Females	Subtotal	Males	Females	Subtotal
65-	14.23	15.79	30.02	12.08	13.46	25.54
70-	13.84	19.10	32.94	13.27	22.77	36.04
75-	8.77	13.06	21.83	9.11	12.08	21.19
80-	4.10	6.24	10.34	4.55	6.34	10.89
85--	1.56	3.31	4.87	0.99	5.35	6.34
Total	42.50	57.50	100.00	40.00	60.00	100.00

Table 3
The distributions of scores of HDS in HBRA and CA

Scores of HDS		HBRA		CA		Total	
		Cases	Rate(%)	Cases	Rate(%)	Cases	Rate(%)
Illiterate	≤16	32	6.24	25	4.95	57	5.60
	>16	356	69.40	341	67.53	697	68.47
Primary sch.	≤20	4	0.78	6	1.19	10	0.98
	>20	108	21.05	125	24.75	233	22.89
Middle sch.	≤24	0	0	0	0	0	0
	>24	13	2.53	8	1.58	21	2.06
Total		513	100.00	505	100.00	1018	100.00

2.3 Scores of HDS

The scores of HDS of 1018 subjects in HBRA and CA are shown in Table 3. The distributions of the scores of HDS in the two areas were similar. There were 67 inhabitants in both areas whose scores were lower than the criteria and made up 6.58% of the total subjects. Of them 36 (7.02%) were in HBRA and 31(6.14%) in CA.

2.4 Prevalence rate of senile dementia in HBRA and CA

Sixty-one cases of senile dementia were diagnosed in both areas. Among them 31 cases were in HBRA and 30 cases in CA. The total prevalence rate of senile dementia in both areas was 5.99%, of which 6.04% in HBRA and 5.94% in CA (Table 4).

Table 4
Prevalence rate of senile dementia in HBRA and CA

	HBRA	CA	Total
Subjects	513	505	1018
Cases	31	30	61
Prevalence rate (%)	6.04	5.94	5.99

Through the Chi-square Test, $\chi^2 = 13.77 \times 10^{-4}$, df=1, $P=0.9452$. No statistically significant difference between two areas was found.

2.5 Clinical classification of cases of senile dementia

All the cases were classified by using DSM III-R. The distribution of types of senile dementia in both areas is shown in Table 5. There were 25 cases of Alzheimer's disease (AD) in HBRA (80.64 percent) and also 25 in CA (83.33 percent). The distribution of clinical classification of cases in HBRA was basically similar to that in CA. Analyzed by sex, the percentage was much higher in females than that in males. Among 31 cases in HBRA, there were 29 cases in females (93.55 percent) while the corresponding number in CA was 25 (83. 33 percent).

2.6 Analysis of factors influencing prevalence rate of senile dementia
2.6.1 Sex and age

Merging the data from two areas, the distribution and the prevalence rate of senile dementia by age and sex are shown in Table 6.

Table 5
Types of senile dementia classified by DSM III-R in HBRA and CA

Type	HBRA			CA		
	Males	Females	Subtotal	Males	Females	Subtotal
Alzheimer's disease	1	24	25	5	20	25
Multi-infarct disease	0	4	4	0	5	5
Non classified	1	1	2	0	0	0
Total	2	29	31	5	25	30

Table 6
Distribution of prevalence rate of senile dementia by age and sex

Age group (years)	Males			Females			Total		
	Subjects	Cases	Prevalence rate (%)	Subjects	Cases	Prevalence rate (%)	Subjects	Cases	Prevalence rate (%)
65-	134	0	0	149	2	1.34	283	2	0.71
70-	138	1	0.72	213	14	6.57	351	15	4.27
75-	91	3	3.30	128	14	10.94	219	17	7.76
80-	44	3	6.82	64	9	14.06	108	12	11.11
85-	13	0	0	44	15	34.09	57	15	26.32
Total	420	7	1.67	598	54	9.03	1018	61	5.99

As shown in Table 6, the prevalence rate of senile dementia increased with age. No cases were found in males over 85 years old, possibly because the sample size of this age group was too small (only 13 inhabitants were interviewed). In the same age truncation, the percentages in female were much higher than those in males.

2.6.2 Education levels

The investigated areas are the countryside in the west of Guangdong Province, China. All of the subjects were peasants or housewives. Among them, illiterates made up 74.07% of the total subjects and the rates of graduation from or attendance of primary school and middle school or beyond were 23.87% and 2.06%, respectively. The relationship between the prevalence rate and education levels is shown in Table 7.

Table 7
The relationship between the prevalence rate of senile dementia and education levels

Education levels	Males			Females			Total		
	Subjects	Cases	Prevalence rate (%)	Subjects	Cases	Prevalence rate (%)	Subjects	Cases	Prevalence rate (%)
Illiterates	171	5	2.92	583	53	9.09	754	58	7.69
Primary sch.	228	2	0.88	15	1	6.67	243	3	1.24
Middle sch.	21	0	0	0	0	0	21	0	0
Total	420	7	1.67	598	54	9.03	1018	61	5.99

From the Table 7, it is noted that the prevalence rates decreased with education levels and reached statistical significance. In the main, the prevalence rate in the illiterate group was much higher than those in the groups who finished or attended primary school and middle school or beyond. $\chi^2=12.54$, df=1, $P=0.0004$.

As noted above, the prevalence rate of senile dementia was associated with the age, sex and education levels in this survey, which is consistent with other studies on senile dementia in the literature[5].

3. DISCUSSION

Based on the results of this survey, no significant difference in the prevalence rate of senile dementia was found between HBRA and CA. There was no evidence showing that the prevalence rate of senile dementia was associated with the high background radiation exposure. No significant difference was found between HBRA and CA in the survey of intelligence of children, Down's syndrome and other studies on mental disorder conducted earlier[6,7]. These results of study on mental functions are basically similar.

As most papers reported, the prevalence rate of serious senile dementia in the world was about 4-5%. Over half of them were Alzheimer's disease (AD). In China, the rate was much lower than this value in earlier reports, but the results were significantly different from each other studies. Based on the report of Zhang Minyuan et al.(Shanghai Mental Health Center, 1990) [5], the prevalence rate of senile dementia (the subjects aged 65 or above) was 4.61% (95%CI 4.03%-5.19%), of which AD made up about 2/3, followed by multi-infarct dementia (MID). It was regarded as close to the actual situation in China. The prevalence rate of senile dementia was 5.99% in our survey and AD made up about 80%; these rates were higher than those in the aforementioned reports. One of reasons causing this difference may be that the investigated areas were the countryside in the west of Guangdong Province and the investigated subjects were peasants or housewife of low education and unfavorable socioeconomic status. It is also noted that the prevalence rate of senile dementia was much higher in old, female and low education level groups, which is consistent with other studies on senile dementia elsewhere [5,8].

HDS was used for screening in the survey. To examine the screening efficiency of HDS, 103 subjects were examined and interviewed directly by using special questionnaire and 11 cases were diagnosed as having senile dementia. Among the samples recruited, there were 11 subjects whose scores of HDS were lower than criteria and 10 cases were diagnosed as having senile dementia in the second stage, and one person was ruled out. There were 92 subjects whose scores of HDS were higher than criteria, of whom one case of MID was diagnosed. It suggested that the screening efficiency of HDS was quite high. The rate of follow-up in this survey was 87.31%. Through the analysis of lost follow-up, the major reason was that the subjects had occasionally gone out. Because the survey was carried out in rural areas and the time of investigation was limited, it was very difficult to follow up everyone who was occasionally out. It may bring about some effects on the results of the survey. But in the main, the results are convincing since the evidences of dementia were collected exhaustively and the diagnoses made in the survey were strictly in accordance to the criteria.

ACKNOWLEDGMENTS

Collaborating investigators: Guangzhou Psychiatric Hospital, China. Drs. S-X. Fang, R-M. Feng, M-J. Jiang and W-C. Zhou. We are grateful to Dr. Y. Lu, the vice-president of Guangzhou Psychiatric Hospital for his helpful technical advise, and would like to acknowledge the contribution of Y.S. Liu and J. Zhang in establishing the database and analysis.

REFERENCE

1. UNSCEAR: Radiation effects of the developing human brain. Sources and effects of ionizing radiation. UNSCEAR 1993 Report to the General Assembly, with Scientific Annexes, Annex H. United Nations, New York, 1993.
2. Tao Z.-F. and Wei L.-X., General conditions and perspective of low dose radiation epidemiological investigations. Chin. J. Radiol. Med. and Prot., 15 (1995) 162-168.
3. Yuan Y.-L., Shen H., Zhao S.-A., et al., Recent advances of dosimetry investigation in high background radiation area in Yangjiang, China. Chin J Radiol Med Prot, 15 (1995) 311-316.

4. Hasegawa, K., et al., The epidemiological study on psychiatric disorder in the early. Research anthology dedicated to Naotake Shinhuku, Prof. Emeritus, Tokyo Jikeikai Medical School, Tokyo, 342-354, 1979.
5. Zhang M.-Y., et al., The prevalence study on dementia and Alzheimer disease. Natl. Med. J. China , 70 (1990) 424-428.
6. Cui Y.-W. et al., Hereditary diseases and congenital malformation survey in high background area. Chin. J. Radiol. Med. Prot., 12 (1982) 55-57.
7. Chen D.-Q. et al., Pilot study on intellectual development of children in high background radiation areas in Yangjiang. Chin. J. Radiol. Med. and Prot., 10 (1990) 387-390.
8. Hiroyuki Sugasaki, Yasuyuki Ohta, et al. An epidemiological study of senile dementia at home in Nagasaki. Acta Med. Nagasaki, Japan., 38 (1993) 211-215.

Advanced cytogenetical techniques necessary for the study of low dose exposures

I. Hayata

National Institute of Radiological Sciences, 9-1, Anagawa 4-chome, Inage-ku, Chiba-shi, 263 Japan

1. SPONTANEOUS INCIDENCE OF DICENTRICS

Chromosome aberrations, dicentrics and rings, are excellent indicators in radiation dosimetry. Since the incidence of rings is much lower than that of dicentrics and small rings are difficult to distinguish from fragments, many researchers score only dicentrics for the dosimetry. Table 1 shows the spontaneous incidence of dicentrics (per thousand cells) obtained by the conventional method. In all the studies over ten thousand cells are investigated. The incidence varies from 0.04 to 3.8 per thousand cells. The difference is more than 100 times, although all the frequencies of dicentrics are obtained from non-irradiated control groups. These results indicate that, in the low dose study, the true incidence of dicentrics is difficult to obtain. According to IAEA technical report No. 260 a consensus is emerging that the background level of dicentrics is about 1 in 1,000 cells[1].

2. THE LOWER LIMIT AND THE NUMBER OF CELLS TO BE SCORED

The lower limit of dose that can be estimated by chromosome analysis is considered to be about two cGy by X- or gamma-ray exposure as stated in the studies by Pohl-Rüling et al. (1983)[21] and Lloyd et al. (1992) [17]. In such a low dose range an enormous number of cells have to be analyzed for obtaining statistically significant data. The number of cells (N) to be scored is given by the equation such as

$$Z_{0.05} = \delta / \sqrt{p(1-p)/N}$$

where $Z_{0.05}$ is the correlation coefficient of 95% significant level, δ is the confidence interval, and p is the incidence. Table 2 gives examples of the number of cells to be scored.

Table 1
Spontaneous incidence of dicentrics

No of subject	Cells counted	Dicentrics (‰)	References
234	23,400	1 (0.04)	Srám et al.(1985) [2]
8	21,570	2 (0.09)	Richardson et al.(1984) [3]
205	22,000	2 (0.09)	Sevankaev et al.(1974) [4]
67	35,500	14 (0.39)	Salassidis et al. (1994) [5]
156	17,344	7 (0.4)	Blackwell et al.(1974) [6]
71	14,164	5 (0.4)	Obe & Herha(1978) [7]
33	55,949	29 (0.5)	Obe et al.(1982) [8]
105	16,267	11 (0.7)	Ivanov et al.(1978) [9]
175	17,500	12 (0.7)	Gundy & Varga(1983) [10]
119	107,792	96 (0.89)*	Tonomura et al.(1983) [11]
170	18,785	19 (1.0)	Muramoto et al.(1988) [12]
23	11,500	13 (1.1)	Léonard et al.(1984) [13]
316	23,300	30 (1.3)	Lloyd et al.(1980) [14]
2	15,218	27 (1.8)	Rohl-Rüling et al.(1986) [15]
4	11,973	24 (2.0)	Lloyd et al.(1988,1992 [16,17]
263	24,409	58 (2.4)	Awa et al.(1978) [18]
44	11,000	31 (2.8)	Heath et al.(1984) [19]
47	24,000	(3.8)	Wagner et al.(1983) [20]

*Average in the 20 to 60 year-old age groups

Table 2
Number of cells to be scored

δ	p = 0.1	p = 0.01	p = 0.001
±10%	3,457	38,832	383,776
±20%	864	9,508	95,944
±50%	138	1,521	15,351

3. THE VALUE OF ONE DICENTRIC IN THE LOW DOSE RANGE

When human lymphocytes are acutely exposed to X- or γ-rays at the dose of 3 - 4 cGy, about one dicentric per 1,000 cells is incurred [17,21]. This means the misjudgment of one dicentric per 46,000 chromosomes results in the difference of

3-4 cGy in dose estimation. The yield of dicentrics per unit dose increases in proportion to the increase of dose. The higher the dose is, the less significant the difference of one dicentric becomes. Thus one misjudgment has a detrimental effect on the dose estimation when the dose is low. Therefore, it is more strictly demanded to avoid misjudgment and overlooking in the study of low dose exposures than in that of high dose exposures.

4. IMPORTANCE OF THE QUALITY OF CHROMOSOME PREPARATION

What is most important in the study of low dose exposures is to make the chromosome slide preparations having good quality for minimizing error in scoring chromosome aberrations. First of all, the spreading of metaphases must be good and the degree of their spreading must be uniform at any area of the slide glass in order to reduce bias in sampling. Then the slide glass must have a large number of analyzable metaphases.

Under the project study of the Nuclear Cross-Over Research supported by the Science and Technology Agency in Japan, we improved chromosome preparations to meet above mentioned requirements that are intrinsic in the study of low dose exposure [22,23]. By the improved methods, we can usually prepare the chromosome slide having over 2,000 metaphases in which over 1,000 cells (at least 5 times more than before) are suitable for the conventional analysis. The methods are briefly described in the sections of 5-7.

5. CULTURING AND HARVESTING THE SEPARATED LYMPHOCYTES

What we examine is the chromosome of a lymphocyte in the sample of the blood culture. For obtaining metaphases more effectively, we separate the lymphocytes from the whole blood and culture them. It allows to collect metaphases in the density that is over five times higher than whole blood culture does [22]. Since there are neither cell debris from red blood cells nor polymorphonucleated granulocytes in the sample of the separated lymphocytes (mononuclear cells), it becomes easy to make well spread metaphases at high density. The separation of lymphocytes (mononucleated cells) from the whole blood can be easily done by the centrifuge using lymphocyte separation medium such as HISTOPAQUE (Sigma Chemical Co.) or LeukoPREP (Becton Dickinson Co.Ltd).

6. ARRESTING AT METAPHASE IN THE FIRST CELL DIVISION

Colcemid and colchicine are the chemicals by which we can arrest cell cycle at the metaphase stage. Fig. 1 shows the rates of metaphases we obtained when they were given at various concentrations from the beginning of the culture [24]. These cultures were set up with the mononucleated cells at the concentration of 5 x 10^5 cells per ml in the RPMI1640 medium containing 20% of fetal calf serum and PHA. The effect of colchicine saturates at the concentration of 0.01É g per ml, while that of colcemid does not become maximum until at the concentration of 0.05 μ g per ml. Colcemid is capable of collecting two times more metaphases than colchicine. Some cells escape from the blockage of cell cycle in the colcemid treatment of 0.05 μ g per ml, but the rate of metaphases in the second cell division after 48 culture is only 0.4%, which is negligible in scoring dicentrics for dosimetry.

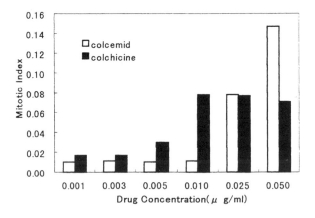

Fig. 1. The effects of concentrations of colcemid and colchicine on mitotic index of human lymphocytes cultured for 48 hours.

Since cell division is influenced by the density of cells, the culture should be set at the fixed cell concentration (5 to 10 x 10^5 cells per ml). The yield of dicentrics is influenced by the quantity of colcemid [24]. Therefore, the dose of colcemid should be accurate. An over-treatment of colcemid produces over-condensed chromosomes which are not appropriate for analysis. When the colcemid of 0.05 μg per ml is given to the culture from the beginning for 48 hours, the majority of metaphases are analyzable.

7. AIR DRYING IN THE WARM AND HUMID BOX

In order to obtain uniform preparation, the air dry should be performed in a constantly warm and humid circumstance without wind [22]. The higher the temperature, the stronger the pressure from the drying fixative solution onto the metaphase cells becomes. The higher the humidity, the longer it takes for drying. If the slide is dried in an appropriately warm and humid air condition, chromosomes spread well. If the temperature and the humidity are too high, the chromosomes are over-dispersed. Advantage of making chromosome slides in the warm and humid box is that you can control the degree of spreading. It leads to the production of the uniformed slide preparation. When a hypotonic treatment is performed in 0.075 M KCl solution at 37°C for 20 min, the optimal conditions of the temperature is 28 to 30°C and of the humidity is 70 to 90%.

8. MICROSCOPE WITH AUTOSTAGE: NECESSITY IN LOW DOSE STUDY

In the conventional way of reviewing chromosomes, especially in low dose exposures, only abnormal cells are usually checked again, because reexamination of all the cells is too time-consuming. In case of high dose exposures, overlooking of the second and/or additional aberrations in one metaphase is the most common error [25]. Such overlooking can be corrected even if only abnormal cells are reviewed. In case of low dose exposures most abnormal cells have only one dicentric. Once a cell is scored as being normal, the cell is not reexamined and overlooking is not corrected by reviewing. This problem can be solved by applying an automated stage with a computer to the microscope, since recording and relocating the position of a large number of metaphases are done mechanically without mistake and reviewing all cells becomes easy. In addition, you can reduce the time of work by separating the task of finding analyzable metaphases under low magnification from that of detecting abnormal chromosomes under high magnification. The overlooking of dicentrics can be minimized if you can concentrate on detecting abnormal chromosomes, not being bothered with changing an objective lens each time for finding a new metaphase.

9. DETECTING DICENTRICS BY THE AUTOMATED IMAGE ANALYSIS

Under the said project we developed an automated system for scoring dicentrics, NIRS-1000 KINETOSCORER [26]. It has ability to perform all the work necessary for detecting dicentrics: loading and/or changing a slide glass, metaphase finding, selection of good metaphases, detection of dicentrics, and tabulation. It also has a program with which a cytogeneticist can review the result. In this system the machine selects candidates of dicentrics and an operator chooses true dicentrics.

The chromosome preparations irradiated by 200 keV X-rays at the doses of 11.3, 23.6, 47.3, and 94.5 cGy were examined with the KINETOSCORER. The threshold level for selecting metaphases was set to be high. The frequencies of dicentrics at those dose points were obtained by counting true dicentrics in the candidates selected by the system. The same preparations were also analyzed by a conventional manual way (the automated stage was used). The result is summarized in Table 3. The dose response relationship is similar in both analyses. The time required for the automated analysis by the KINETOSCORER was 154 hours 26 minutes and that for reviewing was 33 minutes. Manual analysis took us 20.5 hours. When the image analysis system was used, the total time was longer, but an operator's working time was extremely reduced.

Table 3
Detection of Dicentric (Dic) : KINETOSCORER vs Manual (M)

Dose(cGy)	Area(cm^2)	Cells detected	Cells selected	Candidate Dic	True Dic(per cell)
11.3	5.53	1,751	108	47	0 (0.000)
〃	(M)*		542		4 (0.007)
23.6	6.62	2,272	124	76	3 (0.024)
〃	(M)*		491		10 (0.020)
47.3	5.76	2,350	101	64	4 (0.040)
〃	(M)*		464		23 (0.050)
94.5	4.32	1,761	145	123	13 (0.090)
〃	(M)*		473		37 (0.078)

* The whole area of a slide

10. FLUORESCENCE IN SITU HYBRIDIZATION (FISH) TECHNIQUE

Pinkel et al. [27] first reported chromosome painting method by in situ hybridization using chromosome-specific DNA probes. Selected chromosomes are stained entirely with fluorescent dye so that the interchange between painted chromosomes and unpainted chromosomes is easily distinguished. If centromeres are stained simultaneously, it becomes possible to distinguish translocation, dicentrics, and fragments one another. The aberration to be detected by the painting method is only a part of the whole genome (usually the one-fifth to the one-third), and three to five times more cells than those analyzed by the conventional method have to be scored for obtaining the frequency per cell [28]. Therefore, our improved method is extremely useful for the chromosome painting because it can even make a chromosome slide which contains over 1,000 metaphases per square centimeter.

When cells are exposed to radiation, translocations and dicentrics are produced in an equal ratio [29 and references therein]. The incidence of translocation does not change for many years after exposure, while that of dicentrics decreases dramatically within a few years post irradiation [30]. Therefore, even when long time has passed after irradiation or in case of chronic exposure, it is possible to estimate the exposed dose using translocation as a marker, but it is difficult to do so using dicentrics. However, in the study of low dose exposure, translocation's high background frequency itself makes it difficult to detect the exact increase in the frequency of translocations. The background frequency of translocations in healthy adults is about ten times higher than that of dicentrics [31,32]. Tucker et al. [32] report that the frequencies of the translocation and dicentrics per 1,000 cells are 7.7 and 0.62. They are equivalent to the frequencies induced by the acute exposure to ^{60}Co γ-rays at the doses of about 20 cGy and about 1.7 cGy, respectively, according to the dose response relationship reported by Lucas et al. [33]. Since the background level of the incidence of translocation that is two times more than the mean (7.7 per 1,000) is not uncommon [34] it may not be possible to estimate the dose below 20 cGy using translocation as a marker unless the incidence prior to the exposure is known. Therefore, in the particular study of low dose exposures, there is no better indicator than a dicentric.

REFERENCES

1. IAEA technical report No. 260, pp41, Internatl. Atom. Ener. Agent., Viena, 1986.
2. R. J. Srám et al., Mutat. Res., 144 (1985) 277.
3. C. R. Richardson et al., Mutat. Res., 144 (1985) 277.
4. K. Salassidis et al., Mutat. Res., 311 (1994) 39.
5. A. V. Sevankaev et al., Genetika, 10 (1974) 114.
6. N. Blackwell et al., Mutat. Res., 25 (1974) 397.
7. G. Obe and J. Herha, Hum. Genet., 41 (1978) 259.
8. G. Obe et al., Mutat. Res., 92 (1982) 309.
9. B. Ivanov et al., Mutat. Res., 52 (1978) 421.
10. S. Gundy and L. P. Varga, Mutat. Res., 120 (1983) 187.
11. A.Tonomura et al., Radiation-Induced Chromosome Damage in Man, T. Ishihara and M. S. Sasaki (eds.), pp605, Alan R. Liss, Inc., New York, 1983.
12. J. Muramoto et al., Ann. Rep. Fukushima-ken Kankyo Igaku Kenkyujo, 2 (1988) 81.
13. A. Léonard et al., Mutat. Res., 138 (1984) 205.
14. D. C. Lloyd et al., Mutat. Res., 72 (1980) 523.
15. J. Rohl-Rüling et al., Mutat. Res., 173 (1986) 267.
16. D. C. Lloyd et al., Int. J. Radiat. Biol., 53 (1988) 49.
17. D. C. Lloyd et al., Int. J. Radiat. Biol.., 61 (1992) 335.
18. A. A. Awa et al., J. Radiat. Res., 19 (1978) 126.

19. C. W. Heath et al., J. Am. Med. Associ., 251 (1984) 1437.
20. R. Wagner et al., Mutat. Res., 109 (1983) 65.
21. J. Rohl-Rüling et al., Mutat. Res., 110 (1983) 71.
22. I. Hayata et al., J. Radiat. Res., 33, Suppl. (1992) 231.
23. I. Hayata, Biotech. Histochem., 68 (1993) 150.
24. R. Kanda et al., J. Radiat. Res., 35 (1994) 41.
25. M. Bianchi et al., Mutat. Res., 96 (1982) 233.
26. A. Furukawa et al., Abst. 1996 Ann. Meet. Atom. Ener. Soci. Jpn. pp559, 1996.
27. D. Pinkel et al., Proc. Natl. Acad. Sci. USA, 85 (1988) 9138.
28. J. N. Lucus et al., Int. J. Radiat. Biol., 56 (1989) 35.
29. R. Kanda and I. Hayata, Int. J. Radiat. Biol. 69 (1996) 701.
30. K. E. Buckton et al., Human Radiation Cytogenetics. H. J. Evans et al. (eds), pp 106, North-Holland, Amsterdam, 1967.
31. M. Bauchinger et al., Int. J. Rad. Biol., 62 (1993) 673.
32. J. D. Tucker et al., Mutat. Res., 313 (1994) 193.
33. J. N. Lucas et al., Health Phys., 68 (1995) 761.
34. M. J. Ramsey et al., Mutat. Res., 338 (1995) 95.

Preliminary report on quantitative study of chromosome aberrations following life time exposure to high background radiation in China

T. Jiang[a], C.-Y. Wang[a], D.-Q. Chen[a], Y.-R. Yuan[b], L.-X. Wei[a], I. Hayata[c], H. Morishima[d], S. Nakai[e], T. Sugahara[e]

[a]Laboratory of Industrial Hygiene, Ministry of Health, Beijing 100088, China

[b]Labor Hygiene Institute of Hunan Province, Changsha 410007, China

[c]National Institute of Radiological Sciences, Chiba 263, Japan

[d]Kinki University, Atomic Energy Research Institute, Higashi-Osaka, Japan

[e]Health Research Foundation, Kyoto 606, Japan

1. INTRODUCTION

Chromosome aberration is the most reliable biological indicator for quantitative effects of ionizing radiation. The study of dose-effect relationship has long been a subject of numerous papers. For acute irradiation in human being, a well-established relationship between unstable chromosome aberration (Dic+Rc) and radiation dose has been shown in the subjects such as A-bomb survivors in Hiroshima and Nagasaki [1]. However, there is still much interest and debate at low dose level [2 - 4]. Many areas with natural high background radiation can be found around the world, such as in India, Iran, Brazil and China. Analysis of Dic+Rc in peripheral lymphocytes of people living in these areas would provide valuable data for the low dose study. There are several reports about the cytogenetic investigation on the inhabitants of high background radiation area (HBRA)[5-8]. Those reports suggested that the Dic+Rc frequencies increased in the inhabitants of HBRA in comparison with their respective controls. In order to obtain accurate quantitative data, we carried out an investigation in HBRA with improved techniques. The present results provided a good evidence for the positive dose-effect relationship in the case of low-dose chronic exposure.

2. MATERIALS AND METHODS

2.1 Study population

The present investigation chose family members as objective samples, which minimized possible biases for the effect of radiation due to biological and environmental modifiers other than radiation. Fifteen study subjects were chosen from six households in HBRA. Among them, 6 were in the grandparental, 3 in the parental and 6 in the child generations, respectively. Thirteen persons were selected from four households in a control area. Of whom, 5 were in the grandparental, 4 in the parental and 4 in the child generations, respectively. All study subjects and the control people were in good health and were native inhabitants either in the HBRA or in the control area and had not received occupational exposure.

2.2 Dose estimate

The data of individual dose estimate were based on the work of Task Group on Radiation Levels and Dose Assessments in terms of effective dose. The cumulated effective doses ranged about 24-240 and 5-51 mGy for people from HBRA and Control Area, respectively.

2.3 Cytogenetic method

Separated lymphocyte cultures were set up according to a new technique developed by Hayata et al.[9] with some modification; i.e. mononuclear cells were separated with Leuco PREP tube from 3 ml blood sample of each subject. The cells were cultured in RPMI 1640 medium supplemented with 20% FCS, 2%PHA and 0.04 mg/ml of colcemid and kept at 36-37 °C for 49-50 h. The preparations obtained were in high mitotic indices and with good chromosome morphology. More than 99.5% of cells were in the 1st division [10]. This provided us with a sufficient number of metaphases at individual level for the low dose chromosome aberration analysis. About 2,000-3,000 metaphases having centromere counts of 46 ± 1 for each individual were analyzed under 100X oil lens. All metaphases with either definite or suspected dicentric, tricentric or centric ring were recorded and confirmed by a third person. Since the microscope was equipped with an auto-stage, the coordinates of all scored cells were recorded. This facilitates the check of all the cells analyzed again.

2.4 Statistical methods

Statistical analyses was performed for the incidence of dicentrics and rings. To compare the difference of Dic+Rc frequencies, two-tail u-test was made. To investigate the cumulative dose effect dependence, linear regression analysis was performed.

3. RESULTS AND DISCUSSION

3.1 Age dependency

A great difference between two areas was found in the age dependent relationship. For HBRA, the Dic+Rc yields increased with age (Fig.1). Even within some of the households, a significant difference of aberration yield could be seen between the oldest and the youngest individuals. This kind of increment, however, was not found within any of the household from control area. Neither was a clear age dependent relation found in the Control (Fig.2).

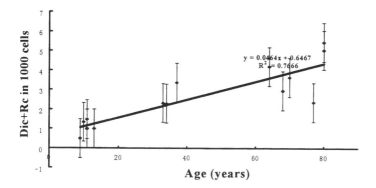

Figure 1. Chromosome aberrations vs. age of individuals from different households in HBRA.

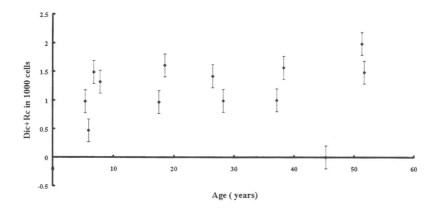

Fig. 2 Chromosome aberrations vs. age of individuals from different households in the control area

A significantly increased mean aberration yields in either middle-age-group or old-age group in HBRA was seen in comparison with those in the similar age groups in the control. For the youngest age group, there was no much difference between the two areas, both of which were consistent with a nominal value of 1 in 1000 cells (Table 1).

Table 1
Frequencies of chromosome aberrations detected in different age groups of inhabitants in HBRA and control area

Area	Mean age (years) ± SD	Cells scored	Dic + Rc in 1000 cells	P
HBRA	79.00 ± 1.4	8505	4.23	
	67.33 ± 3.1	9155	3.60	<0.001
	34.66 ± 1.7	7812	2.56	<0.05
	10.83 ± 1.2	12418	1.05	
CA	61.80 ± 5.6	9829	1.22	
	31.25 ± 4.0	12029	1.25	
	8.75 ± 0.8	10222	1.08	

Tonomura et al.[11] suggested that the yield of Dic+Rc was affected by the age of subjects. Some researchers suggested that the age dependency is most probably caused not only by the biological effects, but also eventually by certain dependency in the cumulative life time exposure to environment mutagens and radiation. Our investigation showed that the unstable chromosome aberrations of inhabitants living in HBRA increased with age. The increment seems mainly caused by the contribution of background radiation over a life time exposure.

3.2 Dose dependency

In the case of HBRA, the older family members had been exposed to higher cumulative dose and at the same time possessed the greater frequency of Dic+Rc. It is clear that the aberration frequencies increased in proportion to the cumulative doses. Results of linear-regression-analysis supported this positive correlation . This was especially true for total Dic+Rc detected (Fig.3). It means that in the case of chronic exposure, total cumulative Dic+Rc must be taken into account for quantitative dose effect study.

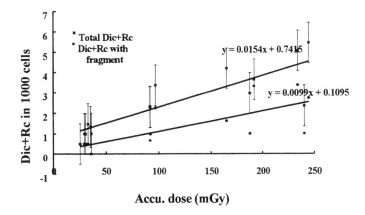

Figure 3. Relationship between chromosome aberration yields and cumulative doses for family members in HBRA.

For quantitative cytogenetic study it is essential to score only the metaphases at the first in vitro division. This is because unstable aberrations tend to be lost through further cell cycles, which will lead to the alteration of the quantitative response. Therefore necessary measures must be taken in the lymphocyte culture to prevent cells from going into the second cell cycle. In general, there is a large individual variation in the duration time of cell cycle. Consequently, it is not sufficient only to set up 48 hour culture time. For this reason, we added colcemid from the beginning of the culture at the optimal concentration of 0.04 (g/ml. This enabled us to cumulate a sufficient number of first mitotic metaphases for analysis.

The studies on cytogenetic effect of high background radiation, such as in Iran and Brazil [5,6], chromosomes were analyzed using lymphocytes cultured for 72 hrs. In another study made in India [7], the lymphocytes were cultured for 64 hrs. As a result, some of the unstable aberrations might have been lost since all of the data in those reports are within the level of controls [12]. In addition, those results [5-8] could only show the difference between two areas, rather than dose effect evaluation. The present study provided data of cumulative dose and aberration frequency at individual level. And an apparent positive dose response relationship in HBRA was demonstrated.

Dicentrics and centric rings are a kind of definitely radiation-induced abnormalities. Compared with stable aberrations, such as translocations and inversions, their special structures made them easier to be recognized under optical microscope with less artifact and lower background noises; hence, they were adopted as an end point in our study. It seems that Dic+Rc can be continuously cumulated over lifetime chronic low dose exposure and can serve as a reliable biological indicator.

REFERENCES

1. Awa, A.A., et al., Chromosome data for A-bomb dosimertry reassessment, In new dosimetry at Hiroshima and Nagasaki and its implications for risk estimates, Proceedings No. 9, National Council on Radiation Protection and Measurements, Bethesda, MD., 15-202, 1988.
2. Lloyd, D.C., et al., Chromosome aberrations in human lymphocytes induced in vitro by low level doses of X-rays, Internl. J. Radiat. Biol., 61 (1992, 335-345.
3. Pohl-Ruling, et al., Effect of low-dose acute X-irradiation on the frequencies of chromosome aberrations in human peripheral lymphocytes in vitro, Mutation Research, 110 (1983) 71-82.
4. Pohl-Ruling, J., Chromosome aberrations of blood lymphocytes induced by low-level doses of ionizing radiation, Advances in Mutagenesis Research,Vol. 2, Edited by G.Obe, Springer-Verlag, Berlin, 155-190 ,1990.
5. Marcello A.et al., Cytogenertic investigation in a Brazilian population living in an area of high natural radioactivity, American Journal of Human Genetics, 27 (1975) 802-806.
6. Fazeli, T.Z., et al., Cytogenetic studies of inhabitants of a high level radiation area of Ramsar, Iran, Proceedings of an international conference on high levels of natural radiation, 459-464 (1990).
7. Kochupillai, N., et al., Down's syndrome and related abnormalities in a area of high background radiation in coastal Karala, Nature, 563 (1976) 60-61.
8. Chen, D. and Wei, L., Chromosome aberration, cancer mortality and hormetic phenomena among inhabitants in areas of high background radiation in China, J. Radiat. Res. Supplement, 62 (1991) 6-53.
9. Hayata, I. et al., Robot system for preparing lymphocyte chromosome, J. Radiat. Res., 33, Supplement, 231-241, 1992.
10. Kanda, R., et al., Effects of colcemid concentration on chromosome aberration analyses in human lymphocytes, J. Radiat. Res., 35 (1994) 41-47.
11. Tonomura, A., et al., Types and frequencies of chromosome aberrations in peripheral lymphocytes of general populations. Radiation-induced chromosome damage in men, Alan R. Liss, Inc., New York, 605-612, 1983.
12. In Biological dosimetry: chromosome aberration analysis for dose assessment. Technical Report Series No. 206, IAEA, Vienna, 1986.

Effect of low dose rates on the production of chromosome aberration under lifetime exposure to high background radiation

Sayaka Nakai [1], T. Jiang [2], D. Chen [2], I. Hayata [3], Y. Yuan [4], N. Morishima [5], S. Fujita [6], T. Sugahara[1] and L. Wei[2]

[1]Health Research Foundation, Kyoto 606, JAPAN

[2]Laboratory of Industrial Hygiene, Ministry of Health, Beijing 100088, CHINA

[3]National Institute of Radiological Sciences, Chiba 263, JAPAN

[4]Labor Hygiene Institute of Hunan Province, Changsha, CHINA

[5]Kinki University, Higashi-Osaka 577, JAPAN

[6]Radiation Effects Research Foundation, Hiroshima 732, JAPAN

1. INTRODUCTION

Study of the biological effects of high background radiation is very valuable for evaluating the radiation risks induced by low level radiation under the condition of chronic exposure to the radiation. The underlying condition of the dose range in a high background radiation area is a level similar to the dose range of a population exposed to an accident of a nuclear plant. In order to estimate risks induced by low-dose radiation, quantitative data providing the dose effect relation is indispensable. For this purpose, highly sensitive and reliable data on the biological effect of radiation is necessary. Tonomura et al. reported that increased frequencies of chromosome aberration in populations were observed with increasing age of the persons examined under a natural environment [1], suggesting that the increase in chromosome aberration frequency could be due to a cumulative dose effect of background radiation. However, the existence of other possibilities, such as a biological aging effect, still remains to be clarified. Thus, we have explored the dose rates effect on chromosome aberration under the condition of lifetime exposure to high background radiation in people of different age groups, in order to clarify the role of background radiation on the production of chromosome aberration in the general population.

2. A STUDY OF INDIVIDUAL DOSE ASSESSMENT

Accurate and precise estimation of individual absorbed does is an essential requirement for a study of basic information relevant to the dose effect relationship. However, a study of individual dose assessment under the condition of lifetime exposure to radiation is a difficult problem, since there are uncertainty inherent in a number of the modifying factors. Recently. we have carried out a cooperative study [2,3] to investigate the nature of various factors concerned using direct and indirect methods in a high background radiation area in China. The data showed that individual dose consisted mainly of indoor dose since people's living time in their own dwellings i.e., occupancy factor, is about 70% in this area, and since the indoor dose rate is usually 2- 3 times higher than the outdoor rate in a high background radiation area. The indoor dose in people's own dwellings is derived from the amount of radionuclide contained in the building materials. Therefore, the individual dose is primarily family dependent. On the other hand, the contribution of the outdoor dose is considered to be a secondary factor, which is reflected to the variation of exposed doses within members of family due to their personal behavior in the radiation environment. While, their living conditions are very homogeneous since they are mostly farmer in this area. Accordingly, although the outdoor activities of persons are thought to be largely age-dependent, however, the data showed that individual dose rates were unexpectedly uniform with respect to the persons of different ages except for babies and very old persons [2] . In fact, the deviation of individual doses within member of families was less than half that of mean doses for different families within same hamlet. Accordingly, it would be safe to assume that the individual dose rates of background radiation will be *operationally constant* throughout individual lifetimes, and could be *family dependent*.

In order to obtain a more accurate dose rate relationship for chromosome aberration, we have directly measured personal doses using a thermoluminescent dosimeter (TLD) for two months. Assuming that the level of dose rate per year could be extrapolated from the observed level per month, we estimated the dose rates for each individual members of 7 families in a high background radiation area (HBRA), as well as for the members of 5 families in a control area (CA). Blood samples were obtained from the three generations in each family. As shown in Table 1, the data demonstrated that the individual dose rates within members of a family were actually uniform as expected. Thus it may be safe to consider that the measured dose rates are almost constant during their lifetimes, and that the estimated dose rates are family dependent. In order to make a quantitative comparison between different cytogenetic data for the dose response relation *in vivo* and *in vitro* studies [1,4,5,6,7,8,9] it is necessary to define a common dose unit in terms of air absorbed dose (mGy) instead of effective dose for cancer study. For this purpose, we have calculated dose rates in term of air absorbed dose

Table 1

Assessment of the dose rates for family members with blood samples taken from people with lifetime exposure to background radiation

Family Code	Measured Dose Rate (mR/Y)[1] Measurement by TLD[2]					n	Mean	SD[3]	Estimated Dose Rate (mGy/Y)[4]
(HBRA group)									
A	478*	424	365	379*	390*	5	407.2	45.1	3.42
	(64)	(59)	(30)	(9)	(36)				
	(M)	(F)	(F)	(F)	(M)				
B	407*	362	371*	388*		4	382.0	62.6	3.21
	(68)	(58)	(33)	(13)					
	(M)	(F)	(M)	(F)					
C	406	369	421			3	398.7	26.8	3.35
	(55)	(34)	(28)						
	(F)	(M)	(F)						
D	429*	404	371*	380	407*	5	398.2	23.1	3.35
	(66)	(53)	(35)	(32)	(10)				
	(M)	(F)	(M)	(F)	(M)				
E	389*	511	537	448*		4	471.3	66.4	3.96
	(80)	(43)	(39)	(11)					
	(M)	(M)	(F)	(F)					
F	489	480*	454*	485	473	5	476.2	13.8	4.00
	(70)	(14)	(80)	(38)	(48)				
	(F)	(M)	(M)	(F)	(M)				
G	447*	450	436*	485	458	5	455.2	18.4	3.83
	(77)	(66)	(10)	(28)	(31)				
	(M)	(F)	(F)	(M)	(M)				

Table 1 continued

Family Code	Measured Dose Rate (mR/Y)[1] Measurement by TLD[2]						n	Mean	SD[3]	Estimated Dose Rate (mGy/Y)[4]
(CA group)										
K	75 (62) (F)	77 (38) (M)	76 (32) (F)	81 (12) (M)			4	77.3	2.6	0.65
L	80 (63) (M)	77* (55) (F)	83* (26) (F)	92* (8) (M)			4	83.0	6.5	0.70
M	86* (70) (M)	70* (37) (F)	107* (10) (F)				3	87.7	18.6	0.74
N	93* (64) (M)	84 (61) (F)	97 (35) (M)	95* (32) (M)	93 (34) (F)	84* (8) (F)	6	91.0	5.6	0.77
P	64* (64) (M)	77* (56) (F)	103 (26) (M)	78* (30) (F)	92* (9) (F)		5	82.8	17.8	0.70

* Indicates person from whom those sample was taken.
1) Data in parenthesis show the age and gender of the person examined.
2) TLD carried by the person. Contribution of cosmic rays is estimated to be 0.23mGy/Y.
3) Standard deviation.
4) For calculation, the following conversion factors are employed; Conversion factors for mR to mGy and conversion with phantom are 0.87 [10] and 0.96 [3], respectively.

(mGy) using appropriate conversion factors [3,10] as shown in Table 1. The data obtained in the present study will lead to more accurate information compared with previous studies for estimation of the dose rate effect relationship at individual level.

3. ANALYSIS OF DOSE-RATES EFFECT RELATION ON CHROMOSOME ABERRATION

In order to obtain more reliable dose-response relationship compared with previous studies, very accurate chromosome data are necessary, since extremely low yields of chromosome aberration are expected at low dose conditions. Accordingly, we have undertaken a very careful examination on the chromosome aberration induced by background radiation as presented in this Conference. In this study, we have used improved cytogenetic techniques [11,12,13] to observe an enormous number of metaphase cells. Moreover, we have studied blood samples taken from members of big families, which has enabled us to minimize the deviation due to individual variations. Thus, we obtained more accurate data compared with previous studies on the yields of dicentrics and rings with and without fragments of chromosome aberration at individual levels [13]. We performed analysis on the dose rates effect relationship using these cytogenetic data combined with present data of individual dose assessment.

In order to assess the biological effect of lifetime exposure to chronic radiation, it is necessary to exclude the possibility of age dependent chromosomal effects due to agents other than radiation, such as a biological aging effect or environmental mutagen effect. For this purpose, we have tried to examine the dose rates effect relationship under the condition of persons of the same age. We have classified three groups according to the ages; the elderly, adults, and children, and verified the aberration frequency at certain ages with these groups. We have plotted yields of dicentric chromosome aberration against the dose rates of the members in a family observed about 3,000 cells in average at individuals. The results are shown in figure 1. It is evident that an apparent dose rates relationship is demonstrated at the elderly group. Statistical analysis shows that the relationship is fitted well to a regression line as a linear model against dose rates ($p<0.001$) as shown in Table 2. This result may indicate that radiation plays a principal role for the production of chromosome aberration in very low dose rates exposure for 60 years under chronic irradiation. If an agent other than radiation is significant, the observed linear dose rates relation could not logically be expected. This means that the so-called *aging effect* other than radiation would be almost negligible under the condition, though the possibility of a certain amount of aging effect could not be excluded. Similarly, a significant dose rates dependency in the 30 years old of adult age group ($p<0.001$) was observed. The effectiveness of the dose rates compared with the elderly group is, however, apparently decreased i.e., about half that of elderly group. These results together

Table 2
Results of linear regression analysis on chromosome aberration($\times 10^{-3}$)

$Y = a + bX$ where X is dose rate (mGy/Y)

Group	Parameter	Estimate	SE*	P-value
Elderly	a	0.47	0.52	----
(60 years)	b	0.96	0.19	<0.001
Adult	a	0.85	0.29	----
(30 years)	b	0.55	0.13	<0.001
Children	a	1.07	0.24	----
(10 years)	b	−0.0003	0.007	>0.1

* Standard error

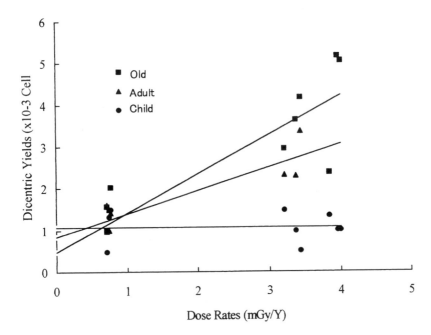

Figure 1. Dose rate effect relationship on the production of dicentrics

with the results of the children group, strongly suggest that the effectiveness of the dose rates is closely correlated with the observed ages relevant to the exposure time to background radiation. As shown in Table 2 and Figure 1, the effect of background radiation did not manifest distinctly in the children group with less exposure time.

4. DISCUSSION AND FURTHER STUDY

Although further intermediate dose rates study is needed, an apparent dose rates effect has been shown in this study. This result indicates that even very low-level radiation plays a principal role on the production of dicentric aberration in a population. Furthermore, the result indicates that the effectiveness of the dose rates on the increased yield of aberration is closely related to the ages of persons under the exposure. The result is consistent with the view that *cumulative doses*, i.e., a simple cross product of the dose rate and the time of radiation exposure has a critical meaning on the yield of dicentric aberration in chronic exposure which is supported to other studies [13,14]. Detailed analysis on the effect of cumulative doses will be published elsewhere.

Another interesting point is that no apparent dose rates relations were demonstrated in the children's group. An action of radiation did not show any measurable increase of chromosome aberration for this underlying condition. Under such conditions, we will be able to estimate the *lowest dose level* at which a chromosomal effect would be apparent. Here practical border lines for the action of low level radiation could be demonstrated. Above the border line, a chromosomal effect of the radiation will emerge significantly, whereas the radiation dose below this line will not be distinct. The estimated cumulative doses of the border line(s) are supposed to be around 50-100mGy in term of cumulative dose under the conditions of chronic irradiation as shown in Figure 2. It should be noted that the linearity of the chromosomal effect can be demonstrated down to doses of about 20mGy at *in vitro* system under the conditions of acute irradiation reported by Lloyd et al. [8]. The level of chromosome aberration corresponding to the respective border line might be determined by two factors; one is the dose-rates effect of radiation, the other is the agent for controlling spontaneous chromosome aberration. In this regard, we suppose that the spontaneous level of chromosome aberration due to agents other than radiation may cause a masking effect, where by the radiation effect could not be detected. These notions could remind us of the concept of *de minimus* dose in a point of view of the radiation protection standards. Thus, the nature of the mechanism(s) controlling certain levels of spontaneous chromosome aberration is of particular interest. For this purpose, further study is needed to develop more advanced techniques.

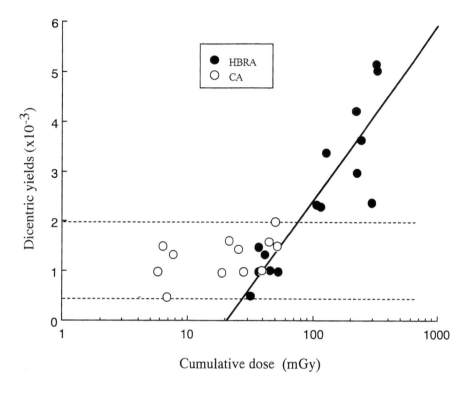

Figure 2. Dose effect relationship on the dicentric yields against cumulative dose.

The present study shows that radiation plays a major role in the production of dicentric chromosome aberration, and that the yield of dicentric aberration seems to be able to accumulate under conditions of exposure to low dose rates background radiation. This characteristic nature of dicentric aberration will provide a reliable indicator to estimate the biological effect of radiation in the general population under the condition of chronic exposure to low dose rates irradiation.

REFERENCES

1. Tonomura, A., Kishi, K. and Saito, F. (eds.), Alan R. Liss. Inc., New York, NY Radiation Induced Chromosome Damage in Man, 605-616, 1986 .
2. Yuan Yongling, Shen Hong,Zhao Shian et al. ,Chinese J. Radiol. Med. Protet. , 15 (1995) 311-316 .

3. Morishima, H.,T. Koga, K. Tatumi, S. Nakai, T. Sugahara, Y. Yuan, Q. Sun and L. Wei., Proceedings of The 4th International Conference on High Levels of Natural Radiation (in press).
4. Awa. A. A., Ohabitaki. K., Itoh. M., Honda. T., Pristine. D.L. and McAonney. M., Proceedings No.9, National Council on Radiation Protection and Measurements., Bethesda. MD. , 15-202, 1988.
5. Evans. H. J., Buckton. K. E., Hamilton. G. E., and Carothers. A., Nature, 77 (1979) 531-534.
6. Pohl-Ruling, J., Advances in Mutagenesis Research. Vol. 2., G. Obe (ed.), Springer Verlag. Berlin , 155-190, 1990.
7. Masao Sasaki., Alan R. Liss, Inc., New York, NY., Radiation-Induced Chromosome Damage in Man ,585-604, 1983.
8. Lloyd, D. C., Edwards, A. A., and 9 others., International Journal of Radiation Biology., 61 (1992) 335-345.
9. Takahashi, E., Hirai, M., Tobari, l., Utsugi, Y. and Nakai, S. ,Mutation Research, 94 (1982) 115-123.
10. ICRU report 43, International Commission on Radiation Units and Measurements, Bethesda. Maryland U. S. A. ,1991.
1 1 . Hayata, I., Tabuchi, H., Fukukawa, A., Okabe.N., Yamamoto, M. and Sato, K., Journal of Radiation Research(supplement) , 33 (1992) 231 -241 .
1 2. Hayata, J. , Proceedings ofThe 4th International Conference on High Levels of Natural Radiation (in press).
1 3. Jiang Tao, Wang Chunyan, Chen Deqing, Yuan Yongling, Wei Luxin, Isamu Hayata, Hiroshige Morishima, Sayaka Nakai, Tsutomu Sugahara., Proceedings of the 4th International Conference on High Levels of Natural Radiation (in press).
14. Chen Deqing, Yao Suyan, Zhang Chaoyang, Dai lianlian, Jiang Tao, Wang Chunyan, Li Shu, Yuan Yngling, Wei Luxin, Sayaka Nakai, Tsutomu Sugahara, Proceedings of the 4th lnternational Conference on High Levels of Natural Radiation (in press).

Cytogenetic findings in inhabitants of different ages in high background radiation areas of Yangjiang, China

Chen Deqing[a], Yao Suyan[a], Zhang Chaoyang[a], Dai Lianlian[a], Jiang Tao[a], Wang Chunyan[a], Li Shu[a], Yuan Yongling[b], Wei Luxin[a], Sayaka Nakai[c], Tsutomu Sugahara[c]

[a]Laboratory of Industrial Hygiene, Ministry of Health, Beijing 100088, China

[b]Labor Hygiene Institute of Hunan Province, Changsha 410007, China

[c]Health Research Foundation, Kyoto 606, Japan

1. INTRODUCTION

Many areas of the world possess unusually high levels of natural radioactivity leading to increased natural radiation exposures of their inhabitants. An increase in chromosome aberrations in peripheral blood lymphocytes of persons living and /or working in an environment with elevated natural background radiation level has been found such as India, Brazil, Iran and China [1-4]. An investigation was carried out by Pohl-Ruling and Fischer amongst the residents and workers in the area of the "radon spa" Badgestein in Austria. The resulting dose-effect curve showed that at very small dose levels and small dose rates the aberration frequencies increase with dose, up to about 300 mrad/yr (α-and γ-blood dose), and additional fractional delivered doses resulted in a leveling-off of the curve into a plateau [5]. However, the yields of unstable chromosomal aberrations in human lymphocytes induced in vitro by X-rays over the dose range 0-300 mGy indicated that the aberration yields significantly in excess of control values are seen at doses > 20 mGy and these were consistent with a linear extrapolation from higher doses[6]. There is much interest and debate on the relationship between biological effect and absorbed radiation doses from elevated natural background radiation. The results of the studies on cytogenetic survey of inhabitants in high background radiation area of Yangjiang are presented and discussed in this paper.

2. MATERIALS AND METHODS

2.1. Choice of subjects

Blood samples were obtained from healthy inhabitants of both sexes in high background radiation areas (HBRA) and control areas (CA), respectively. An attempt was made to approximately match subjects from HBRA and CA for factors such as age, smoking habits and medical X-exposure. According to age and residence history of inhabitants the candidates of donor were selected from demographic data stored in a computer. Before collection of peripheral blood the candidates were interviewed and taken physical examination to select qualified blood donors. All of the donors were farmers, native-born and lived there ever since. Persons who had received any occupational exposure or significant medical dose of radiation or who had recently received diagnostic X-irradiation in half a year or more than two diagnostic x-ray examinations in their life-time were excluded.

2.2. Cumulative exposure

The sources of high background are the natural radionuclides such as uranium, thorium and radium in soil and construction materials in high contents. The environmental radiation levels for each blood donor were measured with dose-rate-meter, including indoors (sitting room, bedroom, kitchen) and outdoors (lane, yard, pond, well, road and farmland or rice field). In order to convert the results of dose-rate meter into annual exposure (or absorbed dose), residential time factor of local inhabitants was investigated. According to each blood donor's age, sex and environmental radiation level, the cumulative dose for donor's lifetime was calculated. In addition, for blood donors and their environment the cumulative exposures were measured with thermoluminescent dosimeters (TLD) for two months. The results from the both measurements are in good correlation. Detailed data were published elsewhere [7]. Table 1 gives average age, smoking habits, diagnostic X-irradiation and accumulated dose for each group.

2.3. Cytogenetic techniques

Heparinized venous blood from each donor was incubated in RPMI 1640 medium supplemented with 20% fetal calf serum, antibiotics and phytohemagglutinin. Colchicine was added into the whole blood microcultures at the beginning of culture period and cultures were incubated at 37°C for 48-54h and harvested by standard procedures, which had more than 99.9% of mitotic cells in the first metaphase [8]. 800-1,300 metaphases were scored from each subject. All types of structural chromosome aberrations were analyzed from coded preparations. Fragments associated with dicentric or ring chromosomes were not included as the chromosome aberration count. A translocation was recorded only when the morphology of the derivative chromosome was clearly indicative of this kind of rearrangement.

Table 1
Status of blood donors

Age groups (years)	No. of donors	No. of smokers	Frequency of X-irradiation	Age (years)+ Mean	Range	Cumulative dose ++ (mGy)
HBRA						
10	10	0	0	10.5	9.3-11.4	24.9±2.1
40	10	7	4	40.9	37.4-43.4	94.1±7.8
55	8	4	3	54.9	51.2-58.8	131.5±7.2
70	10	8	3	70.9	69.1-74.2	161.4±12.6
CA						
10	10	0	0	10.6	10.2-11.1	6.4±0.5
40	10	6	4	40.0	38.7-42.3	23.9±1.3
55	7	4	5	54.0	49.7-57.2	34.4±3.9
70	10	7	0	70.3	68.3-73.4	44.5±1.2

+ Exposure time is equal to age.
++ For one medical diagnostic X-irradiation 0.6mGy is added.

Numerical abnormalities were considered only when hyperdiploidy was observed.

Statistical analyses were carried out using the Student's test and linear regression.

3. RESULTS

Table 2 shows the cytogenetic results obtained from inhabitants in HBRA and CA. The frequencies of the different chromosome-type aberrations were slightly higher in almost all age-groups of HBRA than in corresponding age-groups of CA, but not significantly so. However, the mean frequencies of total number of aberrations in HBRA showed a significant difference compared with CA, such as dicentrics with fragment (0.77 $^0/_{00}$ vs. 0.28 $^0/_{00}$, P<0.01), translocations (2.34 $^0/_{00}$ vs. 1.65 $^0/_{00}$, P<0.05) and total structural chromosome-type aberrations (4.69 $^0/_{00}$, vs. 3.15 $^0/_{00}$, P<0.001). The frequencies of each structural chromosome-type aberration in HBRA were likely to increase roughly with the increase of age (or accumulated dose).

Although there are only four experiment points in this investigation, an attempt was made to fit total structural chromosome-type aberrations to the linear model. For ages the best linear fit was:

$$Y_{HBRA} = 2.41 \times 10^{-3} + 5.43 \times 10^{-5} A \quad (r = 0.98) \quad (1)$$
$$Y_{CA} = 2.05 \times 10^{-3} + 2.75 \times 10^{-5} A \quad (r = 0.82) \quad (2)$$

For accumulated doses the best linear fit was

$$Y_{HBRA} = 2.35 \times 10^{-3} + 2.37 \times 10^{-5} D \quad (r = 0.99) \quad (3)$$
$$Y_{CA} = 2.03 \times 10^{-3} + 4.46 \times 10^{-5} D \quad (r = 0.85) \quad (4)$$

Table 2
Chromosome aberrations in inhabitants of different age groups in HBRA and CA

Age groups (years)	Cell scored	No. of aberrations (/10^3 cells)					Hyper-diploidy (n=47)
		Dic+CR +f	Dic+CR -f	AF	Tr	Total	
HBRA							
10	11,910	$^b_5R_{(0.42)}$	1 (0.08)	10 (0.84)	17 (1.43)	33 (2.77)	1.13
40	10,200	7 (0.69)	$_5R_{(0.49)}$	13 (1.27)	$^a_{25}$ (2.45)	50 (4.90)	1.04
55	10,000	10 (1.00)	$^c_2$ (0.20)	15 (1.50)	$^a_{28}$ (2.80)	55 (5.50)	1.76
70	9,700	10 (1.03)	1 (0.10)	19 (1.96)	28 (2.89)	58 (5.98)	2.96
Total	41,810	32 (0.77)**	9 (0.22)	57 (1.36)	98 (2.34)*	196 (4.69)***	1.52
CA							
10	11,700	1 (0.09)	2 (0.17)	12 (1.03)	16 (1.37)	31 (2.65)	0.69
40	11,090	$^a5(0.45)$	2 (0.18)	10 (0.90)	10 (0.90)	27 (2.43)	0.92
55	7,330	4 (0.55)	2 (0.27)	7 (0.95)	14 (1.91)	27 (3.68)	1.18
70	9,200	1 (0.11)	1 (0.11)	12 (1.30)	25 (2.72)	39 (4.24)	9
Total	39,320	11 (0.28)**	7 (0.18)	41 (1.04)	65 (1.65)*	124 (3.15)***	1.11

Dic= dicentric, CR= centric ring, +f or -f = with or without fragment, Tr= translocation, AF = excess acentric fragment, acentric ring and minute. R: the data include a centric ring. a: indicates that the data include a cell with two same type aberrations, b: a cell with three same type aberrations; c: a cell with a tricentric. *: u=2.196, P<0.05; **: u=3.00, P<0.01; ***: u=3.498, P<0.001.

The number of cells with hyperdiploidy was slightly higher in the different age-groups of HBRA than in that of CA, but without statistically significant difference.

When smoking habits were taken into account for each age-group, the frequencies of the different chromosome-type aberrations in most age-groups were slightly higher in smokers than in non-smokers; only for the mean frequency of total number of aberrations in 55-years-old group of HBRA, the difference between smokers and non-smokers was significant (7.01 $^0/_{00}$ vs. 4.08 $^0/_{00}$, P<0.05) (see Table 3).

Table 3
Comparison between smokers and nonsmokers for chromosome-type aberrations(number/10^3 cells)

	10 years-old		40 years-old		55 years-old		70 years-old	
	Smokers	Non-smokers	Smokers	Non-smoker	Smokers	Non-smokers	Smokers	Non-smokers
HBRA								
Subjects	10	7	3	4	4	8	2	
Cells	11910	7050	3150	4850	5150	7800	1900	
Dic+CR	0.50	1.00	1.59	1.86	0.58	1.28*	0.53	
AF	0.84	1.28	1.27	1.44	1.55	2.05	1.58	
Tr	1.43	2.69	1.90	3.71	1.94	3.08	1.05	
Total	2.77	4.97*	4.76	7.01*	4.08*	6.41*	3.16	
CA								
Subjects	10	6	4	4	3	7	3	
Cells	11700	6600	4490	4000	3330	6550	2650	
Dic+CR	0.26	0.45	0.89	1.25	0.30	0.31*	0	
AF	1.03	1.06	0.67	0.75	1.20	1.07	1.89	
Tr	1.37	0.91	0.89	2.50	1.20	2.60	3.77	
Total	2.65	2.42*	2.45	4.50	2.70	3.97*	5.66	

* P<0.05, for 7.01 vs. 4.08, 4.97 vs. 2.42, 6.41 vs. 3.97 and 1.28 vs. 0.31

4. DISCUSSION

Although most studies on chromosome aberrations in persons living and/or working in an environment with elevated natural radioactivity showed higher levels of asymmetrical aberrations than controls, no relationship was found between radiation dose and yield of aberrations (for reference see Pohl-Ruling, 1989, 1990, Pohl-Ruling and Fischer, 1983) [9-11]. Pohl-Ruling and Fischer (1979) [12] investigated the chromosomal aberration in the lymphocytes of sap workers exposed to enhanced natural radiation level, and found that a plot of the yield of total aberrations against dose-rate showed a linear relationship between effect and dose up to dose-rates of 2 mGy per 6 months, followed by a plateau at higher dose rates.

In the present study, the mean frequencies of total number of aberrations in HBRA showed a significant increase compared with CA, especially for dicentrics with fragment, which may be interpreted as an effect of elevated natural radiation present in the area, because an attempt was made to approximately match for age, smoking habits, while all subjects were farmers. Furthermore, linear regression analysis showed the correlation between the incidences of the total structural chromosome-type aberrations and accumulated doses, either in HBRA or in CA. It is evident that if anything meaningful is to be seen in men exposed to low levels of radiation, a large nummer of cells are needed to be examined.

The effect of age is equivocal. No significant change in frequency with age has been observed in the study by Bender et al. (1986) [13]. However, Tonomura et al. (1983) [14] investigated the frequencies of chromosome aberrations in lymphocytes of general populations, and suggested that the frequency of dicentric aberrations increases linearly with age. The present data of the total frequencies of chromosome-type aberrations do support an age effect. The relationship between the chromosome aberration and increasing age was similar to that between the chromosome aberration and accumulated dose. This may suggest that the increase is due to an increase in accumulated exposure to environmental natural background radiation.

The effect of smoking on chromosome aberration has been largely ignored in previous studies on inhabitants living in elevated natural background radiation areas. The influence of smoking has not been universally demonstrated, but the present study seems to support that cigarette smoking has an effect on chromosome-type aberrations(Tawn and Binks, 1989) [15].

In our study, the frequencies of hyperdiploidy in inhabitants of HBRA were slightly higher than those in CA, without statistically significant difference. However, we find influence of age on the incidence of such aberrations. The frequencies of hyperdiploidy in age-groups increase significantly with age, either in HBRA or in CA. According to Barquinero et al. (1993) [16], the frequency of hyperdiploidy in hospital workers occupationally exposed to low levels of ionizing radiation (1.6-42.71mSv) was significantly higher than that in controls. They, however, did not find any influence of age on the incidence of such aberrations, probably due to the narrow range of age (only 30 through 50 years).

When compared with inhabitants in CA, the inhabitants of HBRA did not show any correlation between the excess frequencies of the different chromosome type aberrations considered and the excess doses received. The influence of factors such as number of cells analyzed, interindividual differences in susceptibility and the activation of DNA-repair enzymes cannot be discarded.

REFERENCES

1. George, K.V, Aravindan, B. Joseph, M.V. Thampi, V.D. Cheriyan, C.J. Kurien, V.R. Shah K. Sundaram, Cytogenetic studies. Abstract of 2nd Special Symposium on Natural Radiation Environment. Bombay, India, 87, 1981.
2. Barcinski, M. do Ceu, A. Abreu, J.C.C. de Almeida, J.M. Naya, L.G. Fonseca, L.E. Castro, Am. J. Hum. Genet., 27 (1975) 802.
3. Fazeli, R.G. Assaei, M. Sohrabi, A. Heidary, R. Varzegar, F. Zakeri and H. Sheikholeslami, Proceedings of an International Conference on High Levels of Natural Radiation, 459, 1990.
4. Chen D.-Q. and Wei L.-X., J. Radiat. Res., Suppl. 2 (1991) 46.
5. Pohl-Ruling and P. Fischer, in: T. Ishihara, M.S. Sasaki (eds), Radiation-Induced Chromosome Damage in Man, Alan R Liss, New York, NY10011, 527, 1983.
6. Lloyd, A.A. Edwards, A. Leonrard, G.L. Deknudt, L. Verschaeve, A.T. Natarajan, F. Darroudi, G. Obe, F. Palitti, C. Tanzarella and E.J. Tawn, Int. J. Radiat. Biol., 61 (1992) 335.
7. Yuan Yongling, Shen Hong, Zhao Shian, et al. Chinese J. Radiol. Med. Prot., 15 (1995) 311.
8. Chen D.-Q. and Zhang C.-Y., Mutat. Res., 282 (1992) 227.
9. Pohl-Ruling, in: G. Obe (Ed.), Advances in Mutagenesis, Vol.II, Springer, Berlin,155, 1990.
10. Pohl-Ruling, Proceedings of the XVth Berzelius Symposium on Somatic and Genetic Effects of Ionizing Radiation, Umea, 103, 1989.
11. Pohi-Ruling and P. Fischer, in: T. Ishihara and M.S. Sasaki (Eds.), Radiation-Induced Chromosome Damang in Man, Liss, New York, (1983) 527.
12. Pohl-Ruling and P. Fischer, Radiat. Res., 80 (1979) 61.
13. Bender, A.A. Awa, A.L. Brooks, H.J. Evans, P.G. Groer, L.G. Littlefield, C. Pereira, R.J. Preston and B.W. Wachholz, Mutat. Res., 196 (1988) 103.
14. Tonomura, K. Kishi and F. Saito, in: T. Ishihara and S.M. Sasaki (Eds), Radiation-Induced Chromosome Damage in Mam, Alan R. Liss, Inc; New York, 605, 1983.
15. Tawn and K. Binks, Radiat. Prot. Dosim., 28 (1989) 173.
16. Barquinero, L. Barrios, M.R. Caballin, R. Miro, M. Ribas, A. Subias and J. Egozcue, Mutat. Res., 286 (1993) 275.

Cytogenetic studies in lymphocytes of healthy inhabitants of Mazandaran a high level natural radiation area in Iran*

R.M.Raie and M.Aledavoud

Department of Biology, Faculty of Science, Tehran University, IRAN.

Chromosomal aberrations (CAs) and sister chromatid exchanges (SCEs) in peripheral lymphocytes of normal human population living in a high level natural radiation area in Iran (Mazandaran) and in one of the low level natural radiation areas in Iran (Tehran) were studied and compared. In all 97 blood samples from healthy inhabitants of Mazandaran (30 samples from Gonbad, 34 samples form Babol and Ghaemshahr and 33 samples from Ramsar, one of the high level natural radiation areas in the world) and 30 samples from Tehran were cultured. The results showed that the highest frequency of CAs belonged to Ramsar and the lowest belonged to Tehran. The mean frequency of total number of CAs per 100 cells for Ramsar is 5.75, while for Tehran is 1.53. The mean frequency of total number of CAs per 100 cells in the three groups residing in Mazandaran are significantly higher than in Tehran. Gonbad 3.2, Ramsar 3.7, Babol and Ghaemshahr 2.26 times higher respectively than Tehran. Similarly the scored SCE per cell in the Mazandaran groups are 1.5, 1.3 and 1.4 times higher respectively than healthy individuals in Tehran. The scored SCE in the healthy inhabitants of Gonbad is 5.172 per cell, which is highly different from that of Ramsar, 4.245 per cell, but no difference was found between the scored SCE for Gonbad in comparison with the scored SCE for Babol and Ghaemshahr which is 4.71 per cell. In addition significant correlation was found only between the length of residence in Ramsar and the mean number of gaps and the total number of CAs/100 cells.

1. INTRODUCTION

CAs including gaps, breaks, dicentric chromosomes, minute chromosomes, ring

---

*This investigation was carried out with finantial support for the Research Deputy of Tehran University under contract No. 513/1/225.

and tetraploidy are the more important outcomes of ionizing radiation exposure of cells and are indicators of genetic damage[1]. SCEs is also considered a general marker of mutagenesis and is often used as an indicator of levels of genetic damage and deficiency of DNA repair mechanisms[2]. In this study we decided to determine and compare the levels of CAs and SCEs in the normal inhabitants of Mazandaran province. Ramsar in Mazandaran province is known as one of the high natural radioactive areas in the world, with radiation levels ranges from 0.08-9 mR/h [3]. In this province Babol and Ghaemshahr have normal level and Gonbad seems to have medium level of natural radioactivity [4]. Normal individuals of Tehran were selected as another control group residing in a region with a low level or natural radiation background [4,5].

The epidemiological studies demonstrated that cancer could be associated with ionizing radiation [1]. The relevance of cytogenetic damage to cancer has been elucidated elsewhere [6]. Therefore to study the possible effect of ionizing radiation on human health, correlation between natural radiation dose and cancer incidence rates in Mazandaran province, a province with high incidence rate of cancer[7] has also been studied. In this respect Mazandaran was divided into three parts. Gonbad has a high incidence rate of cancer, while Ramsar has medium, and Babol+Ghaemshahr has low risk rate[7]. The results are presented and discussed in this paper. In addition correlation between the length of residence in these areas and the level of CAs and SCEs was studied.

2. MATERIALS AND METHODS

Two ml of blood from the individuals of the selected groups were collected and transferred to Tehran within 24 hours. A standard culture technique [8] with slight modification in hypotonic solution (3.9 g/l of KCl) was employed. At least two whole blood cultures were setup for each sample and for each technique. Scoring of CAs was done on 100 cells of each individual. The abnormalities were separately recorded as gaps, breaks, dicentric, minute, tetraploid and other. The classification of "other" included fragments, acentric chromosomes, iso gaps and ring. The data is presented as the average number of each type of aberration per 100 cells of each individual in each of the grouping. Statistical analysis of the data was performed using the Student's t-test and Pearson's correlation coefficient. The level of significance is 0.05 and 0.01.

3. RESULTS

The data on SCEs and CAs are presented in table 1 and 2 respectively. A comparison of the total CAs and SCEs for all selected groups is shown in the form of histograms in Fig.1. Figure 2 shows the percentage of individuals in each group whose investigated chromosomes showed at least one case of each type of

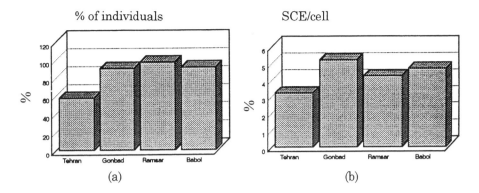

Fugure 1. A comparison of the total number of chromosomal aberrations (a) and SCEs(b) for all groups

aberrations. It is clear that the percentage of such individuals among all subgroups of Mazandaran province is significantly higher than those of Tehran for all the aberration types. The northern subgroups significantly differed from each other only in the parameters of gaps, breaks and minutes. The normal of Tehran had significantly lower incidence of all aberrations as compared to the northern subgroups.

4. DISCUSSION

A more significant finding of the present study is the high incidence of various CAs among healthy inhabitants of Mazandaran. The chromosomal aberration

Table 1
Frequencies of SCE/cell in the studied groups

Samples	n	Ave. No of SCE/cell	Std. deviation
Healthy Individuals from Mazandaran			
Gonabad	25	5.2	1.6
Babol and Ghaemshahr	30	4.7	0.8
Ramsar	30	4.2	1.0
Normals of Tehran	30	3.2	0.6

n = No. of Individuals

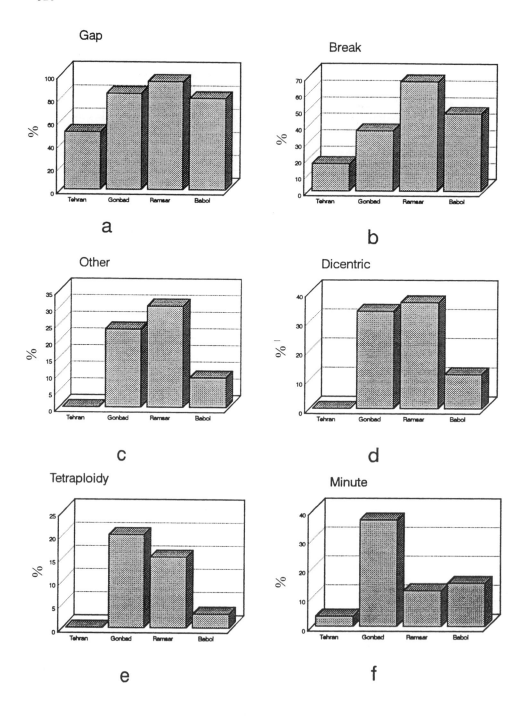

Figure 2. The percentage of individuals from each group whose investigated chromosomes showed at least one aberration of identified type

Table 2
Frequencies of CAs/100 cells in the studied groups

Sample	n.	Gap	Break	Minute	Dicen.	Ter.	Others	Total
Healthy Individuals from Mazandaran								
Gonbad	30	1.86	0.5	0.56	0.5	0.5	0.7	4.63
		1.6 [a]	0.7	1.0	1.1	1.1	2.5	5.5*
Ramsar	33	3.1*	1.2*	0.15	0.6*	0.27	0.4	5.75*
		1.8	1.4	0.4	1.1	0.8	0.8	3.7
Bobol &	34	2.5	0.6	0.15	0.12	0.03	0.1	3.47
Ghemshahr		2.1	0.8	0.4	0.3	0.2	0.3	3.44*
Noramal	30	1.3	0.2	0.03	0	0	0	1.53
of Tehran		1.6	0.5	0.2				1.8

n; Number of individuals   a; Each number of lower position shows standard deviation. *Significantly higher than control group P<0.05

levels in the northern subgroups were 2.3 - 3.7 times higher than their normal counterparts in Tehran. This is in agreement with the results reported by Fazeli et.al. [5] for Ramsar. Within Mazandaran province the mean number of CAs/100 cells for Ramsar and Babol + Ghaemshahr differ significantly (P<0.01). In addition it was found that in Gonbad %90, in Ramsar %97, in Babol + Ghaemshahr %91.2 and finally in Tehran %56.7 of the individuals studied had CAs (Fig.1a). An immensely high level of environmental radiation exist in Ramsar seems to be relevant for the significant higher rate of chromosomal breaks, dicentric chromosomes and gaps (Fig.2b,d and a respectively) compared to other areas of Mazandaran. This has been confirmed by the study of the correlation between the length of residence in Ramsar and the total number of CAs and the mean number of gaps (correlation coefficients are 0.543 and 0.515 respectively). No correlation between the length of residence in other areas of Mazandaran and the number of SCE and CAs was found. The mean scored SCEs/cell for northern subgroups were 1.3 - 1.5 times higher than normal individuals of Tehran (p<0.01). The high incidence of SCEs in Ramsar in comparison with Tehran is in disagreement with the results reported by Fazeli et.al.[5]. Within Mazandaran province the mean number of scored SCE/cell in

Gonbad was higher than Ramsar (p<0.05). The distribution of age, older or younger than 30 yeears, and sex in the major two groups was similar. The mean frequency of SCEs and CAs in women and in individuals older than 30 years was higher, but no correlation between SCE and CAs in regard to sex and age was found. These results suggest that geographical differences in natural radiation level in northern subgroups does not cause significant increase in the incidence of SCE in these groups in comparison with each other, while Ramsar with high level of natural radioactivity showed significant increase (p<0.01) in gaps and the total number of CAs/100 cells.On the other hand significant increase in CAs and SCEs in northern subgroups are consistent with the increase of the incidence of cancer in these areas. Therefor it can be concluded that CAs and SCEs is higher in population who are genetically susceptible to chromosomal damage and probably are consistently exposed to high levels of chromosomal damaging agents such as ionizing radiation(e.g. in Ramsar) and chemical carcinogens. The significance of high level of chemical carcinogens such as Ni, one of the carcinogenic trace elements, and low level of Mn and Mg, two anticarcinogenic trace elements has been studied and confirmed in these areas[9].

REFERENCES

1. Alpen, E.L. "Radiation Biophysics", Prentice-Hall International (eds.), N.J., 283 - 299, 1990.
2. Carrano, A.V.et.al, Nature, 271 (1978) 551 - 553.
3. Sohrabi , M., Proceeding of ICHLNR, Ramsar, 39 - 47, 1990.
4. Sohrabi, M.et.al., Proceeding of ICHLNR, Ramsar, 425 - 432, 1990.
5. Fazeli, T.Z. et.al., Proceeding of ICHLNR, Ramsar, 459 -463, 1990.
6. Heim, S. and Mitelman, F. " Cancer Cytogenetics", 2nd.ed. Wiley- Liss Inc. U.S.A., 9 - 33, 1995.
7. Nadim, A.,et.al. "Global Geocancerology", Chapter 16, 241-252, 1985.
8. Venitt, s., Parry, J.M., "Mutagenecity testing - A Practical approach." IRL Press., 1984.
9. Azin, F.and Raie, R.M., et.al., "Correlation between the levels of certain carcinogenic and Anti - carcinogenic trace elements and Esophageal Cancer in Northern Iran (In press).

VI Radiation hormesis

The radiation paradigm regarding health risk from exposures to low dose radiation

Tsutomu Sugahara

Health Research Foundation,
Pasteur Building 5F, 103-5 Tanaka-Monzen-Cho, Sakyo-ku, Kyoto 606, Japan

1. INTRODUCTION

Health effects of ionizing radiation on human populations have been studied most extensively on Japanese A-bomb survivors. But since the studies are all on the population exposed to a single exposure to high doses, now there are numerous studies in the literature of human populations to low doses and at low dose rates. Studies on populations in high background radiation areas to be discussed in this conference are one of them. The studies are unique in the sense that the population under study have lived for many generations in areas of high background radiation while other populations under study such as radiation workers or people living in contaminated area with fallout from nuclear weapon tests or nuclear facility accidents have been exposed to radiation for limited time. Radiation sources are stable thus can be measured repeatedly to confirm individual doses.

Such studies, of course, have many difficulties in methods and interpretations as will be discussed in this conference. In spite of these difficulties it is an important finding that the populations who have lived for generations in the high background area in China and India do not show noticeable difference in cancer mortality but show an significant increase in the incidence of chromosome aberrations in peripheral blood lymphocytes in adults. We have to consider it seriously in estimating health risks at low doses.

Here, I would like to discuss this problem in the context of radiation paradigm for radiation protection.

2. A BRIEF HISTORY OF RADIATION PARADIGM

Radiation protection standard was based on a threshold model of radiation effects as usual biological responses until 1950. After ICRP recommendation in

1950 which applied this model excluding genetic effect many genetisists appealed against the recommendation and published their recommendation on genetic effects of radiation which was based on a linear model for mutation induction by radiation. Leukemia and life-shortening by radiation were added to this recommendation based on their somatic mutation hypothesis and were included in the next recommendation. Since 1958, ICRP has adopted the linear model for its radiation protection standard though main hazards of radiation now placed on cancer induction but not on genetic effect and life-shortening. These processes were discussed previously [1, 2].

Cancer risk at low doses has been estimated by a linear extrapolation from the data at high doses, mostly those of A-bomb survivors. As mentioned in the introduction the radiations from A-bomb were a single instantaneous ones and quite different from daily exposures to radiation workers and population in general. Data on the population in high background area are appropriated in this aspect but have not been accepted for direct risk assessment.

Recent radiobiological studies have revealed many new insights at low dose range which are different from the results obtained previously at high doses and by short term observations. Adaptive responses are observed after 1-10cGy irradiation and genetic instability after many cell divisions. Some biological responses are better understood by assuming whole cells as targets rather than DNA as a target of hit. Furthermore many epigenetic changes have been observed in radiation responses as reviewed previously [3]. In summary we have now many contradictory data and new biological findings to the present linear model of radiation-induced carcinogenesis as schematically shown in Fig. 1.

3. CHARACTERISTICS OF RADIATION-INDUCED CARCINOGENESIS

Before presenting an alternative model to the current one, i.e., the mutation directly induced by radiation initiates cancer, the observations obtained so far on radiation-induced carcinogenesis are analyzed briefly. The current model for the whole processes from the initiation to clinical cancer has been well summarized by Mendelsohn [4] in his paper entitled "A simplified reductionist model for cancer risk in Atomic bomb survivors" as follows: In view of the increasing evidence for multiple oncogene and suppressor gene changes in human cancer, as well as the evidence that human cancer rate is often proportional to age to the power of 6 or so, it is postulated that the radiation has contributed one and only one oncogenic mutational event to the radiation induced cancers. The radiation induced cancers should therefore display a cancer rate versus age relationship that has a power of n-1, where n is the power for the corresponding background cancers. It is shown that this is precisely what is happening in the collective solid cancer incidence of the atomic bomb survivors.

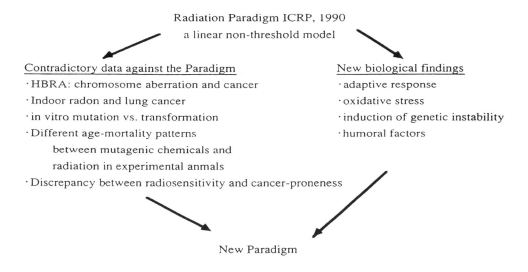

Fig. 1 Contradictory data and new findings so far obtained

This is apparently very persuading model. but as he himself mentioned in his discussion the model has biologically serious problems. It assumes that all solid cancers have similar multistage behavior, and that the stages are independent of order, representing gene mutations which have roughly identical sensitivity, rate over time, and hit number. In addition it assumes one mutation per exposure irrespective of radiation dose per exposure. These assumptions are quite unlikely in the sense of complex biological processes such as carcinogenesis and randomness of radiation-induced mutations. Very low dose rate effectiveness factor so far observed is contradictory to somatic mutation by radiation which has been known to have a clear dose rate effect.

Now characteristics of radiation-induced cancer are simply summarized as follows:
(a) The latency for solid cancers is characteristically longer than that for leukemia, and the age-distribution of the induced cancers tends to resemble that of "naturally" occurring cancers of corresponding types [5].
(b) Radiosensitivity of somatic cells and radiation-induced cancer susceptibility does not correlate well as previously assumed. [6]
(c) Virus-related malignancies such as adult T-cell leukemia and cervical cancer are not affected significantly by radiation [7, 8].
(d) Significant increase of malignancies after a single irradiation can be observed above 20 cGy [9].
(e) Cancer cells have multiple mutations which would be induced along a definite sequence [10].

The current model as summarized by Mendelsohn fits some of those characteristics but not all.

4. EPIDEMIOLOGICAL STUDIES ON POPULATION EXPOSED CHRONICALLY TO LOW DOSE RATE RADIATIONS

As referred in the introduction briefly there have been published many epidemiological studies on populations exposed chronically to low dose rate radiations as shown in Table 1. Dose and dose rate are different in different populations. Thus by combining these data we can cover a wide range of dose and dose rate. Unfortunately, however, none of them are mature enough to reach a definite conclusion on the cancer risk of low dose radiations.

Table 1
Epidemiological studies on populations exposed to low, low dose rate radiations

Populations	Main studies in
Nuclear facillty workers	U.K., U.S.A, Japan
Populations under fallouts of experimental nuclear detonatlons	SCOPE-Radtest 1994-96
Effects of Chemobyl accident	IAEA, WHO, OECD and others
Population near nuclear facllitles	U.K, U.S.A.
High natural background area	China, India and so on
Medical X-ray workers	Japan, China, U.S.A.
Air plane cabin crew	Finland, Canada, U.S.A.
In door radon	Sweden, U.K., U.S.A., China

Among them the populations in high background radiation area are unique in the sense that the radiation source and population there are stable enough to make dose estimation and mortality survey as accurate as possible. In other populations we often have to make a dose reconstruction based on memory and model cases. Even in high background cases we have a lot of problems as discussed in this Conference. But by expanding the dose range and population size available and also by standardizing research protocol and methods adopted through the collaboration of many countries we can improve the accuracy of measurement and statistical power of the studies.

In Yangjiang, China and Kerala, India rather large epidemiological studies are going on. As far as I know, the results in the two area are quite similar in the sense that the incidence of chromosome aberrations in peripheral blood lymphocytes were observed in the adult in high background radiation area while no increase in cancer mortality (China) or morbidity (India) are observed. This apparently contradictory data to the current somatic mutation theory on carcinogenesis may be due to the difference in statistical power in the two studies or may really indicate the independence of radiation-induced somatic mutation on cancer. In the latter case, the incidence of chromosome aberrations can be used as a biological dosimeter but not a predictor of cancer. The above-mentioned data and their interpretation are referred in Fig. 1 as one of the contradictory data against the current paradigm.

In addition, such epidemiological studies would have a great impact on the perception of general public on radiation safety.

5. PROPOSAL OF AN ALTERNATIVE MODEL

Since, at low doses, it is now impossible to obtain cancer risk directly from epidemiological data, a biological model of carcinogenic processes must be adopted for extrapolation from high doses to low doses. Currently a DNA-damage-mutation model has been adopted widely. Here I will propose an alternative model to this.

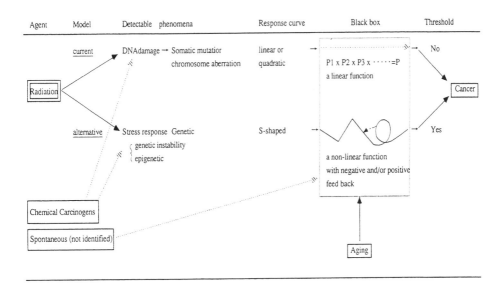

Fig. 2 Current and alternative models for radiation carcinogenesis

We recently know that at low doses radiation induces stress response such as adaptive response, signal transduction, and/or epigenetic changes. These are often followed by genetic instability which results in chromosome instability, delayed mutations and gene amplification, all related to carcinogenic processes. At present I have no idea what will happen after these changes until finally cells become malignant. I have to say we have a black box here. Probably spontaneous carcinogenic process may be incorporated in genetically unstable cells. The unstableness thus induced by radiation, probably, do not accelerate the carcinogenic process but enhances the process. The model comparing current and the alternative ones is schematically shown in Fig. 2.

In this figure detectable phenomena include DNA damage, somatic mutations and chromosome aberrations as well as stress response, genetic and epigenetic, which are followed by genetic instability. It is different in the two models how these phenomena connect with the black box. In the current model a simple linear function is assumed in the black box but in the alternative one a rather complex non-linear function is assumed. This corresponds to my criticism to Mendelsohn's assumption as mentioned previously. Similarly Little [11] proposed an alternative model in which the initial critical step in radiation carcinogenesis is the induction of genetic instability.

In the alternative model we can easily understand the biological meaning of the increased incidence of chromosome aberrations at high background area without the increase in cancer mortality, i.e., the increase in chromosome aberrations is not directly related to cancer but only related to radiation exposure. At very low dose rate no stress response may occur and no cancer may be induced. There are reports that adaptive response can be observed in normal cells but not in malignant cells [12,13]. These are well correspond to the inefficiency of radiation to virus-related cancers as mentioned above.

It is quite natural to assume a non-linear function in the black box since in many biological systems such as immunological and hemopoietic systems we always find very complicated feedback regulations. Chemical carcinogens especially genotoxic chemicals seem to be different from radiation which is a rather weak mutagen. Chemicals appear to fit the Mendelsohn's model well and result in the shortening of latency as shown in NCRP report No.96.

Unsolved problems in this model are spontaneous carcinogenesis and incorporation of aging process in the model. The both are fundamental problems in biology. I hope that studies on radiation carcinogenesis would contribute to the understanding of this kind of fundamental problems.

6. PARADIGM AND ITS SHIFT

The paradigm for radiation protection,i.e., the biological bases how we can estimate health effects of low dose irradiation as encountered in daily life and in radiation works , is one of the principal problems in environmental safety. Now we have many biological data which seem to be contradictory to the current paradigm. Unfortunately, however, we have no alternative model approved by scientists and accepted by the public. In this presentation I have proposed a model as mentioned above to promote the shift of the paradigm. My model is quite premature and incomplete. But I hope this would stimulate the discussion on the current paradigm.

In another aspect I am suspicious in the notion that even natural background radiation is harmful according to the current paradigm, because human beings have been evolved for many years under continuous exposure to the background radiations. They must be adapted to this condition. Studies on the populations in high natural background radiations may contribute to elucidate the tolerable limit of radiations for human being under the adaptation. All studies so far reported have demonstrated a threshold for the induction of stress response. The proposed model supports my suspicion.

Extrapolation to unknown region ,i.e., low dose level, for risk assessment has always been done from high dose level but the extrapolation from low natural radiation level up to somewhat higher level should be tried as well. In this case homeostasis of biological systems in question and its limit should be studied. Unfortunately, however, there have been published few papers concerning carcinogenesis and homeostasis. Recently C.S.Potten and his colleagues [14] studied the spontaneous and radiation-induced apoptosis in the gastrointestinal tract in mice and have concluded that it is possible that the spontaneous apoptosis observed in intestinal epithelia is not the consequence of infidelity of normal DNA replication or due to the occasional damage induced by background radiation or other potentially cytotoxic damage which imposed p53-responsive DNA double strand breaks. Incidence of apoptosis independent on the existence of p53 gene is the main reason of this interpretation . Further study is urgently needed to clarify the role of natural radiations in homeostasis.

General public has a fear to radiation. The risk at very low doses of radiation has never been demonstrated but only estimated based on the present paradigm. ICRP always emphasizes that the current paradigm has been adopted for protection . But the agency has never taken care of public perception on the above statement, especially when the agency applied the protection standard to the public. The agency insisted the safety but as a countervailng risk the public has had a fear to radiation which could not be supported by scientific data. I hope the studies on populations living in high background radiation area would be a good example of no health hazard by low dose irradiation.

7. FUTURE SCENARIOS AS A CONCLUSION

Kuhn,T.S. who first proposed the word, paradigm, in science in 1962 clearly wrote how difficult it was for the paradigm to shift. We are really experiencing now what he said. According to the method of futurology I would like to propose three scenarios again [15].

(1) Preserving the present non-threshold linear model

It is very hard to expect a paradigm shift because most of radiation protectionists are conservative and the present model is simple and easy for them to understand assuming the same mechanism of radiation effect throughout all dose range.

The dose limit will be reduced step by step based on new reports of epidemiological studies which suggest the higher risk of cancer than previous one. Movement against nuclear energy will be elevated by the public who insists on zero risk.

(2) Complete amendment of the present protection standard

A new paradigm that a threshold exists in radiation-induced carcinogenesis as I proposed here. But it will take a long time for a com-pletely new paradigm to be established. Together with biological studies on radiation-carcinogenesis epidemiological studies at high background radiation area and on nuclear facility workers will contribute to establish a threshold level.

However, since it is practically very difficult to verify that there is no effect at all at very low doses, a long debate will continue among scientists who support a new model and who insist on the old one. Anyway, it will take a long time before the new paradigm will be accepted.

(3) Compromised amendment

For safety standard, the present paradigm will be adopted for protection of radiation personnel. But lower risk than present one will be accepted at very low doses near natural background level. For calculating a population dose (man-Sv) a LQ model which will give a very low risk at very low doses will be adopted. Thus, a lower population dose will be estimated than that of the present one. The role of indoor radon in the etiology of lung cancer will be less evaluated. In this case the study on high background radiation area will contribute to the estimation of risks at low dose rate.

Note: After completing this paper as above, I have obtained some important informations concerning the effects of adaptive response. Watanabe et al. reported in this Conference that the preirradiation of 2 cGy reduced the extent of genetic instability observed after challenging dose of several Gy in mouse m5s cells. Azzami et al.(Radiat. Res. 146, 369, 1996) reported that the irradiation of 0.1 or 1.0 cGy and 24 hr incubation reduced the incidence of spontanous transformation in mouse C3H10T1/2 cells. These results indicate that the results of stress response at low doses are different in different cells and conditions. Thus a simple model as I suggetsted cannot explain the complicated phenomena such as carcinogenesis. It is strongly requested to establish low dose radiobiology as a new paradigm based on all the results so far obtained at cGy order radiations. On the contrary the current radiobiology seems to be based on the studies at Gy or higher order of radiations.

REFERENCES

1. T.Sugahara, J.Radiat.Res., 35 (1994)48.
2. T.Sugahara and M.Watanabe, Radiobiological Concepts in Radiotherapy, D. Bhattacharjee and B.B.Singh (eds.) (1995) 1.
3. T.Sugahara and M.Watanabe, !nt.J.Occup.Med.&Toxicol., 3 (1994)129.
4. M.L.Mendelsohn, RERF CR, 1-96 (1996)
5. NCRP Report No.96 (1986)
6. K,Sankaranarayanan and R.Chakraborty, Radiat.Res. ,143 (1995)121.
7. Y.Shimizu, H.Kato H, W.J.Schull. Radiat. Res. ,121 (1990)120.
8. D.A.Pierce, Y.Shimizu, D.L.Preston, M.Veath and K.Mabuti. Radiat.Res. , 146 (1996) 1.
9. Y.Shimizu, H.Kato, W.J.Schull and K.Mabuti. Low Dose Irradiation and Biological Defence Mechanisms T.Sugahara, L.Sagan and T.Aoyama (eds.), (1992) 71-74.
10. E.R.Fearon and B.Vogelstein, Cell, 61 (1990) 756.
11. J.B.Little, Radiation Research 1895-1995, Vol.2, (1996)597.
12. M.S.Sasaki, Int. J. Radiat. Biol. ,68 (1995) 281.
13. K.Ishii and M.Watanabe, Int. J. Radiat. Biol. , 69 (1996) 291.
14. A.J.Merritt, C.S.Potten, C.J.Kemp, J.A.Hickman, A.Balmain, D.P.Lane and P.C.Hall, Cancer Res. , 54 (1994) 614.
15. T.Sugahara, 100 Years of X-rays and Radioactivity, D.D.Sood et al.(eds.) (1996) 367.

Cellular and molecular basis of the stimulatory effect of low dose radiation on immunity*

Shu-Zheng Liu

MH Radiobiology Research Unit, Norman Bethune University of Medical Sciences, 8 Xinmin Street, Changchun 130021, China

It has been observed in human populations and animal studies that low dose radiation (LDR) could stimulate the immunological responses. The up-regulation of immunity following LDR involves a series of cellular and molecular reactions as well as their systemic regulation. The studies in our laboratory and elsewhere in recent years have convinced us that whole-body irradiation (WBI) with X- and -rays in the dose range within 0.2 Gy has definite positive effect on the immune system which can be considered as beneficial to the organism. In the present paper evidence will be given for this statement.

1. CELLULAR REACTIONS IN THE IMMUNE ORGANS AFTER LDR

1.1. Reactivity of T lymphocytes

In an area of high background radiation in Yangjiang, Guangdong, China with radiation exposure rate 3 times as high as the adjacent control area of Enping County, it was first observed in 1979 that the reactivity of the T lymphocytes of peripheral blood to *in vitro* stimulation with phytohaemagglutinin (PHA) was up-regulated (Fig. 1) and the observation was confirmed in 1982 when a repeat examination was made on blood samples from the inhabitants of the same area. Meanwhile, the percentage of the suppressor T cells (T_S) was also counted with a surface receptor method (the Tcells) and no significant difference was found between the samples from the high background radiation group and those from the control. The level of unscheduled DNA synthesis (UDS) of the lymphocytes was assessed by measuring ^{3}H-TdR incorporation in the presence of hydroxyurea and was found to be higher in the blood samples from the inhabitants of the high background radiation area, indicating an increased capability of ecision

---

*This work was supported by grants from NSFC

repair of DNA damage (Fig. 2) [1,2]. These early observations in human populations have led to a series of experimental studies in animal models [3].

WBI of C57BL/6J mice with doses from 0.5 to 6.0 Gy caused a dose-dependent depression of the reaction of the splenic lymphocytes to both concanavalin (Con A) and lipopolysaccharide (LPS) with the D_{37} being 1.60 and 0.65 Gy, respectively, indicating that the B lymphocytes have a much higher radiosensitivity than that of the T lymphocytes. owever, WBI of mice with doses within 0.2 Gy caused stimulation of the reactivity of the lymphocytes to the mitogens, with more marked up-regulation of the response to Con A than to LPS. There has been reports proving a stimulation of the reactivity of rat splenic lymphocytes to Con A 4h following WBI with 50 mGy[4]. It is interesting to note that *in vitro* irradiation of mouse splenocytes could not completely replicate what was observed after WBI. The stimulatory effect of *in vitro* LDR on the reaction of splenic lymphocytes to Con A was very slight, and the reaction of these lymphocytes to LPS showed a dose-dependent depression after *in vitro* irradiation.

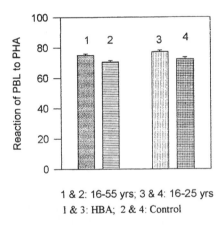

Fig. 1. Increased reactivity of peripheral blood lymphocytes from the inhabitants of high background area (HBA)

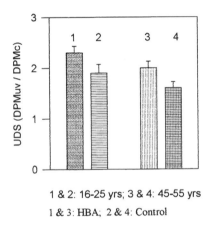

Fig. 2. Enhanced unscheduled DNA synthesis of peripheral blood lymphocytes from the inhabitants of HBA

1.2. Antitumor cytotoxicity

The antitumor activity of the natural killer (NK) cells was measured using ^{125}I-UdR labeled YAC-1 cells as the tumor targets. The cytotoxic effect was found to be markedly enhanced after WBI of mice with 50, 75 and 500 mGy X-rays with an E:T ratio of 200:1. It was reported that examination of the blood samples of the AB survivors in Japan using ^{51}Cr-labled K562 cells as the tumor targets at an E:T ratio of 20:1 showed a significant increase in antitumor cytotoxicity in the individuals exposed at the age above 15 years to doses of 0.01 to 1.00 Gy[5]. The antibody-dependent cell-mediated cytotoxicity (ADCC) was assessed using S180 cells incorporated with ^{125}I-UdR as the tumor targets showing significant stimulation of this killer (K) cell activity after WBI with 75-100 mGy X-rays. The specific cytotoxic T lymphocyte (CTL) activity in mice bearing Lewis lung cancer against these cancer cells *in vitro* was stimulated after WBI of the tumor-bearing mice with 75 mGy X-rays.

1.3. Antibody formation

The plaque-forming cell (PFC) reaction was examined using the monolayer liquid phase technique with sheep red blood cells (SRBC) as the antigen. This reaction is highly radiosensitive with a D_{37} value of 0.95 Gy. WBI of mice with low doses, however, caused stimulation of the PFC reaction of the splenic cells which was most marked after a dose of 75 mGy X-rays at a dose rate of 12.5 mGy/min and after a dose of 72 mGy of -rays at a dose rate of 68 μGy/min. Chronic irradiation of mice in a ^{60}Co -field with a dose rate of 15 μGy/min (6h/d and 6d/week) also caused stimulation of the PFC reaction of the spleen at cumulative doses of 32.4 and 65 mGy. Under the above conditions the PFC reaction was depressed when the single exposures exceeded 0.1 Gy and the chronic exposures exceeded 0.39 Gy.

1.4. Macrophage activity

The macrophages (M) have two categories of immune functions, one being nonspecific and not MHC-restricted and the other being antigen-specific and MHC-restricted. Of the former category the phagocytic and digestive activities as well as the secretion of IL-1 were examined after LDR. It was found that the phagocytosis and digestion of the chicken red blood cells (CRBC) by the peritoneal M's were enhanced after WBI of mice with 50-75 mGy X-rays. At the same time the secretion of IL-1 by the splenic M's was stimulated in the course of one week after WBI with 75 mGy X-rays with most prominent changes occurring on days 1-2. The cooperative effect of the M on T lymphocytes was assessed in a coculture system in which macrophages from the spleen of irradiated or sham-irradiated mice were mixed with splenic lymphocytes from the irradiated animal and the proliferative response of the T lymphocytes was measured by ^{3}H-TdR incorporation in the presence of Con A. It was found that WBI with 75 mGy X-rays caused significant enhancement of the cooperative effect of the macrophages on the T cells.

1.5. Cytokine secretion

As mentioned above the secretion of IL-1 by the M's was stimulated after LDR. The increased production of IL-1 provides a maturation signal to the helper T lymphocyte (HTL) in the thymus and up-regulate its release of colony-stimulating factors (CSF) (Fig. 3) which in turn give stimulatory signals to the macrophages resulting in a positive feedback loop of enhancing effect on the HTL. The latter will increase its production of IL-2 upon antigenic or mitogenic stimulation. It was found that IL-2 production by HTL was up-regulated after LDR together with increased expression of its receptors (IL-2R) thus giving rise to signals for the clonal expansion of the T cells. At the same time the secretion of -interferon (IFN-) was also enhanced by LDR(Fig. 4). In the high background radiation area it was also found that there was an increased frequency of IL2-producing cells in the peripheral blood in the inhabitants in comparison with those of the control area[6]. The increased secretion of IL-2 and IFN- are important regulatory factors promoting the cytotoxic effect of the CTL and NK cells on the tumor cells.

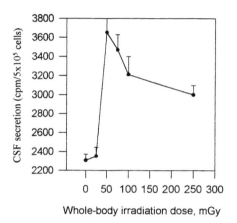

Fig. 3. Increased secretion of CSF by mouse thymocytes after LDR

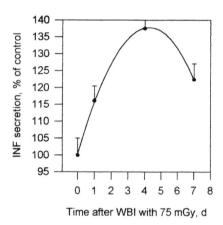

Fig. 4. Stimulated secretion of IFN-γ by mouse splenic lymphocytes after LDR

1.6 Thymocyte maturation and proliferation

The changes in the thymus are of utmost Importance in the understanding of the up-regulation of T cell functions after low LDR. Flow cytometric analysis showed that WBI of mice with 75 mGy X-rays caused a slight increase in percentage of the CD4/CD8 double negative thymocytes, up-regulation of the expression of the TCR/CD3 complex on the thymocytes and increase in the proportion of S phase cells, suggesting stimulation of cell renewal, maturation and proliferation within this central immune organ of the T cells. Meanwhile, the spontaneous incorporation of ^{3}H-TdR into the thymocytes and the number of thymocytes were found to be higher than the sham-irradiated control giving further evidence to the enhanced process of cell proliferation in the thymus. All these changes in the thymus after LDR may give rise to increased supply of mature T cells to the periphery in response to challenge[3].

1.7 Targets of LDR in the immune system

The immune system as a whole is ranked as highly radiosensitive among various tissues in the body. However, the cellular constituents of the immune organs are very heterogeneous in response to ionizing radiation. In the order of radiosensitivity the B cells rank first followed by T_S, T_H and M, and in the thymus the CD4/CD8 double positive thymocytes have the highest radiosensitivity followed by the CD4/CD8 negative, CD8 and CD4 single positive thymocytes. It was speculated that LDR would preferentially damage the T_S cells resulting in a rise in the T_H/T_S ratio which would lead to a preponderance of the helper function and thus immunologic stimulation [7]. However, flow cytometry with double immunofluorescence and functional tests did not give support to such an assumption[3, 8, 9]. It was observed that following WBI with 75 mGy X-rays the most radiosensitive double positive thymocytes did not change and the ratio of CD4+ to CD8+ cells only increased with doses above 1 Gy. Another speculation is the "altruistic cell suicide" or increase of apoptosis of the more primitive thymocyte subset by LDR leading to an increase of proliferation of the more mature ones and thus stimulation of immunity[10]. But up to the present there have not been any firm experimental data supporting this hypothesis, and, on the contrary, flow cytometric analysis of the apoptotic bodies and fluorescence analysis of the DNA fragmentation of the thymocytes showed that only doses above 0.2 to 0.5 Gy caused definite increase of apoptosis of mouse thymocytes and lower doses might even cause a decrease of thymocyte apoptosis(Fig. 5)[3, 11]. Therefore, the speculations of the preferential deletion or damage of the relatively more radiosensitive cell elements by LDR with secondary increase in function or number of the others have not gained experimental support. Data presented in this paper would rather suggest that the target cellular elements of LDR might be the T_H and M (including the dendritic cells) with their functional stimulation. Further evidence will be given at the molecular level (vide infra).

2. MOLECULAR BASIS OF THE CELLULAR CHANGES

The molecular mechanism of immunoenhancement after LDR has only been studied recently, so its complete picture has to be clarified by further work. The existing data point to the significance of facilitation of the signal transduction process in the immune cells after LDR. Here only a sketchy outline will be given as follows.

2.1. Up-regulation of expression of TCR/CD3 on the thymocytes

It has been disclosed that after WBI of mice with 75 mGy X-rays the expression of the TCR and CD3 molecules on the thymocytes increased with time in parallel within the first 24h(Fig. 6) together with a gradual increase in percentage of thymocytes bearing high density TCR and CD3 molecules(not shown in the figure). It is known that the TCR/CD3 complex is the key receptor of the T cells receiving external stimuli. The rearrangement of the TCR gene and its expression on the T cells, just as that of sIg on the B cells, is the molecular basis for the T cells to

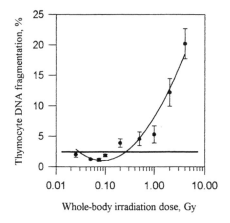
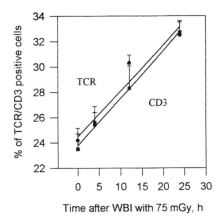

Fig. 5. Dose-effect curve of apoptosis of mouse thymocytes (14h after WBI)

Fig.6. Expression of TCR(dot) and CD3 (square) molecules on thymocytes after LDR

recognize the infinite variety of antigens. The CD3 molecule functions to transduce the signals received by TCR. The up-regulation of the expression of TCR/CD3 complex after LDR is a sign of expedited maturation of the T cells on one hand and provides the basis of facilitation of signal transduction into the cytoplasm in response to antigenic and mitogenic stimulation on the other.

2.2. Mobilization of intracellular free Ca^{2+} and activation of protein kinase C

Following WBI of mice with 75 mGy X-rays the intracellular free Ca ions ($[Ca^{2+}]_i$) in the CD4 and CD8 single positive T subsets in both the thymus and spleen was increased, but in the thymic T subsets its increase to the level above the sham-irradiated control only occurs after Con A stimulation while in the spleen its basal level was higher than the control before mitogenic stimulation. The time course of $[Ca^{2+}]_i$ mobilization in response to anti-CD3 monoclonal antibody after LDR was found to coincide with that of the up-regulation of the TCR/CD3 complex on the cell surface after LDR. Meanwhile, the time course of the opening time and probability of the Ca2+-activated K^+ channels of the thymocytes in response to Con A after LDR was also found to parallel that of both the $[Ca^{2+}]_i$ mobilization and TCR/CD3 expression[12]. These data indicate the dependence of the changes of ion channels on the surface and those of signal molecule Ca^{2+} in the cytoplasm on the up-regulation of the expression of TCR/CD3 molecules of the thymocytes after LDR.

The total activity of PKC in the splenic homogenate as well as in the separated T and B cells was markedly increased with that in the T cells reaching its peak 12h after WBI with 75 mGy X-rays and that in the B cells on day 4[3]. Flow cytometry with immunofluorescence showed that the expression of protein kinase C (PKC), 1 and 2 in mouse thymocytes was significantly up-regulated 12h after LDR (Fig. 7). The mobilization of $[Ca^{2+}]_i$ and activation of PKC are two important links in the cascade of signal transduction leading to the induction of the immediate early genes in the T cells.

2.3. Induction of c-fos gene and expression of Fos protein

Dot blot hybridization showed that WBI of mice with 75 mGy X-rays led to a rise of the mRNA level of c-fos in both the thymus and spleen[3]. Further analysis of the induction of the c-fos gene and expression of the Fos protein in the immune organs using in situ hybridization and immunohistochemistry, respectively, disclosed the time sequence and localization of these molecular changes after LDR. The up-regulation of c-fos mRNA in the thymus and spleen in response to LDR reached its peak at 1h and 2h, respectively, and returning to the basal level within 12h, followed by increased expression of the Fos protein which began at 2h, peaking at 24h and returning to the control level within 72h[13]. It should be pointed out that the induction of the c-fos gene and the expression of the Fos protein were most marked in the dendritic or inderdigitating cells and the large lymphocytes.

2.4. Expression of Bcl-2 and the ratio of Bcl-2/BAX

Immediately following the increased expression of Fos protein the expression of Bcl-2 protein in the immune organs was up-regulated, reaching its peak at 12-24h (Fig. 8), and the Bcl-2/BAX ratio was higher than control 24h after WBI with 75 mGy X-rays (Fig. 9). It is known that Bcl-2 serves as a "brake" in the signal pathway of apoptosis activated by different genotoxic agents and the homodimer of BAX protein has the action of promoting apoptosis. When Bcl-2 forms a heterodimer with BAX protein, the pro-apoptosis action of the latter is alleviated. Thus the above mentioned experimental results of a rise of the Bcl-2/BAX ratio would lend support at the molecular level to the decrease of thymocyte apoptosis after LDR [3, 11, 13].

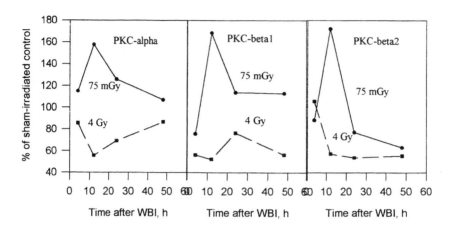

Fig. 7. Changes in PKC, PKC1 and PKC2 in mouse thymocytes after WBI

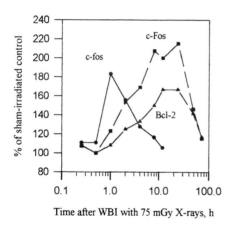

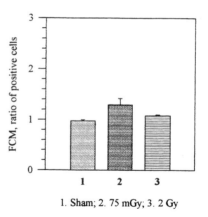

Fig. 8. Induction of c-fos and expression of c-Fos and Bcl-2 proteins in the thymus after LDR

Fig. 9. Bcl-2/BAX ratio changes after low dose versus high dose WBI with X-rays

2.5. Expression of p53

Increased expression of the wild type (wt) p53 is a critical step in the molecular pathway of radiation-induced apoptosis. After WBI of mice with 2 Gy X-rays the expression of wtp53 was markedly increased[13]. It has been shown that WBI with 2 Gy X-rays also led to an increased rate of apoptosis of the thymocytes expressed as increased DNA fragmentation and apoptotic bodies[3, 11, 15]. Wtp53 is generally recognized as a "guardian of the genome", recently also called "cell cycle watchman"[14], since wtp53 suppresses the G1 to S phase transit which would allow a chance of repair of DNA damage caused by high dose genotoxic agents before DNA replication, thus preventing the passage of possible gene mutations to the daughter cells. WBI of mice with 2-4 Gy X-rays caused increased expression of wtp53 and G1 arrest while 75 mGy X-rays down-regulated the wtp53 expression in the thymocytes (Fig. 10). This down-regulation of wtp53 expression in the thymocytes after LDR is not necessarily accompanied with cell cycle changes in the opposite direction to those observed after high dose radiation and even a slight increase in the percentage of G1 cells with concomitant slight decrease in percentage of S phase cells may be observed [15].

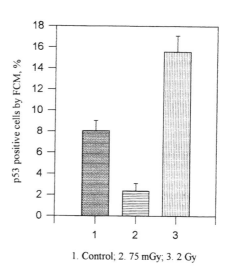

Fig. 10. Flow cytometric analysis of mouse thymocytes 24h after WBI with low and high doses of X-rays using indirect immuno-fluorescence with monoclonal antibody against wtp53 as the first antibody and FITC-conjugat-ed goat-anti-mouse-IgG as the second antibody

1. Control; 2. 75 mGy; 3. 2 Gy

The significance of such observations can only be realized with parallel studies on other molecular changes to disclose the temporal relationship of the induction of different genes after different doses of radiation. It can only be stated at present that low versus high dose radiation may cause entirely different changes in the thymocytes at the molecular level which may open a new avenue for further exploration.

2.6. Activity of transcription factors

The binding of transcription factors to certain DNA sequences is an important mechanism of initiation of gene induction. Data on the effect of ionizing radiation, especially low dose radiation, on the binding activity of transcription factors are scanty[16-18]. It was pointed out by Woloschak et al that genes induced by high dose radiation might be a part of the apoptosis pathway, while genes induced by low dose radiation might have important influence on cell survival[19]. We examined the activity of several transcription factors (NFkB, CREB, AP1, OCT1 and GRE) in the nuclear protein extract from mouse immune organs after LDR using the gel mobility shift assay. It was found that 4h after WBI with 75 mGy X-rays the binding activity of NFkB and CREB was markedly up-regulated, rising to 6- and 4.3-fold of control, respectively, in the thymic extract and 7- and 5-fold of control, respectively, in the splenic extract. The binding activity of AP1 was only slightly increased (about 2-fold of control). No changes were observed for the other transcription factors examined [13].

Data listed above indicate that WBI with low dose X-rays could induce a series of molecular changes, beginning from the expression of surface receptors on the T cells to cytoplasmic and nuclear signals, finally resulting in T cell activation and proliferation. The signal cascade initiated by the up-regulated expression of the TCR/CD3 molecules may only represent one aspect of the signal transduction pathways concerned, and the activation of other pathways simultaneously or in sequence may also exert their effect in concert. For example, it has been found that WBI with low doses could increase the expression of Ras protein and activate the process of phosphorylation of a number of proteins, suggesting that the Ras-Raf-MEK-MAPK cascade may also be involved in the activation of transcription factors. This area certainly deserves further exploration.

3. SYSTEMIC REGULATION

The immune system is under the regulatory influences of the nervous and endocrine systems, with the most prominent effect exerted by the hypothalamic-pituitary-adrenocortical (HPA) axis. There exists a bi-directional regulation between the HPA axis and the immune system. The effect of ionizing radiation on the functions of HPA axis depends on the radiation dose. High dose radiation causes accentuation of the function of the HPA axis manifested as increased secretion of ACTH and corticosteroid hormone, while LDR causes a down-regulation of the HPA axis with lowering of serum ACTH and corticosterone (CS) levels in mice and decreased urinary output of 17-hydroxy-corticosteroids in humans [20, 21]. Normally the HPA axis exerts a tonic suppression on the immune system, especially the thymus. The down-regulation of the HPA axis after LDR would lessen the degree of such tonic suppression on the immune system which may be an important promoting factor in the LDR-induced stimulation of the immune reactions. We also found that WBI of mice with low dose X-rays caused a reduction of the hypothalamic content of met-enkephalin and leu-enkephalin. In situ hybridization showed a lowered transcription of the POMC gene in the arcuate nucleus of the hypothalamus, suggesting that the LDR-induced lowering of serum ACTH and CS might originate from the radiation action on the vegetative nervous system. Interestingly, the transcription of the

POMC gene in the immune organs took a completely opposite direction to that in the hypothalamus in the course of 12h after WBI with 75 mGy X-rays. These observations disclose the molecular mechanism of the lowering of serum ACTH and CS on one hand and indicate the down-regulation of the function of the HPA axis being accompanied with stimulation of immune functions on the other. It is known that the cellular elements of the immune system (M's and T and B cells) express ACTH receptors. The up-regulation of POMC transcription level in the immune organs after LDR might be the result of a positive feedback reaction to the lowering of serum ACTH content.

4. CONCLUDING REMARKS

The studies on the induction of adaptive responses by LDR, including stimulation of DNA damage repair and enhancement of immunologic reactions, not only promote theoretical explorations in radiation biology, but also raise the possibility of practical application of the experimental data in clinical trials as well as in re-evaluation of the existing themes or paradigms in radiation protection. The mostly concerned point to the public and academic spheres is the cancer risk of low level exposures to environmental radiation. The present hypothesis on risk assessment of carcinogensis by low level radiation is chiefly based on assumption of the low dose effects by extrapolation from the dose-response relationship established with observations from the effects of high or medium doses with high dose rates. The present concept holds that carcinogenesis arises from genetic changes in single cells with characteristics of multi-stage development. Previous data from *in vitro* studies on cell transformation suggested that the dose-effect relationship of induction of malignant changes in certain cell lines may conform with the no threshold hypothesis, but a recent report shows that malignant transformation of C3H10T1/2 cells at low doses may be significantly lower than the spontaneous transformation level in the control [22]. And it should be pointed out that *in vitro* transformation is not equivalent to cancerogenesis. It is a long process with complicated cellular and molecular changes, which are subjected to systemic modulation, to go through genetic mutation via precancerous clonal formation to end in the appearance of clinical cancer. And many details of this process are not yet fully understood. The primary action of the immune system is to protect the body from infection and tumor growth, and enhancement of immunologic surveillance would exert a suppressive effect on cancer development. The stimulatory effect of LDR on immunity has to be assessed by challenge experiments. It has been found that WBI with low dose X-rays could retard tumor growth and prevent tumor metastasis, and pre-exposure of mice to LDR could prevent the carcinogenic effect of high dose radiation[23, 24]. Preliminary trials of total body or half body low dose irradiation have been introduced to the clinic in the treatment of cancer. Such regimens have been found to increase the efficacy of chemotherapy of non-Hodgkin's lymphoma [25]. We have observed in mice that such promoting effect of LDR on cancer chemotherapy may be closely related to its stimulatory effect on anticancer immunity.

In the high background radiation area cancer mortality rate in the population has been found to be slightly lower than the control and the difference is statistically significant in the age group of 40-70 years [26]. It may be true that the statistical power is not strong enough for epidemiological survey to reach a definite conclusion of whether or not a threshold for cancer induction exists with low level exposures to radiation, but work along this line could possibly accumulate valuable scientific data to shed light on this important and much concerned problem.

REFERENCES

1. S.Z. Liu, P.X. Xiao, S.Y. Ma, et al., Chin. J. Radiol. Med. Protect , 2 (1982) 64-70.
2. S.Z. Liu, G.Z. Xu, X.Y. Li, et al., Chin. J. Radiol. Med. Protect, 5 (1985) 129-140.
3. S.Z. Liu, Radiation Hormesis with Low Level Exposures. Beijing, Science Press, 212-316, 1996.
4. S. Hattori, Int. J. Occup. Med. Toxicol., 203-217, 1994.
5. O. Zhou, M. Akiyama, Y. Kusunoki, et al., Chin. J. Radiol. Med. Protect, 11 (1991) 20-23.
6. J. Yao, Y.R. Cha, Z.X. Lin, ISBEELES '93, Changchun, China, 64.
7. R.E. Anderson, I. Lefkovits, G.M. Troup, Contemp Top Immunobiol 11 (1980) 245-274.
8. S.J. James, T. Makinodan, Int. J. Radiat. Biol., 53 (1988) 137-152.
9. Z. Du, L. Su., J. Radiat. Res. Radiat. Proc., 12 (1994) 59-60.
10. S. Kondo, Int. J. Radiat. Biol., 53 (1988) 95-102.
11. S.Z. Liu,.Y.C. Zhang, Y. Mu, et al. Mutat. Res., in press.
12. C.W. Zhang, G. An, G.G. Zhong, et al., Chin. J. Radiol. Med. Protect , 16 (1996) 154-156.
13. S.Z. Liu, H. Wan, S.L. Chen, J. Norman Bethune Univ. Med. Sc., 22 (1996) 553-559.
14. T. Jacks, R.A. Weinberg. , Nature, 381(1996) 643-644.
15. S.Z. Liu,.Y.C. Zhang, Y. Mu, et al. J Norman Bethune Univ. Med. Sc., 21 (1995) 551-558.
16. A.V. Prasad, N. Mohan, B. Chandraseka, et al. , Radiat. Res., 138 (1994) 367-372.
17. S.P. Singh, M.F. Lavin. , Mol. Cell. Biol., 10 (1990) 5279-5385
18. B. Teale, M.F. Lavin. ,Radiat. Res., 138 (1994) S52-S55.
19. G.E. Woloschak, C-M. Chang-Liu, J. Panozzo, et al. , Radiat .Res ., 138 (1994) S56-S59.
20. S.Z. Liu. , J. Norman Bethune Univ. Med. Sc., 4 (1978) 1-5.

21. S.Z. Liu, Y. Zhao, Z.B. Han, et al. , Chin .J. Radiol. Med . Protect , 14 (1994) 11-14.
22. E.I. Azzam, S.M de Toledo, G.P. Raaphorst, et al., Radiat. Res ., 146 1996) 169-172.
23. X.Y. Li, Y.B. Chen, F.Q. Xia, et al. , Chin. J. Radiol. Health, 5 (1966) 21-23.
24. X.Y. Li, Y.J. Li, H.Q. Fu, et al. ISBELLES '93, Changchun, China, p103.
25. Y.Takai, S.Yamada, K. Nemoto, et al. in: Sugahara T, Sagan LA, Aoyama T.(eds.), Low Dose Irradiation and Biological Defense Mechanisms. Amsterdam, Elsevier Science Publishers, 113-116, 1992.
26. L.X. Wei, J.Z. Wang. , Int. J. Occup. Med.Toxicol.,3 (1994) 195-201.

A survey of long-term and low dose radiation effect on SIL-2R of residents

Ziyi Zeng, Yuyou Cheng, Gang Zeng, Wenting Lu, Zhing Li and Shue Zhang

Guizhou Provincial Institute of Radiation Protection, Guiyang 550004, P.R. China

At a mine site a large number of residential houses for staff and workers were built on the slag which contains radioactive nuclides. The residents received long-term and low dose external radiation. In order to observe the radiation immune function, we carried out a survey on the soluble interleukin-2 receptor of the residents. The results are as follows.

1. METHOD

1.1. Survey subjects

By means of random sampling 31 females living on the slag were investigated (the survey group). Their average residing time is 23.8 years. A control group of 35 females lives at a distance of 1 km from the survey group, their living quarters have no slag. The average residing time is 24.4 years.

1.2. Radiation doses of environment

The indoor and outdoor external radiation doses are $(53.99 \pm 26.12)(10^{-8}$ Gy/h and $(65.33 \pm 27.26) \times 10^{-8}$ Gy/h, respectively. The average individual effective dose equivalent for the survey group is 3.75 ± 10^{-3} Sv. These for the control group are $(12.34 \pm 1.69) \times 10^{-8}$ Gy/h, $(11.46 \pm 1.09) \times 10^{-8}$ Gy/h and 1.05×10^{-3} Sv, respectively. The houses of the two groups are constructed by brick and the ground made by cement.

1.3. Experimental method

The SIL-2R was determined by enzyme-labeled immunosorbent assay (ELISA). The reagent was provided by Beijing Bangding Biomedical Co. of Military Medical Academy. The method is briefly described as follows: The enzyme labeling plate is coated by single ant-SIL-2R, and sealed with serum albumin. Add the sample. There are two parallel wells and add the reference standard for each plate. After incubation, add consecutively ant-SIL-2R and enzyme labeled antibody. Finally add substratum to develop color. According to the reference criterion OD value, the content of SIL-2R in the sample is calculated.

2. RESULTS

There is statistically significant difference in the levels of serum SIL-2R between the survey and control groups under 50 years of age, while no significant difference between the two groups over 51 years of age (Table 1).

Table 1
The levels of serum SIL-2R in the residents (ku/l)

Group		No. of cases	Average age	X±S	P
Under 50 of age	Survey group	16	37.9	158.06±106.89	<0.05
	Control group	7	42.1	277.57±124.33	
Over 51 of age	Survey group	15	53.7	198.00±128.58	>0.05
	Control group	28	56.4	138.89±68.43	

These results are close to those reported by Bai [1] and Zhang [2], 193.70 ± 92.55 ku/l and 241.85 ± 58.47 ku/l, respectively.

3. DISCUSSION

Since Robin et al.[3,4] first reported in 1985 that the lymphocytes of peripheral blood activated *in vitro* have SIL-2R expression not only on their cell membrane but also in the substrate of the culture. The soluble interleukin-2 receptor is an important immuno-inhibiting factor and has important immno-regulating function *in vivo*. The serum SIL-2R level may reflect the base level of immune system activation under physiological condition. Especially, the negative feedback regulation of SIL-2R immune response aroused great interest by scholars. The assay of the content of SIL-2R in the peripheral blood may make clear not only the immune condition but also the seriousness and interrelation of the disease.

Based on the environmental monitoring doses, the ratio of indoor- to outdoor-external radiation doses for the survey and control groups are 4.4:1 and 5.7:1, respectively. Although the residents of survey group received a long-term and low external radiation, the results of SIL-2R assay were within the limits of national normal reference values. The SIL-2R level in the survey people of 51 years of age is higher than that in the control group of the same age, but there is no significant difference (P>0.05). The SIL-2R value in the control group under 50 years of age is higher than that in the survey group of the same age and the difference is significant (P<0.05) because one case in the control group has a high SIL-2R value (480 ku/l), which may affect the average value and the test of significance.

It was confirmed that the effect of low dose irradiation on the whole body may enhance the immune reaction and the excitement of immune system, but the information about whether or not low dose (external radiation affect the SIL-2R of residents is not available in the literature at home. The results of our investigation may be attributed to the following causes: 1) the dose level is not high enough to cause the change of SIL-2R; 2) the small number of surved people cannot reflect the minor difference between the two groups. The effects of low dose (external radiation on SIL-2R of residents needs further study.

ACKNOWLEDGMENT

This work was supported by research grant from Guizhou Provincial Health Department and we thank vice-chief technician Dianxi Luo for his kind technical help.

REFERENCES

1. Bai Nan, Kong Xiantao and Jiang Yunwei, Clin. Lab. Sci., 12 (1994) 30.
2. Zhang Jingxiu and Zuo Dapeng, Chin. Med. Lab.Sci., 19(1996) 54.
3. Liu Shuzheng, Xu Guizheng, et al. ,Chin. Radiol. Med. Prot., 5(1985) 124.
4. Liu Shuzheng. ,Chin. Radiol. Med. Prot. , 15 (1995) 217.

Immune competence and immune response to virus in high background radiation area, Yangjiang, China

J.-M. Zou, J. Yao, N.-G. Chen and Y.-R. Zha

Guangdong Institute of Prevention and Treatment of Occupational Disease, 165 Xingangxi Road, Guangzhou 510310, China

1. INTRODUCTION

Immune system constitutes one of the most important defense mechanisms against the establishment and growth of cancer induced by various environmental agents, including ionizing radiation. It has been determined that large or moderate dose radiation induces inhibition of immune system, but the immune effects of low-dose ionizing radiation are still disputed. Therefore, it is important to clarify biological effects of low-dose and low-dose rate radiation. Epidemiological investigation on high background radiation can provide very important data on the effect of long-term low level exposure to radiation on human beings. While cancer mortality study was conducted by the High Background Radiation Research Group (HBRRG), immunological parameters have been studied for inhabitants in high background radiation area (HBRA) and the nearby control area (CA). The results before 1990 indicated that the total number of peripheral blood lymphocytes(PBL), and the relative and absolute number of T cells were lower in inhabitants from HBRA than in those from CA, but the reaction of T cells to antigen was higher. This suggests that long-term low level radiation may have a stimulatory effect on immune system[1-2]. The present studies were designed to further study immune competence and immune responses to virus in HBRA inhabitants on the basis of previous study.

2. MATERIALS AND METHODS

2.1 Frequency of interleukin-2 secreting cells

The subjects were healthy males, 25 from HBRA and 27 from CA. Their families had lived there for generations. They were divided into three age groups (<20, 20-50, >50 years old) .The interleukin-2 secreting cells (IL-2SC) were examined as an index of peripheral blood lymphocyte (PBL) immune functions, by using the monoclonal antibody APAAP method.

2.2 Anti-EB virus antibody levels in human being

Sixty-one inhabitants from HBRA and 60 inhabitants from CA were selected as donors for blood serum examination. They were all healthy and local inhabitants, being divided into two groups based on their age (10-29, 30-84). By using indirect immunoenzymatic assay, the levels of anti-VCA (viral capsid antigen) immune globulin A antibody (VCA/IgA), anti-VCA immune globulin G antibody (VCA/IgG), anti-EA (early antigen) immune globulin A (EA/IgA) and anti-EA immune globulin G (EA/IgG) were adopted as criteria of evaluation.,

2.3. Infection rate of hepatitis B virus (HBV)

Twenty middle school students, including 10 male subjects and 10 female subjects, aged 15-17 years, having not been inoculated against anti-hepatitis B vaccine, were selected as donors of blood from HBRA and CA, respectively. The evaluation indexes were hepatitis B surface antigen (HBsAg), hepatitis B surface antibody (HBsAb) and hepatitis B core antibody (HBcAb), using ELISA method.

3. RESULTS AND ANALYSIS

3.1. Frequency of interleukin-2-secreting cells

Considering the influence of age on immune function, the results were divided into three groups (Table 1).

Table 1
Comparison of frequency of IL-2SC between HBRA and CA

| Age group | HBRA | | CA | |
(years)	No. of analyzed	Frequency(%)	No. of analyzed	Frequency(%)
<20*	9	20.11±1.97	9	17.02±3.90
20-50**	9	19.89±1.75	9	17.17±2.37
>50***	7	20.71±0.86	9	15.78±1.58
Total***	25	20.20±1.61	28	16.63±2.76

* $P>0.05$, ** $P<0.05$, *** $P<0.01$

Generally, the results showed that the frequency (%) of PBL IL-2SC activated by PHA and TPA was significantly increased in subjects from HBRA, compared with those from CA. In analyzing the results in different age groups, the frequency of IL-2SC was higher in every age group of HBRA than in the same age groups of CA. However, there was no significant difference between teenagers of

HBRA and CA; the reason may be that there was one case had higher frequency of 26.2% than those of others in CA. If excluding that case, there was significant difference between them in this age group.

3.2. Anti-EB virus antibody levels

Because the incubated EB virus in body is most reactivated after 30 years old [3], the results were divided into two groups (10-29 years and 30-84 years). The results of VCA/IgA, VCA/IgG and EA/IgG are shown in Table 2, Table 3 and Table 4, respectively.

Table 2
Levels of serum VCA/IgA antibody

Age (years)	Titer geometric mean (GMT)		Rate of detection(%) when dilution ≥ 1:5					
			HBRA**			CA**		
	HBRA*	CA*	Subjects	No. of detected	%	Subjects	No. of detected	%
10 − 29	4.1	3.3	27	7	25.93	21	3	14.29
30 − 84	5.2	4.8	34	13	38.24	39	13	33.33
Total	4.8	4.3	61	20	32.79	60	16	26.67

* $P>0.05$, ** $P>0.05$

Table 3
Levels of serum VCA/IgG antibody

Age (years)	Titer geometric mean (GMT)		Rate of detection(%) when dilution ≥ 1:640					
			HBRA**			CA**		
	HBRA*	CA*	Subjects	No. of detected	%	Subjects	No. of detected	%
10 − 29	111.7	53.8	27	4	14.81	21	1	4.76
30 − 84	192.2	104.4	34	8	23.53	39	6	15.38
Total	151.2	82.8	61	12	19.67	60	7	11.67

* $P<0.05$, ** $P>0.05$

Table 2 and Table 3 showed that all the subjects were infected by EB virus. The levels of VCA/IgA in two areas and the levels of VCA/IgG in CA were similar to the levels of those in Guangdong Province, but the levels of VCA/IgG in HBRA were higher than those in Guangdong province[3-4]. The levels of antibodies, including VCA/IgA and VCA/IgG, were higher in HBRA than in CA, but there was significant difference only in the geometric mean of titers of VCA/IgG between HBRA and CA. The rate of positive titer detection of VCA/IgA(≥1:5) in the two areas were significantly higher than those in Guangdong Province (6.25%)[5] (P<0.05).

Table 4
Distribution of titers (reciprocals) for serum EA/IgG antibody

Age (years)	HBRA							CA				
	No. of subjects	2.5	5	10	20	40	GMT	No. of subjects	2.5	5	10	GMT
10−29	27	3	1	1	0	0	3.8	21	3	0	0	2.5
30−84	34	1	4	3	0	1	9.4	39	3	1	3	5.0
Total	61	4	5	4	0	1	5.8	60	6	1	3	4.1

In this test, the levels of EA/IgA in subjects were very low (titer<1:2.5); only EA/IgG antibody is shown in Table 4. The results indicated that the levels of EA/IgA antibody were negative in two areas, but there was a tendency that the titers for serum EA/IgG antibody in HBRA were higher than those in CA, the difference being not significant statistically.

3.3. Infection rate of hepatitis B virus (HBV)

In China, the positive rate of HBsAg ranges from 16.% to 27.12%, and it is higher in countryside than that in city[6]; the infection rate of HBV is around 58.2%, and it is higher in teenagers than in adults[7]. In our case, the infection rate of HBV and the positive rate of HBsAg in both HBRA and CA were similar to the rates in China; the positive rates of HBsAg and HBcAb were the same or similar between HBRA and CA; both the infection rate of HBV and the positive rate of HBsAb were higher in HBRA than in CA with no significant differences between them (Table 5.)

Table 5

Areas	No. of subjects	HBsAg		HBsAb		HBcAb		HBV*	
		Positive cases	%	Positive cases	%	Positive cases	%	Positive cases	%
HBRA	20	5	25.0	13	65.0	5	25.0	17	85.0
CA	20	5	25.0	9	45.0	4	20.0	13	65.0

* Subjects were infected by HBV with one of the following reactions positive: HBsAg, HBsAb and HBcAb.

4. DISCUSSION

4.1. Immune competence of inhabitants in HBRA

Immune reaction is the sum of interaction of many subset lymphocytes, and some interactions are mediated by cellular decomposed products. IL-2 is one of the most important interleukins in regulating the immune system and mediating a variety of immune functions. Changes in IL-2 activity may reflect immune levels of an individual. Therefore, it is very important to examine IL-2 activity in different individuals so as to look into their immune functions and to explore the mechanism of immune regulation. Usually, IL-2 activity examination by means of PBL IL-2SC frequency *in vitro* by mitogen and antigen could not reflect the real activity of in vivo IL-2SC. In this study, however, we used human IL-2 monoclonal antibody to examine IL-2SC frequency at cellular level. This may more objectively reflect the immune status in vivo. Animal experiments have also shown that low-dose radiation can significantly enhance IL-2SC production by splenocytes[8]. Data presented here indicated that the inhabitants in HBRA had a higher IL-2SC frequency than those in CA. This may be related to enhancement of T cell function after long-term exposure to low level radiation.

4.2. Immune response to virus after exposure to high background radiation

EB virus is a herpes virus closely related to nasopharyngeal carcinoma (NPC). it is usually inoculated in B cells, giving rise to EB virus-specific memory T cells mediated cellular immunity[9]. VCA/IgG antibody is thought to be an index when one is infected by EB virus, and antibodies to EA are the markers indicating that inoculated EB virus is reactivated. Epidemiological investigation and clinic data have demonstrated that the levels of IgA antibodies to anti-EB-virus (including VCA/IgA and EA/IgA) are specific to NPC[10]. The results in this study showed that there was higher positive rate of VCA/IgA in two areas, indicating that there was higher incidence of NPC in them, and the levels of anti-EB virus were higher in HBRA than in CA, but the mortality of NPC in HBRA was not significantly higher than that in CA based on epidemiological data[11]. This study suggests that the inhabitants in HBRA who continuously exposed to low level irradiation may have an increased reactivity of EB virus-specific memory T cells, and consequently an increased immunity against EB virus.

Hepatitis B virus (HBV) is one of the potential predisposing agent of liver cancer. Usually, HBsAg and HBcAb play an etiologic role in development of liver cancer, and HBsAb plays a neutralizing role in eliminating HBV. Data in this study showed that the infection rate of HBV in HBRA was higher than that in CA, so the mortality of liver cancer in HBRA should be higher than that in CA upon the above theory, but real results of epidemiological data demonstrated that the mortality of liver cancer in HBRA was lower than that in CA[11]. The preliminary findings give a hint that the inhabitants in HBRA may have an increased reactivity of T cells.

Based on the theory of immunosurveillance inspect, the immunity system plays an important role in controlling cancer incidence, but the immune function is affected by many confounding factors. Recent data have demonstrated that the immune function is increased in both human beings and animals after exposure to low level radiation for a long time. The tendency of reaction of T cells was observed on some immunologic parameters in previous and present studies. It is possible that there exists a hormetic response after exposure to low level radiation, which needs further study.

REFERENCE

1. Liu Shuzheng et al., Chin. J. Radiol. Med. Prot., 2 (1982) 64.
2. Liu Shuzheng et al., Chin. J. Radiol. Med. Prot., 5 (1985) 124.
3. Liu Yuxi, et al., Chin. Int. J. Cancer, 5 (1983) 337.
4. Luo Yan, et al., Screening for EB virus in general population(report of 11832 cases detected). The 6th academic conference of nasopharyngeal carcinoma in China,4 (1998) 1.
5. Chen Aimei, et al., Cancer, 3 (1984) 54.
6. Si Chongwen, et al., Progresses and consults for virus hepatitis. The Union Publish House of Peking Medical College and Peking Union Medical College ,158, 1993.
7. Li Wen, Chin. Int. Med. J., 33 (1994) 273.
8. Liu Shuzheng, Chin. J. Radiol. Med. Prot., 7 (1987) 2.
9. B. Rickinson, Cellular immunology responses to virus infection. In The Epstein-Barr Virus (M. A. Epstein and B. C. Achong Eds), pp. 75-125. William Heinmann Medical Books, London, 1986.
10. Zhong Jianming, et al., Tumor and Protection, 12 (1989) 54.
11. He Weihui, et al., Chin. J. Radiol. Med. Prot., 5 (1985) 109.

Stimulatory effect of *in vivo* exposure to enriched uranium oxyfluoride on DNA excision repair in mouse lymphocytes

Z. S.Yang, S. P. Zhu and S. Q. Yang

Department of Radiation Medicine, Suzhou Medical College, Suzhou 215007, P.R.China

1. INTRODUCTION

DNA is very sensitive to many environmental substances such as ionizing radiation or chemical compounds which might induce cancer, teratogeny and mutation. We have previously observed that enriched uranium oxyfluoride ($^{235}UO_2F_2$) could induce DNA single and double strand breaks. DNA is an important target molecule for ionizing radiation [1]. Radiation-induced DNA strand breaks are repaired efficiently in proliferating mammalian cells [2], non-dividing cells e.g. hepatocytes [3], cerebellum cells [4] and various lymphocytes [5]. The DNA repair is independent of the stage of cell cycle [6]. So it is labeled as unscheduled DNA synthesis (UDS). The DNA excision repair was not affected by DNA synthesis inhibitor such as hydroxyurea [7]. Non-repair or incorrect repair of DNA damage will affect the function of cells. The effect of various damaging agents on UDS in mammalian cells have been extensively studied. Extensive as the application of $^{235}UO_2F_2$ is, the biological effect of $^{235}UO_2F_2$ which is closely related to man's living environment, especially its effect on DNA excision repair function in immune cells, has not been reported yet. In the present investigation, ultraviolet light (UV)-induced UDS, by incorporation *in vitro* of ^3H-TdR into DNA of isolated spleenic lymphocytes of mice, was measured to study the effects of different doses of $^{235}UO_2F_2$ on the DNA excision repair capability of mouse lymphocytes.

2. MATERIALS AND METHODS
2.1. Experimental animals

BALB/c male mice, 7-8 weeks of age and 21 ± 3 g of weight were used at the beginning of the experiment. These animals were obtained from the Center of Experimental Animals of Suzhou Medical College.

2.2. Internal contamination doses of $^{235}UO_2F_2$

18.9 percent of abundance of $^{235}UO_2F_2$ (60 mg/ml) were diluted with injected water as working solution. BALB/c mice were injected through the tail vein with 0, 0.001, 0.01, 0.1, 1, 20, 100 and 500 µg $^{235}UO_2F_2$/kg weight.

2.3. Assay of UV-induced UDS in spleen lymphocytes

Spleens were removed with aseptic technique on day 12 after injection and teased apart to release spleenic cells in RPMI 1640 medium. After centrifugation at 2000 r/min. for 10 min., the spleenic cells were resuspended at 1×10^7 cells/ml in RPMI 1640 medium containing penicillin (100 IU/ml), streptomycin (100 µg /ml), 10 percent fetal calf serum and 10^{-2} mol hydroxyurea [8]. From each spleenic cell suspension, each of five samples (0.1 ml each) was dispensed respectively into two pieces of 24-well microplate. The duplicate plates were preincubated in a CO_2 incubator at 37°C for 30 min., and 37 kBq ^{3}H-TdR was added to each well. 100 µl of 200 µg /ml PHA was added to No.3-4 wells per spleen suspension per plate. One of the two plates was irradiated with 20 J/M^2 of UV-rays. The other plate was not irradiated by UV-rays as controls. Then the duplicate plates were incubated at 37°C under 5% CO_2 for 120 min. At the end of incubation, the cells were immediately collected on glass fiber filters with a cell autoharvester. The filters were dried and counted in the Beckman LS 6800 liquid scintillator. The average count-per-minute values of five samples of the controls were subtracted from those of the corresponding UV-irradiated samples. The results were expressed as cpm/10^6 cells.

3. RESULTS

3.1. Effect of $^{235}UO_2F_2$ on UV-induced UDS of spleen lymphocytes

The effects of different doses of $^{235}UO_2F_2$ on UV-induced UDS in spleenic lymphocytes are shown in Table 1. It can be seen from Table 1 that UV-induced UDS values were significantly increased (P<0.05 or 0.01) when the mice were exposed to $^{235}UO_2F_2$ at doses of 0.1-20 µg /kg body weight. The results showed that low doses of $^{235}UO_2F_2$ had obvious stimulatory effect on DNA excision repair capacity.

3.2 Effect of $^{235}UO_2F_2$ on UDS in spleenic lymphocytes unirradiated by UV

The effects of $^{235}UO_2F_2$ on UDS in spleenic cells which were not irradiated by UV are shown in Table 2. The UDS in the spleenic cells of the experimental groups showed a pronounced increase (P<0.05, 0.01 or 0.001) on the 12th day after internal contamination of $^{235}UO_2F_2$. The results indicated that $^{235}UO_2F_2$ *in vivo* had continuous damaging effect on the spleenic cells, thereby leading to enhancement of DNA excision repair capacity.

Table 1
Effect of $^{235}UO_2F_2$ on UV-induced UDS in spleenic lymphocytes on day 12 after injection

Doses (μg/kg)	Cpm/10^6 cells[a]	Relative value	P value
0	43.50±55.07	1.00	
10^{-3}	133.43±48.52	0.93	>0.05
10^{-2}	215.48±49.65	1.50	>0.05
10^{-1}	280.00±38.44	1.95	<0.01
1	244.75±44.97	1.71	<0.05
20	225.50±38.60	1.57	<0.05
10^2	173.50±60.82	1.21	>0.05
5×10^2	124.00±69.30	0.86	>0.05

a: M±SD (n=5)

A half logarithmic curve and corresponding curve equation of UV-induced UDS in the spleenic cells after *in vivo* $^{235}UO_2F_2$ contamination were made according to the data in Table 1 (Figure 1).

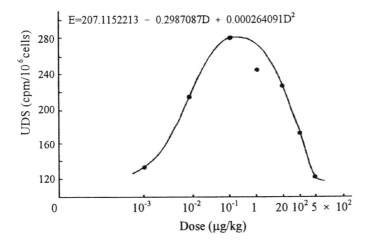

Figure 1. Responsive curve of UV-induced UDS in spleenic cells on day 12 after *in vivo* $^{235}UO_2F_2$ contamination.

Table 2
Variation of UDS in spelling lymphocytes exposed to $^{235}UO_2F_2$ on day 12

Doses (μg/kg)	Cpm/10^6 cells	Relative value[a]	P value
0	183.50±20.51	1.00	
10^{-3}	250.08±60.35	1.36	<0.05
10^{-2}	240.88±37.10	1.31	<0.05
10^{-1}	269.00±10.46	1.47	<0.001
1	232.25±14.86	1.27	<0.01
20	236.00±42.15	1.29	<0.05
10^2	235.13±26.65	1.28	<0.01
5×10^2	264.00±45.59	1.44	<0.01

a. M±SD (n=5)

3.3 Effect of PHA on UV-induced UDS in spleenic lymphocytes after $^{235}UO_2F_2$ internal contamination

Mitogen PHA could induce *in vitro* a differentiation and proliferation of lymphocytes, and led to the increase of semi-conservative DNA replication in the cells. There were higher UV-induced UDS values in the cycling lymphocytes than in resting ones [9]. In order to approach the effects of PHA on UV-induced UDS in spleenic lymphocytes after exposure to different doses of $^{235}UO_2F_2$, the variation of UV-induced UDS in PHA-stimulated, but non-proliferating (hydroxyurea-treated) spleen lymphocytes on day 12 after $^{235}UO_2F_2$ internal contamination was investigated (Table 3). The results showed that the UV-induced UDS values were significantly decreased in PHA-stimulated, but non-proliferating cells than in corresponding unstimulated cells (P<0.05 or 0.01).

4. DISCUSSION

4.1 Stimulatory effect of $^{235}UO_2F_2$ on *in vivo* UV-induced UDS in spleenic lymphocytes

$^{235}UO_2F_2$ is a high-LET (radioactive nuclide. The soluble $^{235}UO_2F_2$ can be rapidly distributed to kidneys, bone, liver and spleen after injection through the

Table 3
Inhibitory effect of PHA on UV-induced UDS in spleenic lymphocytes on day 12 after $^{235}UO_2F_2$ internal contamination

Doses (µg/kg)	cpm/10⁶ cells[a]		PHA	P value
	non-PHA	PHA	non-PHA	
10^{-3}	133.43±48.52	36.25±35.71	0.27	<0.05
10^{-2}	215.48±49.65	38.67±20.28	0.18	<0.01
10^{-1}	280.00±38.44	181.67±66.25	0.65	<0.05
1	244.75±44.97	156.57±29.48	0.64	<0.05
20	225.50±38.60	146.34±40.70	0.65	<0.05
10^2	173.50±60.82	66.00±28.28	0.38	<0.05

a. M ± SD (n=5)

tail vein. The results of Table 1 suggested that low-doses of 0.1-20 µg $^{235}UO_2F_2$/kg weight could increase DNA excision repair capacity, i.e. had significant stimulatory effects on UV-induced UDS in the spleenic lymphocytes.

In a previous study, we found that the ^{3}H-TdR incorporation of PHA-stimulated spleenic T lymphocytes was increased significantly (P<0.05) after intake of $^{235}UO_2F_2$ at a dose of 0.1 µg /kg weight as compared with the control [10]. Tuschl also found that a population potentially receiving elevated background levels of chronic low-dose exposure to high-LET radiation by inhalation of ^{222}Rn and its daughters revealed an enhanced repair capacity for DNA damage inflicted by a second insult, e.g. UV-irradiation [11]. These results indicated that low doses of $^{235}UO_2F_2$ had an obvious stimulatory effect on DNA replicative synthesis and UDS in spleenic lymphocytes, and the stimulatory effect occurred only at a limited range of $^{235}UO_2F_2$ [12].

The extent of UDS after irradiation is correlative with the excision-repair of DNA base damage and is dependent upon several cellular factors: the amount of DNA damaged, the amount of DNA per cell, the specific activity of cellular thymidine pools, the average size of the repaired regions, the number of repaired sites per unit length of DNA [13] and the levels of ATP [14] and NAD$^+$ [15]. The stimulatory effect induced by low dose of irradiation may activate related compound enzymes, especially DNA polymerase [9,11,16] and poly-(ADP-ribose) polymerase [17].

4.2 The continuing effect of $^{235}UO_2F_2$ on DNA damage and excision repair function in spleenic lymphocytes

UV-induced UDS values of spleenic lymphocytes were still higher on day 12 in the $^{235}UO_2F_2$ groups than in the control group (P<0.05,0.01 or 0.001) (Table 2). The results suggested that the DNA damage and excision repair of spleenic cells had not yet recovered to normal owing to the continuing effect of *in vivo* $^{235}UO_2F_2$. The results were different from the biological effect of external irradiation. The X or (ray irradiation-induced DNA strand breaks are repaired in lymphocytes almost completely within 12 hours [18]. The overwhelming majority of DNA single-strand breaks are repaired within 1 hour [14]. Non-UV-irradiated cells of both exposed persons and controls did not show any incorporation of label for the autoradiograph [11]. However, except for the excretion of 70-80% $^{235}UO_2F_2$ by urine after 24 hours of intake, the remainder of the $^{235}UO_2F_2$ excreted very slowly and continuously produced DNA damage. The average count-per-minute values of non-UV-irradiated cells of both $^{235}UO_2F_2$ groups and controls were subtracted from ones of corresponding UV-irradiated cells. Owing to the identical dose of the UV-irradiation, the extent of the second damage for DNA induced by UV was the same. For this reason, the UV-induced UDS values represented the effect of $^{235}UO_2F_2$ on UDS in spleenic lymphocytes, indicating that the low doses of $^{235}UO_2F_2$ could induce the increase of the DNA excision repair capacity in the spleenic lymphocytes.

4.3 The inhibitory effect of PHA on the UV-induced UDS in the spleenic lymphocytes after $^{235}UO_2F_2$ internal contamination

The level of UV-induced DNA repair synthesis was approximately tenfold higher, and DNA strand breaks were repaired more rapidly in PHA-stimulated lymphocytes than in PHA unstimulated cells [9,19]. When human peripheral lymphocytes were irradiated by X-rays after various intervals of PHA stimulation, UDS generally increased during the first 6 hours, but then tended to decrease again [20]. Lymphocytes from old mice showed a lower level of UDS than lymphocytes from young mice. After *in vitro* treatment with nicotinamide, a precursor of cellular NAD$^+$, Con A activated, but non-proliferating (hydroxyurea-treated) lymphocytes from old mice displayed a level of DNA repair similar to that of lymphocytes from young animals [15]. At present investigation, the level of UV-induced UDS was lower in PHA-stimulated, but non-proliferating lymphocytes than in PHA-unstimulated lymphocytes in the group receiving the same dose. The results suggested that mitogen PHA had an obviously inhibitory effect on UV-induced UDS in spleenic lymphocytes after $^{235}UO_2F_2$ intake. Owing to treatment with hydoxyurea, PHA could only be combined with the corresponding acceptor of the lymphocytes and stimulated the lymphocytes, but could not proliferate the cells. On the contrary, the effect of PHA on lymphocytes might partly consume ATP in the cells. The repair of DNA strand breaks requires ATP [11,14]. A loss of ATP might be one of the reasons for the reduction in DNA repair capacity.

REFERENCES

1. Zhu, Chin. J. Radiol. Med. Prot., 12 (1992) 232.
2. Ormerod and U. Stevens, Biochim. Biophys. Acta., 232 (1971) 72.
3. Ono and S. Okada, Int. J. Radiat. Biol., 25 (1974) 291.
4. Wheeler and J.T. Lett, Proc. Natl. Acad. Sci. USA, 71 (1974) 1862.
5. Tempel, Radiat. Environ. Biophys., 29 (1990) 19.
6. Sawada and S. Okada, Radiat. Res., 41 (1970) 145.
7. Young, Cancer Res., 27 (1967) 526.
8. Evans and A. Norman, Radiat .Res., 36 (1968) 287.
9. Hamlet, Int. J. Radiat. Biol., 41 (1982) 483.
10. Z-S. Yang, Chin. J. Radiol. Med. Prot., 14 (1994) 147.
11. Tuschl, Radiat. Res., 81 (1980) 1.
12. Planel, Health Phys., 52 (1987) 571.
13. Koval, Mutat. Res., 49 (1978) 431.
14. CH. Gartner, CH. Sexauer and U. Hagen, Int. J. Radiat. Biol., 32 (1977) 293.
15. Licastro and R.L. Walford, Mech. Ageing Dev., 35 (1986) 123.
16. S-Z. Liu, Chin. J. Radiol. Med. Prot., 9 (1989) 247.
17. Marini, Int. J. Radiat. Biol., 58 (1990) 279.
18. Spiegler and A. Norman, Radiat .Res., 39 (1969) 400.
19. Freeman and S.L. Ryan, Mutat. Res., 194 (1988) 143.
20. Gundy and M.A. Bender, Radiat. Res., 97 (1984) 519.

VII Adaptive response and genetic effects

Adaptive response in human beings living in the area contaminated by chernobyl accident

I. Pelevina, G. Afanasjev, A. Aleschenko, V. Gotlib, O. Kudrjachova, L. Semenova, and A. Serebryanyi

Institute of Chemical Physics Academy of Sciences, Institute of Biochemical Physics Academy of Sciences, Kosygina st. 4, V-334, Moscow,117977, Russua

1.INTRODUCTION

The adaptive response (AR) - the decrease of damaging action of genotoxic agents in large doses after the preliminary action of the same or other agents in low doses. AR is perhaps the universal protective reaction that was formed in the all livings during the evolution. AR is observed in bacteria, plant, animals, in cultured cells *in vitro*, on stimulated human blood lymphocytes etc. [1-5]. The ability to AR is genetically determinant perhaps, and that was proved with the alkylating agents, where AR is connected with the functioning of the gene ada and the absence of this gene prevented the cell to development the AR [6,7]. It is well known the induction of AR by ionizing radiation. The proposed mechanisms in this case are the induction and expression of genes, stimulation of repair, detoxification of free radicals, cell cycle delay as the response of the genomic damages, that have been caused by low level radiation; so, the damages induced by the subsequent large irradiation have been repaired better[8-10].

In man AR have been discovered in 1984 [1] on the human blood lymphocytes and up to day have been confirmed on many objects [2-5] It was shown that people vary in their ability to AR; and there are some individuals that haven't capacity to AR; some times the preliminary treatment in low level dose result in the increase not the radioresistance but the enhancement of radiosensitivity [11-13].

It seems that population aspects of AR problem is very important because of it have to be known if all the people can induce the AR, if this property can be formed during the process of the oncogenesis, if it is the professional influence on the AR, influence of the style of life, stress and other ecological situation (conditions).

This problem became very actual because of now a lot of people are living and working in the conditions of elevated level of radiation. There are many investigations in which have been studied the late effects of this elevated irradiation for the health of the people; but as usually there have been registered only the mean for the whole population of risks assessment. For example, the study of 96,000 workers on the nuclear industry in USA, England, Canada have shown that the absorbed during the life dose of 0.1 Sv result in the probability increase of death from leukemia (in the mean) on the 22%. But it still remained unknown how the risk of the appearance of pathological events can be evaluated for the each individual. So that the problem of the individual risk estimation and the detection of persons the reaction of which to damage or stress is different from the whole population, from the mean of the whole population,espesially in the direction of sensitivity enhancement is a very actual problem for the science up to date. The deviation from the mean reaction can justify to genomic defects, disturbances of damages repair, changes in the organism homeostasis. These all permit to suppose, that there are beings with increased sensitivity to unfavorable factors. The problem is very actual for different areas of medical science because there is the possibility of individual risks evaluation of late effects for each individual.

2.MATERIAL AND METHODS

The level of lymphocytes with micronuclei have been scored in Moscow donors and people living in Bryansk region with the degree of contamination 15 - 40 Ci/km^2. The doses that liquidates have been obtained were not higher then 25 cGy.

The scheme of experiments was usual. Human lymphocytes have been stimulated by PHA-P and cultivated for 72 h. In 24 h after beginning of cultivation culture was irradiated in G1-phase of the cell cycle in the adaptive dose of 0.05 Gy and 5 h after in the challenge dose of 1.0 Gy. Cytochalasin B (0.006mg/ml) was added after 48 h of cultivation. The frequency of micronuclei (MN) was expressed on the 1000 binuclear cells, the 2000 - 3000 cells were scored on each slide and the χ^2-test was used for estimation of validity of differences between variants.

3.RESULTS AND DISCUSSION

It was investigated 9 groups of people: the adults from Moscow; 4 groups of children 3 - 7 age old living in different districts of Moscow and Moscow region; liquidators - the people that took part in the liquidation of Chernobyl disaster consequences (obtained doses 250 mSv and below); 2 groups from the towns of Vishkow and Klincy Brjanskja region - region with the degree of contamination up to 40 Ci/km^2. The all number of investigated people was 349.

When the comparison was made (χ^2 - test) of individual effects by the irradiation in conditioning (0.05 Gy) and challenged dose (1 Gy) the next results have been obtained: 1. the two groups doesn't differ and χ^2 is equal or is near to the 0; 2, 3 - the groups differs from each other but AR or radiosensitivity enhancement are statistical insignificant (< 3.84); 4, 5 - and there is significant AR or radiosensitivity enhancement.

Because of it is very different to carry out a strong border between the group 1 and groups 2 and 3 in this paper these groups were joined in one pool group. Selected by us criteria by which we can take every person to one or other group is not very strong (not below 95% confidence limit). If we shall select the 90% probability, the part of donors from the group with the nonsignificant effect can pass to the subpopulation with significant AR or to the subpopulation with radiosensitivity enhancement.

It will be noted that the value of (hi-square test depends on the number of cells scored, so that in the most cases there was 2000 cells register that not all people have ability to AR was known earlier, but the radiosensitivity enhancement after the radiation in conditioning dose has been rare observed [11-13].

The reasons of AR absence may be different - genetically determinant property of individuals (perhaps the need of other conditioning dose); the decrease of the protein synthesis rate, so the time for AR induction will be greater (or smaller); the presence of constitutive adaptive proteins, that were induced earlier (by radiation or chemicals, because there is cross effect between these agents) and after the additive irradiation the protein synthesis have not been induced [14].

Special attention must be made to the group of persons with the synergistic effect, when after conditioning irradiation and then challenged dose (0.05 + 1 Gy) the effect is significant higher then after irradiation in one dose of 1 Gy. Perhaps these individuals must be take to the groups of risk. To the same group perhaps we must take the persons which have effect of radio sensitivity enhancement with confidence limit greater then 85 - 90 %, but smaller then 95%. Our proposition is that in adults this effect after conditioning dose can be explained by genetically determinant defects (deficiency) in the synthesis of protective proteins or repair proteins. In children this effect can depend on that these systems haven't been formed yet in the process of oncogenesis.

The persons from the group of risk can be characterized by the enhanced probability of the increased sensitivity to the damaging action of radiation, some chemicals, enhanced probability of malignant transformation and other late effects.

It is known that in the repair process of genomic damages after many chemical agents and radiation the same repair systems take part; so that we can propose that individuals with enhanced radiosensitivity would be more sensitive to the chemical environmental factors. The adults in the group of risk must have a limit to the some kind of professional work. On the children from the group of risk there must be pay especially attention, because they have to be limited in the number of rentgenological investigations, the use of soadults (Table 2) are discovered the heterogeneity of this population. Because of this cohort was divided on two cohorts - persons living in Moscow; and persons living on contaminated area and liqidators. This mean that low level radiation induce changes in the population

Table 1
The distribution of children living in different districts of Moscow and Moscow region on the type of blood lymphocytes response *in vitro* on conditioning irradiation in the dose of 0.05 Gy and subsequent irradiation in challenged dose of 1 Gy 5hr thereafter

Group	The number of investigated persons	With adaptive response				With enhanced radiosensitivity			
		significant number	%	nonsignificant number	%	significant number	%	nonsignificant number	%
1	41	10	24.4	13	31.7	8	19.5	10	24.4
2	15	2	13.3	4	26.7	4	26.7	5	33.3
3	38	7	18.4	4	10.5	14	36.8	13	32.2
4	23	4	17.4	6	26.1	7	30.4	6	26.1
Total	117	23	19.6	27	23.1	33	28.2	34	29.0

structure, expressed in the decrease of the frequency of persons which lymphocytes were not able to AR and simultaneously in the increase of persons frequency with radiosensitivity enhancement. For example, in comparison with Moscow in Klincy and Vishkow the part of people with the significant AR decreasing is 38.7 and 36.7, and in Moscow 56.8 ($P < 0.1$). The more pronounced decrease is observed in liqidators - to 26.3% ($P < 0.1$). Simultaneously there is change in the number of another group observed, and in some cases these differences are significant. These differences are very important because of the all other parameters (the normal distribution, the mean frequency of spontaneous and acute irradiated cells with MN are the same, doesn't differ mode, standard deviation) weren't differ in Moscow, Slinky and Vishkow [15].

The data of Table 1 and Table 3 also justify to the influence of radiation on population structure, because there are differences between the children from Moss chronic low dose irradiation result in the genomic disturbances, decreasing of protective mechanisms, repair ability, and in the depression of protein synthesis, which are need to AR induction (repair enzymes, poly-ADP-ribosylation, p 53, PKC, transcript factors etc.).

Another conclusion in the field we can make by the comparison of the data table 1 and 2. Amongst the children the significant AR is observed in 19.5% cases, and amongst the adults - in 56.8%. This difference is statistically significant. In contradictory the part with the statistically significant enhancement of radiosensitivity in children is higher (28.2% and 2.3%). This fact can mean that the ability to AR induction is formed with the age.

Table 2
The distribution of adult beings living in Moscow, contaminated with radionuclides city Klincy and Vishokow and liquidators on the type of blood lymphocytes response *in vitro* on the conditioning irradiation in the dose of 0.05 Gy and subsequent irradiation in challenged dose of 1 Gy 5h thereafter

Group	The number of investigated persons	With adaptive response				With enhanced radiosensitivity			
		significant		nonsignificant		significant		nonsignificant	
		number	%	number	%	number	%	number	%
Moscow	44	25	56.8	16	36.4	1	2.3	2	4.6
Klincy	62	24	38.7	20	32.3	8	12.9	10	16.1
Vishkow	49	18	36.7	19	38.8	2	4.1	10	20.4
Liquidators	19	5	26.3	6	31.6	3	15.8	5	26.3

Table 3
The distribution of children living in contaminated city Novozibkow on the type of blood lymphocytes response *in vitro* on the conditioning irradiation in the dose of 0.05 Gy and subsequent irradiation in challenged dose of 1 Gy 5h thereafter

Group	The number of investigated persons	With adaptive response				With enhanced radiosensitivity			
		significant		nonsignificant		significant		nonsignificant	
		number	%	number	%	number	%	number	%
Novozibkow	58	3	5.2	21	36.2	15	25.8	19	32.8

So that from the results obtained we can conclude that populations investigated are heterogenous by the AR ability - there are people in which AR has been inducted and the people with the lack of this property, and in population there are beings in which the conditioning irradiation result in increasing of radiosensitivity. AR, probably, can form during oncogenesis. The environmental factors, as radiation in low doses change the population structure by the indication of AR ability, and this ability to AR in individual persons may be the base for the distinguish the group of risk.

REFERENCES

1. G. Oliviery, J. Bodycote and S. Wolff, Science, 223 (1984) 594.
2. B. Demple, Ann. Rev. Genet., 25 (1991) 315.
3. S.Z. Lui, L. Cai and J.B. Sun, Int. J. Radiat, Biol., 62 (1992) 198.
4. S.Z. Lui, W.H. Lui and J.B. Sun, Health Phys., 52 (1987) 579.
5. L. Cai and S.L. Liu, Int. J. Radiat. Biol., 58 (1990) 187.
6. D. Shadley and G. Dai, Environ. Mol. Mutag., 21 (Suppl. 22) (1989)
7. S. Wolff, G. Oliviery and V. Afzal in: G. Obe and A.T. Natarajan (eds). Chromosomal Aberrations. Basic and Applied Aspects. Springer Verlag, Berlin, 1990.
8. J.D. Shadley, Radiat. Res.,138 (1994) 9.
9. J.D. Shadley and S.Wolff, Mutagenesis, 2 (1987) 95.
10. T.Ikushima in: Chromosomal Aberrations. Springer Verlag, Berlin, 1989.
11. A. Bosi and G. Oliviery, Mutat. Res., 211 (1989) 13.
12. G. Oliviery and A. Bosi in: G. Obe and A.T. Natarajan (eds). Chromosomal Aberrations. Basic and Applied Aspects. Springer Verlag, Berlin, 1990.
13. J. Hain, R. Jaussi and W. Burkart, Mutat Res., 283 (1992) 137.
14. S. Wolff, V. Afzal and K. Wienue, Int. J. Radiat. Biol., 53 (1988) 39.
15. T.J. Bailey. Statistical methods in Biology. Publ. Foreign Liter. Moscow, 1962, p. 75.
16. I. Pelevina, G. Afanasjev, V. Gotlib, A. Alferovich, M. Antochina, N. Ryabchenko, A. Saenko, I. Ryabtsev and I. Ryabov, Radiat. Biol., Ecology, 33 (1993) 85.
17. I. Pelevina, B. Nikolayev, V. Gotlib, G. Afanasjev, L. Koslova, A. Serebryanyi, D. Terechenko, V. Tronov and E. Chramtsova, Radiat. Biol., Ecology, 34 (1994) 805.

Investigation on the adaptive response to ionizing radiation induced by high background exposure to spa workers

Yao Suyan[1], Jiang Tao[1], Junsuke Yamashita[2] and Yukio Takizawa[3]

[1]Laboratory of Industrial Hygiene, Ministry of Health, Beijing 10088, China

[2]Department of Center Radioisotope, Akita University School of Medicine, Akita 010, Japan

[3]National Institute for Minamata Disease, Minamata, Kumamoto 867, Japan

1. INTRODUCTION

In recent years, more and more scholars are being attracted to the adaptive response (AR) field. Authors revealed that beneficial effects induced by low-dose ionizing radiation, for instance, decreased the incidence of some malignant tumors and extended the life span [1,2]. But the results are different. ICRP and UNSCEAR require more information, particularly on the mechanism [3,4]. The investigation on the AR in the lymphocytes of spa workers exposed to enhanced natural levels of external γ-radiation and internal radiation from inhaled radon decay products is not enough[5].

This study attempts to explore whether a chronic low-dose exposure to radon rendered the lymphocytes of Tamagawa spa workers more resistant to a high gamma ray exposure applied *in vitro*.

2. MATERIALS AND METHODS

2.1 Study population

Tamagawa spa in Akita, Japan is an enhanced natural background area of radon and its decay products. The radon content in the air is different from site to site. The ratio of the average level of radiation in the spa and in the control area was 1.85 (a range of 1-3.5)to 1 (according to observed accumulative absorbed dose). The workers had been exposed to accumulative absorbed doses of 0.04-0.13 mSv/month. Three healthy workers in Akita spa (natural background radiation) were used as controls.

2.2 Cell culture

Slide preparations were made according to standard methods. The cultures were kept at 37°C in a 5% CO_2 incubator for 54h. Colcemid was added 4h before fixation.

2.3 Irradiation and analysis

The occupational exposure was termed as the first adaptive dose.

A second adaptive dose of 0.01 Gy (0.05Gy/min) Co-60 γ-rays was applied in G$_0$-phase (before PHA-P stimulation) or in G_1-phase(5h after stimulation). 1.5 Gy of the challenge dose of Co-60 γ-rays (0.2Gy/min) was given in G_2-phase (at 48h after stimulation).

The number of chromosome-and chromatid-type aberrations was determined. For the adaptive response, the number of chromatid-type was scored. The results were analyzed statistically by t-test.

3. RESULTS

3.1 Lymphocytes exposed to the first adaptive dose

The results obtained from each experiment are listed in Table 1-2.

Table 1
The frequencies of chromosome aberrations in lymphocytes

Group	No. of scored cells	Chromosome-type (per cell)	Chromatid-type (per cell)
Tamagawa spa	1700	12 (0.0071)*	50 (0.029)
Akita spa	1284	2 (0.0016)	55 (0.043)

* $P<0.01$

The frequencies of chromosome-type aberrations in lymphocytes of workers in the spa were markedly higher than those of control ($P<0.01$).

The results of chromosome aberrations induced by 1.5 Gy irradiation *in vitro* in G_0 cells were summarized in Table 2. One of 9 subjects exposed to occupational doses of irradiation showed the AR, that is to say, frequency of chromatid-type aberrations reduced significantly compared with the values in controls.

Table 2
Chromosome damage induced by 1.5 Gy γ-rays in G_0 lymphocytes pre-exposed to occupational irradiation

Donor	Occupational irradiation (mSv/m)	Challenge dose (Gy)	No. of scored cells	Total chromatid-type observed	Observed per cell	AR*
Control	-	1.5	300	76	0.25	
A	0.06	1.5	100	45	0.45	-
B	0.09	1.5	49	14	0.29	-
C	0.04	1.5	100	57	0.57	-
D	0.08	1.5	100	13	0.13	+
E	0.06	1.5	100	34	0.34	-
F	0.09	1.5	100	18	0.18	-
G	0.09	1.5	100	24	0.24	-
I	0.13	1.5	88	44	0.50	-
J	0.05	1.5	100	23	0.23	-

* Compared with the controls (no significant difference between frequencies in controls).

3.2 Experiments with a second adaptive dose carried out in G_0/G_1 phase of lymphocytes

The data on AR in lymphocytes of 9 workers exposed to the second adaptive dose in vitro are shown in Table 3. The cells in the G_0-phase, pre-exposed to the second adaptive dose of 0.01 Gy, in 6 of them exhibited significant reductions in the incidence of chromatid-type aberrations induced by 1.5 Gy challenging 3.2 Experiments with a second adaptive dose carried out in G_0/G_1 phase of lymphocytes

The data on AR in lymphocytes of 9 workers exposed to the second adaptive dose in vitro are shown in Table 3. The cells in the G_0-phase, pre-exposed to the second adaptive dose of 0.01 Gy, in 6 of them exhibited significant reductions in the incidence of chromatid-type aberrations induced by 1.5 Gy challenging γ-rays. The decrease in G1 cells with the same condition was observed in 5 of them.

The incidence of chromosome damage was found to be similar in G_0 and G_1 cells pre-exposed to the second dose, suggesting that under the conditions tested the second adaptive dose offered additional protection against chromosome damage induced by the challenging dose.

4. DISCUSSION

Numerous publications indicate that the AR can be induced in human lymphocytes by low-dose ionizing irradiation, but it is not detectable for every

Table 3
Chromosome damage induced by a challenge dose of 1.5 Gy γ-rays in human lymphocytes pre-exposed to 2 adaptive doses**

Donor	Adaptive dose-2 (Gy)	Challenge dose (Gy)	No. of scored cells	Total chromatid-type	Observed per cell	Expected per cell*	AR
A	-	-	200	6	0.03		
	-	-	100	45	0.45		
	0.01(G$_0$)	1.5	100	14	0.14	0.48	+
	0.01(G$_1$)	1.5	100	16	0.16	0.48	+
B	-	-	200	5	0.03		
	-	-	49	14	0.29		
	0.01(G$_0$)	1.5	65	5	0.08	0.32	+
	0.01(G$_1$)	1.5	100	17	0.17	0.32	+
C	-	-	100	2	0.02		
	-	-	100	57	0.57		
	0.01(G$_0$)	1.5	100	16	0.16	0.59	+
	0.01(G$_1$)	1.5	100	25	0.25	0.59	+
D	-	-	200	3	0.02		
	-	-	100	13	0.13		
	0.01(G$_0$)	1.5	100	25	0.25	0.15	-
	0.01(G$_1$)	1.5	100	10	0.10	0.15	-
E	-	-	200	3	0.02		
	-	-	100	34	0.34		
	0.01(G$_0$)	1.5	100	14	0.14	0.36	+
	0.01(G$_1$)	1.5	100	32	0.32	0.36	-
F	-	-	200	10	0.05		
	-	-	100	18	0.18		
	0.01(G$_0$)	1.5	100	33	0.33	0.23	-
	0.01(G$_1$)	1.5	100	21	0.21	0.23	-
G	-	-	200	10	0.05		
	-	-	100	24	0.24		
	0.01(G$_0$)	1.5	100	20	0.20	0.29	-
	0.01(G$_1$)	1.5	100	15	0.15	0.29	+
I	-	-	200	8	0.04		
	-	-	88	44	0.50		
	0.01(G$_0$)	1.5	100	23	0.23	0.54	+
	0.01(G$_1$)	1.5	100	16	0.16	0.54	+
J	-	-	300	2	0.01		
	-	-	100	19	0.19		
	0.01(G$_0$)	1.5	100	9	0.09	0.20	+
	0.01(G$_1$)	1.5	100	13	0.13	0.20	-

*Expected values are the sum of those induced by first adaptive dose and challenge dose

** The adaptive dose-1 is the occupational irradiation dose

donors [6]. Our result was consistent with this fact. The individual variability in the AR to ionizing radiation was proved again. Bai Yongli et al. reported that a 0.5 cGy dose which in itself did not induce the AR, did so when given twice within the same cell cycle. It was the accumulative effect of two low doses of irradiation in inducing the AR in human lymphocytes [7]. Fan et al. showed that two adaptive low doses (1 cGy x-rays) did not offer any additional protection against the chromosome aberrations induced by the challenge dose, i.e., that there exists a saturation dose [8]. The present investigations could not give sufficient evidence of the adaptive response in occupational low dose of radon exposure *in vivo*, did so when given the second adaptive dose within the same cell cycle (at G_0-phase) *in vitro*. It is possibly due to low dose of radon not reaching the adaptive dose. Some investigations indicate that the adaptive doses which induce AR of chromosome damage by low-dose ionizing radiation are different between types of radiation: 0.5-20 cGy for x-rays , 1-20 cGy for γ-rays and 1-10 cGy for ^{3}H-TdR. Optimal reduction in chromosome aberrations is obtained at 1 cGy dose [9].

Generally speaking, the pre-treatment of human lymphocytes at G_1-phase with low dose ionizing radiation in vitro can induce the AR of chromosome aberrations, but such is not the case for the lymphocytes at G_0-phase. Our results showed that the sensitivity of cytogenetic AR seems similar in human lymphocytes at G_0 and G_1 phases. The cytogenetic AR induced by low level ionizing radiation mainly rises from X, β and γ-rays with moderate or high energy, ranging 8-10 kev/μm of low-LET, but less for high-LET α-rays. The AR was found to be present in cells pre-exposed to low-dose of X-rays when the challenging dose of radon was given [10]. Barquinero et al. showed that the occupational exposure to radiation induces the AR in lymphocytes of 12 workers [11].

ACKNOWLEDGMENTS

We wish to express our thanks to Drs. Yasuo Watanabe and Shinnsaku Kagaya for their help with the calculation of dose estimations. This work was supported by the Department of Public Health, Akita University School of Medicine, Dr. Chen Deqing and Mrs. Toshiko Shibata.

REFERENCES

1. Liu Shuzheng et al. ,Chin. J. Radiol. Med. Prot., 5 (1992) 299.
2. Chen Deqing ,Chin. J. Radiol. Med. Prot., 4 (1986) 269.
3. United Nations UNSCEAR 1994 Report, New York, 1994.
4. ICRP, ICRP Publication 60, Pergamon Press, Oxford, 1990.
5. Yao Suyan, Yukio Takizawa, Junsuke Yamashita et al. , J. Radiat. Res. Radiat. Proc., 11 (1993) 242.
6. Bosi A. and Olivieri G., Mutat. Res., 211 (1989) 13.

7. Bai Yongli and Chen Deqing, J. Radiat. Res. Radiat. Proc., 12 (1994) 34.
8. Fan S. ,et al., Mutat. Res., 243 (1990) 53.
9. Ikushima T., Mutat. Res., 227 (1989) 241.
10. Wolff S., et al. , Mutat. Res., 250 (1991) 299.
11. Barquinero J. F. , et al., Int. J. Radiat. Biol., 67 (1995) 187.

Techniques for detecting the genetic effects of the atomic bombings applicable to the study of the genetic effects of natural radiation

C. Satoh[a], M. Kodaira[a], J. Asakawa[a], N. Takahashi[a], R. Kuick[b], S.M. Hanash[b] and J.V. Neel[c]

[a]Department of Genetics, Radiation Effects Research Foundation, 5-2 Hijiyama Park, Minami-ku, Hiroshima 732, Japan

[b]Department of Pediatrics, University of Michigan Medical School, Ann Arbor, Michigan 48109, USA

[c]Department of Human Genetics, University of Michigan Medical School, Ann Arbor, Michigan 48109, USA

1. INTRODUCTION

A pilot study, at the DNA, level has been conducted to detect the potential genetic effects of atomic bomb radiation [1-3]. Because the techniques being used for the detection of radiation induced germline mutations are also applicable to the study of the genetic effects of high levels of natural radiation, data obtained in our study and the characteristics of the techniques will be discussed. For the detection of deletion/insertion/rearrangement (D/I/R) mutations, believed to predominate among radiation-induced mutations, we chose two types of DNA as targets, that is, minisatellites and single copy sequences. Southern blotting technique and two-dimensional gel electrophoresis (2-DE) of DNA [4, 5] are employed for the examination of the former and the latter, respectively.

2. MUTATIONS AT THE MINISATELLITE LOCI

Minisatellites [6] or VNTRs (variable number of tandem repeats) [7] are tandem-repetitive elements dispersed throughout the human genome that show substantial allelic variation in the number of repeat units. Most of the minisatellite loci are not functional, and spontaneous mutation rates at some of the loci are more than 1000 times higher than those at functional gene loci. Two groups of investigators [8-10] have reported increased mutation rates at these loci in the offspring of male mice irradiated by γ rays. However, the results based on mouse data are not consistent with regard to the doubling dose and radiation-sensitive stage and further study is necessary to provide guidance for humans.

2.1. Samples and probes

We examined 50 exposed families having 64 children and 50 control families having 60 children. These families are a subsample of 1000 families from Hiroshima and Nagasaki consisting of father, mother, and all available children for whom permanent cell lines are being established by using Epstein-Barr virus transformation of peripheral B-lymphocytes. Among the 64 children in the exposed families, only one child had parents who were both exposed. Thus, of a total of 128 gametes that produced the 64 children, 65 gametes were derived from exposed parents and 63 were from unexposed parents, the latter being included in a group of 183 unexposed gametes used for calculating mutation rates. The average parental gonadal dose for the 65 gametes was 1.9 sievert (Sv). We used a value of 20 as the relative biological effectiveness for the integration of the two types of radiation released by the atomic bombs (predominantly gamma and a small neutron component).

DNA samples were extracted from cell lines or peripheral lymphocytes. For screening purposes, DNA samples extracted from the cell lines were used. For the confirmation of mutations, DNA samples extracted from granulocytes or lymphocytes that had not been treated with Epstein-Barr virus were used.

DNA samples were digested with restriction enzymes (*Hinf*I or *Alu*I) and the resulting digests were electrophoresed on an agarose gel. Separated DNA fragments were transferred to nitrocellulose filters. The filters were sequentially treated with six ^{32}P-labeled single-locus minisatellite probes (λTM-18, ChdTC-15, Pc-1, pλg3, λMS-1 and CEB-1). For the detection of DNA fingerprints, a probe (15.1.11.4) that includes the core region from DNA fingerprint probe 33.15 was used. After hybridization with the probes, the bands of fragments on a filter were visualized by autoradiography.

2.2. Mutations detected at six minisatellite loci and in DNA fingerprints

We compared the bands of children with those of their parents and identified mutant bands when bands with identical lengths were absent in both parents. The loci that can be detected with the six probes are denoted as *λTM-18*, *ChdTC-15*, *Pc-1*, *pλg3*, *λMS-1* and *CEB-1*. Because each probe detects a single locus with multiple-length alleles and because heterozygosities for these six loci range between 70% and 97%, each allele in the children can be traced back to one parent.

An example of a mutation detected by the CEB-1 probe is shown in figure 1. In figure 1A, neither bands of the rightmost son in the pedigree is identical to any of his father's bands, although one of the son's bands shows the same mobility as one of his mother's bands. Thus, we consider that this mutant allele is probably derived from his father's allele of approximately 2.1 kb. Because his father was exposed, this mutation was classified as one that had occurred in the exposed gamete. After examination of 1488 alleles in the 124 children, we detected a total of 28 mutations at the three loci (1 at the *pλg3*, 12 at the *λMS-1* and 15 in *CEB-1*) and none at the other three loci. In all of the cases, it was possible to determine in which parental gamete the mutation had occurred.

Six mutations were detected in a total of 390 alleles originating from the 65 exposed gametes and 22 mutations were detected in 1098 alleles from the 183 unexposed gametes. The mean mutation rate per locus per gamete was 1.5% in the exposed parents and 2.0% in the unexposed parents. We observed no significant difference in mutation rates in the children of the exposed and the unexposed parents [1-3].

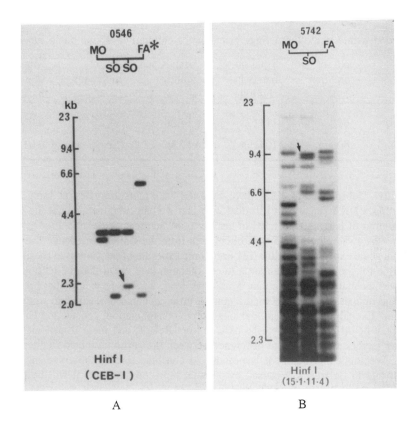

Figure 1. Mutations detected at the *CEB-1* locus (A) and in the DNA fingerprint (B). Arrows indicate mutant alleles. The asterisk (*) indicates the exposed parent.

In a preliminary study, we examined the 124 children and their parents with a multilocus probe (15.1.11.4) which can detect many minisatellite loci simultaneously and detected 25 mutations in the DNA fingerprints of 25 children [11]. An example of a mutation detected in a family is shown in figure 1B. Contrary to the bands detected with single-locus probes, the fingerprint bands of the children were sometimes difficult to trace back to one parent, and we scored 12 and 13 mutations, respectively, for the exposed and the control families (Table 1). Because each child of an exposed family had only one exposed parent, except for one child, fewer than 12 mutations were derived from the exposed gametes. Thus, the exposed and the unexposed gametes did not differ significantly in mutation rates at the minisatellite loci constructing the DNA fingerprint.

Table 1
Mutations detected in the DNA fingerprints of the children of atomic bomb survivors

Group	No. of children	Total no. of bands in children	Bands per child	No. of mutations	Mutation rate per band
Exposed	64	1111	17.36	12	0.011
Control	60	1041	17.35	13	0.012

Recently, Dubrova et al. [12] have reported that mutation rates at minisatellite loci constructing a DNA fingerprint identical with that we had examined, and at the *CEB-1* loci, in 79 children of parents who lived in heavily polluted areas of Belarus after the Chernobyl accident were twice that of 105 control children from the United Kingdom. They suggested that initial acute exposure to iodine-131 or chronic exposure of the parents to caesium-137 was responsible for the increased mutation rates, although they noted that the individual dose of ^{137}Cs "was estimated to be less than 5 mSv per year, a value far below that predicted from mouse and human data." They used a control group composed of children from the United Kingdom. Excluding post-Chernobyl radioactive contamination, genetic and environmental factors may be different between the parents from Belarus and those from United Kingdom and these factors may cause the difference between the germline mutation rates in the two groups. We think that before it is possible to conclude that the observed increase in the mutation rate in Belarus was caused by parental radiation exposure, it will be necessary to select new control Belarus families having children born before the accident [11]. However, there is the possibility that the effects of chronic internal exposure on the highly variable minisatellite loci may be different from those of chronic external exposure to the mouse specific loci. Considering this possibility, it is worthwhile to examine germline mutation rates of inhabitants in areas with high levels of natural radiation.

3. TWO DIMENSIONAL GEL ELECTROPHORESIS (2-DE) OF DNA

To detect D/I/R mutations in single copy sequences, we have introduced a new two-dimensional gel electrophoresis (2-DE) approach developed by Hatada et al. [13] By choosing three restriction enzymes, *Not*I, *Eco*RV, and *Hin*fI, approximately 2000 DNA fragments (0.3-2.0 kb) from a single DNA sample can be separated on a polyacrylamide sheet gel and without using probes, they can be visualized as spots by autoradiography [4, 5]. For the detection of the fragments, we labeled *Not*I sites with ^{32}P which are frequent in the unmethylated "CpG island". This strategy is thought to assure that a high proportion of visualized fragments originate from active genes. In principle, this system will detect two types of genetic variations, that is, nucleotide substitutions at the cut site for the three enzymes used to digest DNA, and variations due to D/I/R events.

The intensity of any spot on the gel is, in the usual case, expected to be determined by two homologous DNA fragments. However, a fresh mutation would usually be detected in a heterozygote with one normal and one mutated allele, and only the normal allele would be at the usual position. Therefore, a quantitative analysis searching for a 50% decrease in spot intensity is required for the detection of the mutation. We have employed computer algorithms which were developed for 2-DE protein gels at the University of Michigan for the analysis of the DNA spot intensity.

We have analyzed DNA samples from three mother-father-child trios [4]. For each sample, gels were prepared in duplicate and autoradiograms were analyzed. Of approximately 2000 spots on the autoradiogram, 774 spots were selected as potential candidates because they were distinct spots. Among them, 482 spots were selected because the average coefficient of variation (CV) (standard deviation divided by mean) for their intensities for the two sets of nine gels was less than 0.12. In a system in which the CV for intensities of the spots is less than 0.12, a spot whose intensity is 50% of the normal value should be detectable. We have detected heterozygotes with normal and deleted alleles using this criterion in our previous screening for enzyme deficiency variants [14] and heterozygous carriers of a deletion in the families of Duchenne muscular dystrophy patients [15]. Thus, the 482 spots are suitable to detect mutations. However, we have detected no mutations among the spots of three children from the three trios.

The mutational yield in the 2-DE study can be calculated only approximately. If we assume that the spontaneous mutation rate is 1×10^{-5}/fragment/generation, we would then expect one mutation per 100 gels from control children (500 diploid fragments scored per gel). Considering this assumed mutation rate, it is not surprising that we detected no mutations in the spots of the three children. This calculation is based on having 500 good spots from the 2000 derived from the *NotI/EcoRV* fragments of 1 to 5 kb separated in the first dimension electrophoresis. We have developed a 2-DE pattern from the 5- to 20-kb *NotI/EcoRV* fragments, which will visualize a new 2000 spots.

Before initiating a study on the 100 human families with the 2-DE technique, we have started a pilot study for the assessment of detectability of germ cell mutations in mice with this technique. We believe that this pilot study will provide basic information for the estimation of the number of children of A-bomb survivors that should be examined in order to obtain statistically significant results. After digestion with *Not*I, *Eco*RV and *Hin*fI, we examined DNA samples of 43 F1 mice (BALB/c) derived from spermatogonial cells irradiated with 5 Gy (X-ray) and 36 control F1 mice born to the same set of 8 pairs of parents before irradiation. Among 1100 spots from a DNA sample on a single gel, 553 spots were suitable for detecting mutations because each of the spots has a CV of less than 0.12. A total of 43,600 spots (23,741 from the exposed group and 19,859 from the control group) were surveyed. One candidate, with intensity 60% of the mean spot intensity, for a mutant was detected among spots from the exposed group [16] and is under study to characterize the mutational event at the molecular level.

We believe that the 2-DE technique will be promising for the detection of germinal mutations.

REFERENCES

1. M. Kodaira, C. Satoh, K. Hiyama, K. Toyama, Am. J. Hum. Genet., 57 (1995) 1275-1283.

2. C. Satoh, N. Takahashi, J. Asakawa, M. Kodaira, R. Kuick, S.M. Hanash, J.V. Neel, Environ. Health Perspect., 104, Suppl. 3 (1996) 511-519.
3. C. Satoh, N. Takahashi, J. Asakawa, M. Kodaira, R. Kuick, S.M. Hanash, J.V. Neel, World Health Statistics Quarterly (1996), in press.
4. J. Asakawa, R. Kuick, J.V. Neel, M. Kodaira, C. Satoh, S.M. Hanash, Proc. Natl. Acad. Sci. USA, 91 (1994) 9052-9056.
5. R. Kuick, J. Asakawa, J.V. Neel, C. Satoh, S.M. Hanash, Genomics, 25 (1995) 345-353.
6. A.J. Jeffreys, V. Wilson, S.L. Thein, Nature, 314 (1985) 67-73.
7. Y. Nakamura, M. Leppert, P. O'Connell, R. Wolff, T. Holm, M. Culver, C. Martin, et al., Science, 235 (1987) 1616-1622.
8. Y.E. Dubrova, A.J. Jeffreys, A.M. Malashenko, Nat. Genet., 5 (1993) 92-94.
9. S. Sadamoto, S. Suzuki, K. Kamiya, R. Kominami, K. Dohi, O. Niwa, Int. J. Radiat. Biol., 65 (1994) 549-557.
10. Y.J. Fan, Z. Wang, S. Sadamoto, Y. Ninomiya, N. Kotomura, K. Kamiya, K. Dohi, R. Kominami, O. Niwa, Int. J. Radiat. Biol., 68 (1995) 177-183.
11. C. Satoh and M. Kodaira, Nature, 383 (1996) 226.
12. Y.E. Dubrova, V.N. Nesterov, N.G. Krouchinsky, V.A. Ostapenko, R. Neumann, D.L. Neil, A.J. Jeffreys, Nature, 380 (1996) 683-686.
13. I. Hatada, Y. Hayashizaki, S. Hirotsune, H. Komatsubara, T. Mutai, Proc. Natl. Acad. Sci. USA, 88 (1991) 9523-9527.
14. C. Satoh, J.V. Neel, A. Yamashita, K. Goriki, M. Fujita, H.B. Hamilton, Am. J. Hum. Genet., 35 (1983) 656-674.
15. J. Asakawa, C. Satoh, Y. Yamasaki, S-H Chen, Proc. Natl. Acad. Sci. USA, 89 (1992) 9126-9130.
16. J. Asakawa, M. Kodaira, N. Nakamura, H. Katayama, S. Funamoto, S. Tomita, J.L. Ohara, D.L. Preston, C. Satoh, The 10th International Mouse Genome Conference, 8-11 October, Paris (1996).

Genetic instability induced by low-dose radiation

M. Watanabe[a], S. Kayata[a], S. Kodama[a], K. Suzuki[a] and T. Sugahara[b]

[a]Laboratory of Radiation and Life Science, Department of Health Science, School of Pharmaceutical Sciences, Nagasaki University, 1-14 Bunkyo-machi, Nagasaki 852, Japan

[b]Health Research Foundation, Kyoto 606, Japan

1. INTRODUCTION

A variety of human cancers have multiple chromosomal abnormalities. Patients afflicted with a range of leukemias and lymphomas show karyotypic abnormalities, many of which include distinct chromosome translocations. Furthermore, the chromosome breakage syndromes such as xeroderma pigmentosum, ataxia telangiectasia, Bloom's syndrome and Fanconi's anemia are all characterized by increased incidence of cancers. The association between chromosomal rearrangements and cancer suggests that chromosomal instability as a phenotype may underlie some fraction of those changes leading to cancer. The importance of genomic changes in tumor progression and their association with cancer underscores the importance of studying the mechanisms by which they arise. We do not know the molecular, genetic and cellular events leading to increased genomic instability.

There is a long history linking radiation exposure and the elevated incidence of cancer. The biological consequences of exposure to ionizing radiation include gene mutation, chromosome aberrations, cellular transformation and cell death. Furthermore, irradiated cells show a tendency toward increased or decreased chromosome numbers (aneuploidy) compared with unirradiated cells. These effects are attributed to the DNA-damaging effects of irradiation resulting in irreversible changes during DNA replication or during the processing of the DNA damage by repair processes. Accordingly, it has been widely accepted that most of these changes take place during the cell cycles immediately following exposure. On the other hand, it has been apparent for many years, however, that delayed reproductive death and delayed chromosome aberration occurred in progeny of cell irradiated with various kinds of radiations[1-3]. Thus, ionizing radiation may induce a transmissible genomic instability producing a variety of cellular effects detected after many cell cycles in the progeny of the irradiated cells.

The reported frequencies of radiation-induced oncogenic transformation in vitro are generally 100-1,000 times greater than those of somatic cell mutation. Oncogenic transformation of cells is a multistep process. Many investigators believe that five or six changes are necessary for this at least and somatic cell mutation is cause of the each change. If such supposition is right, frequency of oncogenic transformation *in vitro* (cancer *in vivo*) becomes the value that multiplied each mutation frequency (10^{-6} per generation). In that event, it is supposed to less than 10^{-25} - 10^{-30}, and frequency of cancer doesn't fit reality. Therefore, the phenomenon that several changes seem to happen in succession during short term needs to be thought about. In this meaning, induction of genetic instability by radiations is attractive as the one candidate.

This paper extends our study of radiation-induced instability in mammalian cells and describes the appearance of several delayed genetical instabilities in the clonal progeny of sublethally irradiated cells. We enforced this study to inspect it whether such phenomenon was universal and whether preirradiation of low dose X-rays modified the frequency of radiation induced genetic instability.

2. MATERIALS AND METHODS

In this study, we used mouse m5s cell and cultured cells in Eagle's MEM medium containing 10% volume of fetal bovine serum. We irradiated cells with X-rays at a dose rate of 0.5 Gy/min. To measure the effects of preirradiation of X-rays, we irradiated cells with 0.02 Gy of X-rays, at 4 hours before challenging irradiation. After irradiation, we inoculated cells into plastic dish at cloning density, and isolated cells from the colony that occurred in dish. We repeated this cloning process twice. At this time point, the cells have already divided 40 times after irradiation at least. Then, we examined the colony forming ability, the frequency of chromosome aberrations and the frequency of mutation in surviving cells. We selected mutants in MEM medium containing 40 mM of 6-thioguanine as previously reported [2].

3. RESULTS

We irradiated cells and cloned 2 times in succession. Then, we examined plating efficiency, the frequency of chromosome aberration and mutation at *hprt* gene.

As the result, a persistently reduced cloning efficiency occurs in many of the cloned progenies of X-irradiated cells (Table 1). Rate of giant cells with multinuclei largely increased in irradiated cell population around 40 cell generations. Colony formation ability of a cell relates to appearance frequency of giant cell closely.

Table 1
Genetically instability appeared as several phenotypes induced in progenies of m5s cells irradiated with X-rays at over 34 cell generations after. Each value is the mean of data from experiments of 14 independent clones.

X-ray Dose (Gy)	Cloning Efficiency	Frequency of Colony with Giant Cells	Frequency of Dicentrics and Breaks	Mutation Frequency ($\times 10^{-6}$)
0	0.84	0.14	0.00083	1.3
0.02	0.86	0.10		
10	0.60	0.29	0.025	8.8
0.02+10*	0.62	0.22		
15	0.27	0.48	0.041	
0.02+15*	0.49**	0.27***	0.020**	

*Cells were preirradiated with 0.02 Gy of X-rays 5 hours before challenging irradiation of 10 and 15 Gy. **P<0.001, ***P<0.02 by Student t-test.

As expected, we observed multiple and complex chromosome alterations at the first passage after irradiation and these anomalies progressively decreased, became quite rare at 3-5 cell generations. Then, these anomalies increased at the later passages (over 34 cell generations) (Table 1). Most strikingly, over 50% of anomalies occurring at the later passages were dicentrics and some cells carrying dicentrics also had chromosome fragment. These results showed that most chromosome lesions directly induced by X-rays are hardly compatible with cell survival and thus disappear after a few cell generations. However, surviving cells acquire a *de novo* chromosome instability leading to the formation of clones with unbalanced karyotypes at late passages. In progenies of irradiated cells, we also found 5-28 times (average 8.8) higher frequency of mutant induction (Table 1).

Preirradiation of low dose X-rays under 0.1 Gy reduced the frequency of this radiation induced genetic instability, such as delayed reproductive death, giant cell formation and chromosome aberrations (Table 1).

4. DISCUSSION

The multistep nature of carcinogenesis sparingly suggests that progression required expression of genomic instability, or mutator phenotype. As stated above, the delayed expression of heritable damage may play an important role in carcinogenesis. We have reported that increased aneuploidy associated with carcinogenesis process [4-6]. In this paper, the delayed reproductive death phenotype also associated with the delayed expression of chromosomal aberrations and mutation at *hprt* gene locus.

Delayed chromosomal instabilities are frequent events occurring in the progeny of irradiated cells. We have analyzed over 18 clones derived from single cell surviving 10 Gy X rays. Fifty percent of these clones contain cells that were expressed genetical instability. The frequency of cells with chromosome aberrations in single colony is 0.025 per cells and this is 30 times greater than that of cells in nonirradiated colony. This indicates a large targeted size for the initiation of delayed chromosome instability. Consequently, the target is probably not a single gene or gene family, and a single mutation is probably insufficient to induce genomic instability. Although the molecular lesion or lesions that initiate genetical instability after cellular exposure to ionizing radiation are not known, several reports suggest that the DNA double strand break may be involved [7].

Recent work in our laboratory has failed to produce any genetically unstable clones by UV irradiation (manuscript in preparation). This means that ionizing radiation specific lesions, i.e., DNA double strand breaks, should be cause of genetical instability. It is generally believed that radiation-induced DNA double strand breaks are repaired rapidly. Cells with persistent unrepaired DNA double strand break should trigger specific cell cycle checkpoints that abrogate further proliferation. Therefore, DNA double strand breaks are efficient at inducing chromosome aberrations, deletion mutations and gene amplification. They constitute a lesion that underlies some fraction of the gross chromosomal change that has the potential to be visible cytogenetically.

Additionally, our work suggests that preirradiation of X-rays at 0.02 Gy restrains an instruction of genetic instability. In the case of chromosome aberrations, preirradiation with 0.02 Gy of X-rays dominantly decreases the frequencies of dicentrics with fragment instead of dicentric without fragment. Dicentrics initiate bridge breakage and refusion cycles associated with chromosomal instability. Dicentrics are generally not compatible with the survival of normal cells and contribute to delayed reproductive death, because this process is frequently accompanied by large-scale genetic loss. In contrast, translocations are generally not cytotoxic, although such events do result in gene rearrangement, which may initiate the cascade of destabilizing events associated with chromosomal instability. Recently increasing evidence suggests that interstitial telomere repeat-like sequences are hotspots from recombination, breakage and chromosome fusion. As a delayed effect of cellular exposure to ionizing radiation, recombination events mediated by interstitial telemeters may contribute to the perpetuation of delayed chromosomal instability. Alternatively, genomic instability may be independent of induced DNA damage and may involve non-nuclear targets, such as those involved in signaling cascades.

This phenomenon has important meaning in risk estimation of the radiations to human being.

REFERENCES

1. M. Watanabe, M. Suzuki, K. Suzuki, K. Nakano, and K. Watanabe, Effect of multiple radiation with low doses of gamma rays on growth ability of human embryo cells *in vitro*. Int. J. Radiat. Biol. 6 (1992) 711-718.
2. M.Watanabe, and K. Suzuki, Expression dynamics of transforming phenotypes in X-irradiated Syrian hamster embryo cells, Mutation Res. 249 (1991) 71-80.
3. M. A. Kadhim, D. A. MacDonals, D.T. Goodhead, S. A. Lorimore, S. J. Marsden, and E.G. Wright, Transmission of chromosomal instability after plutonium alpha-particle irradiation, Nature 355 (1992) 738-740.
4. K. Suzuki et al., Multistep nature of X-ray induced neoplastic transformation in golden hamster embryo cells: Expression of transformed phenotypes and step-wise changes in karyotypes, Cancer Res. 49, 2134-2140, 1989.
5. K.Suzuki, N.Yasuda, F.Suzuki, O.Nikaido, and M. Watanabe, Trisomy of chromosome 9q: Specific chromosome change associated with tumorigenicity during the progress of X-ray-induced neoplastic transformation in golden hamster embryo cells, Int. J. Cancer 44 (1989) 1057-1061.
6. M. Watanabe, K.Suzuki and S.Kodama, Karyotypic changes with neoplastic conversion in morphologically transformed golden hamster embryo cells induced by X-rays, Cancer Res. 50 (1990) 760-765.
7. W.P.Chang, and J.B. Little: Evidence that DNA double-strand breaks initiate the phenotype of delayed reproductive death in Chinese hamster ovary cells. Radiat. Res., 131 (1992) 53-59.

Experimental study of the genetic effects of high levels of natural radiation in South-France

M. Delpoux[1], A. Léonard[2], H. Dulieu[3] and M. Dalebroux[4]

[1]Groupe de Recherches Pluridisciplinaires sur l'Environnement d'Albi de l'Université Paul Sabatier, Toulouse, France

[2]Université Catholique de Louvain, Belgium

[3]Université de Bourgogne, Dijon, France

[4]European Union, Bruxelles, Belgium

A survey made at the ground level in South-West France (Fig. 1) reveals that the gamma radioactivity ranges generally from 0.001 to 0.030 x 10^{-2} mGy/hr and that dose rates as high as I x 10^{-2} mGy/hr are not uncommon.

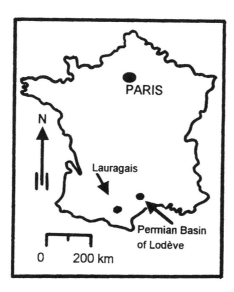

Figure 1. Geographical situation of uranous zones in South France

The Hérault Mission of the French Atomic Energy Commission discovered that the dose rate can reach 10 x 10^{-2} mGy/hr at some "hot spots" located in the Pernian Bassin of Lodève and in the Laurageais. The ionizing radiations originate from the two natural uranium isotopes (99.3 % of ^{238}U and 0.7 % of ^{235}U) and from their 28 radioactive daughter products.

Plants that grow on the most radioactive substrates are apparently normal. However, some differences in germination rates and size of the plantlets, changes in color flowers, etc., transmissible to the progeny suggested, that mutagenic events had been produced by site components. To be able to discriminate the effects of the high natural radioactivity from the influence of other environmental factors, experiments were performed with a strain of Tobacco (*Nicotiana tabacum* L.) carrying genetic markers for leave color.

When evaluating the biological effects of high natural radioactivity on persons and animals living in those regions, one of the main difficulties encountered is the exact determination of the doses received and, in addition, to find appropriate controls. In order to overcome those difficulties, we performed observations on laboratory animals maintained in captivity at a "hot spot" in the bassin of the river Lodève. In the hut where the animals were placed, the γ dose-rate amounted to about 8 x 10^{-2} mGy/h.

1. EXPERIMENTS WITH A PLANT MARKER : THE α_1^+/α_1 α_2^+/α_2 SYSTEM OF TOBACCO (*Nicotiana tabacum* L.)

1.1. The genetic system [1 -5]

It consists of a double heterozygote α_1^+/α_1 α_2^+/α_2 of *xanthi* variety. The plants carrying this genetic system has greenish-yellow leaves. Either spontaneously or under the action of chemical or physical mutagens, the genetic composition of the system can be modified by mutations in α_1^+ op α_2^+ by reversion in α_1 or α_2, reversions being the most common event. In growing cells, each reverted cell will yield a clone that appears as a green spot in the palisade tissue of the greenish-yellow leaf. Since 1,000 to 2,000 cells can be analyzed per mm^2 of leaf; this marker system is particulary appropriate for detecting effects of low and very low doses of radiation with very good statistical precision.

The genetic effect of a given dose, can be expressed as average reversion rate per cell cycle on the basis of the number of reverted cells or, equivalently, of reverted leaf area. It has been showed that, for a given individual, there exists a simple relation between the totai leaf area S and the reverted area Sg on one hand, and the reversion rate on the other :

$$p = 1 - \left[\frac{S - Sg}{S}\right]^{1/t}$$

where t is the number of cell cycle that actually took place during the chronic irradiation:

$$t = \frac{\log N}{\log 2} - 7$$

N is the total number of cells observed that corresponds to the total leaf area S with cell density d (number of cells per unit area), so that :

$$N = S \times d$$

Several experiments were performed and the main results were as follows :

1.2. Results [6-10]
The main results of the numerous experiments performed with this system can be summarized as follows:

1.2.1. First experiment in the Permian Bassin of Lodève (Fig.2)
To study the reversion rate per cell cycle, one hundred plants, divided in two

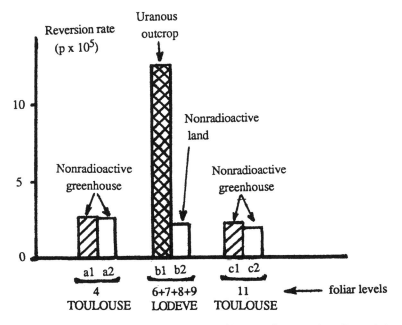

Figure 2. Reversion rates under different conditions of natural radioactivity (Mas d'Alary site)

sets reversion rates in the two sets of 50 pots were maintained in a greenhouse on the University campus in Toulouse and allowed to grow up to four or five leaves. No difference was found between the reversion rates in the two sets.

Thereafter, one set (set 1) was placed at the uranous site of Mas d'Alary where the dose rate ranged between 3 and 10 x 10^{-2} mGy/hr. The second set (set 2), was placed in the vicinity over a substrate yielding a very low dose rate of 0.030 x 10^{-2} mGy/hr considered as a control. Both sets, were allowed to grow up to nine to ten leaves. Whereas, a very large and significant increase in reversion rate was observed in the set 1, no difference was found between the control (set 2) and the plants grown in Toulouse.

Finally, the plants were taken back in Toulouse and placed under the same conditions as before their transfer to Lodève. The plants were allowed to grow a few more leaves. No difference was detected with respect to reversion rates between the latter leaves and the ones produced in the greenhouse at the start of the experiment.

It clearly appears from these data that the reversion rate in the plants maintained in the uranous site, was about five times larger than that of the average control.

1.2.2. Second experiment in the Permian Basin of Lodève (Fig. 3)

To study the dose effect relationship four lots of Tobacco plants were cultivated

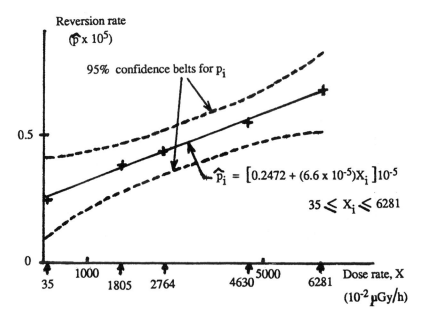

Figure 3. Dose-effect relationship over the Mas d'Alary uranous outcrop in the Permian Basin of Lodève

in areas yielding the following dose rates :

1.805, 2.764, 4.630 and 6.281 x 10^{-2} mGy/hr.

A fifth lot was placed in the vicinity over a substrate yielding the very low dose of 0.035 x 10^{-2} mGy/hr considered as a control. As previously, the response of the genetic system was expressed as reversion rate per cell cycle. The shape of the response curve was found to be linearly dependent of the dose.

1.2.3 Third experiment in Lauragais (Fig.4)

The study was perforryled at the uranous site of Lagravette where the dose rates are much smaller than in the second experiment above, ie:

0.065, 0.106, 0.179, 0.291 , 0.366 and 0.590 x 10^{-2} mGy/hr.

A threshold in terms of reversion rate per cell cycle was observed as long as the dose rate was no much larger than that from a normal natural background. For the two latter higher dose rates, the reversion rate increased significantly .

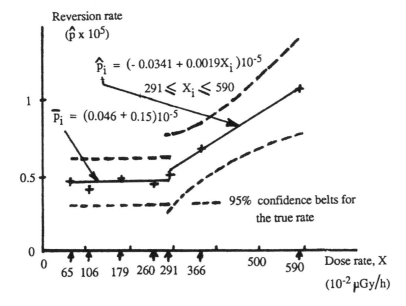

Figure 4. Dose-effect relationship over the Lagravette uranous outcrop in the Lauragais

1.3. Conclusions

1. This marker is very appropriate for detecting *in situ* the genetic effects of low and very low doses of natural radioactivity.
2. There exists a threshold dose rate for the mduction of this type of mutations in Tobacco leaves.
3. From the background to about 0.3×10^{-2} mGy/hr (2.63×10^{-2} Gy/year), we have a plateau.
4. From 0.3×10^{-2} mGylhr (2.63×10^{-2} Gy/year), the response is linearly dependent on the dose.

2. EXPERIMENTS WITH ANIMALS

To study the production of chromosome aberrations in mammalian somatic cells, male laboratory rabbits were kept, for 20 months, in plastic cages placed on the floor and gamma-ray dose determined by individual fluoride dosimeters put around the neck of each rabbit. Blood samples were taken at 4 months intervals to study the occurrence of structural chromosome aberrations in peripheral blood lymphocytes. The effects on male fertility were evaluated on groups of 50 three month old male mice of the BALB/C strain put also in plastic cages with two lithium fluoride dosimeters in each cage. Due to the sensitivity of the mice to low temperatures, the exposure was restricted to the summer period and was repeated with new animals during three successive years. After the exposure, each male mouse was mated in our laboratory to one three month old non-irradiated virgin female of the same strain for a 6 month period. Litter size was recorded at birth and the offspring sexed at weaning 20 days later. To correlate the effects of low doses of ionizing radiation with possible changes in germ cell populations, testicular damage was evaluated at the end of the main period by recording the weight of the testes and by histological examination of 100 carefully selected median cross sections of testicular tubules from one testis of each animal for the presence of spermatogonia, spermatocytes, spermatids and spermatozoa. This method provides mainly a qualitative analysis and allows only a semi-quantitative asssessment because germ cells may be numerous or searce in a given cross section. Meiotic preparations were made from the second testis of each animal and one hundred spermatocytes at the diakinesis-first metaphase stage of meiosis were analysed for the presence of reciprocal translocations.

Control rabblts and control mice maintained near the radioactive site under comparable conditions were handled and sampled in the same exposed animals.

2.1. Results

During the 20 months the exposed rabbits received from 36,300 up to 130.750 mGy according to the locations of their cages in the hut whereas the doses received during the same period by the controls arrounded to 2,400 mGy (Table 1).

Table 1
Dose(10^{-2}mGy) or γ-radiation recieved by the rabbits

Treatment	Rabbit n°	Duration of exposure (months)				
		4	8	12	16	20
Control	1	65	102	152	207	254
	2	40	95	140	195	240
	3	50	110	158	205	235
Irradiated	4	26,500	52,750	70,750	101,750	130,750
	5	22,300	44,300	57,300	76,800	95,800
	6	22,500	43,250	56,000	73,500	91,500
	7	13,000	26,500	38,700	51,200	*
	8	7,800	14,800	20,500	28,500	36,300

*Rabbit 7 died after 19 months

Figure 5 shows that the yield of structural chromosome anomalies typical of an exposure to ionizing radiations (chromosome fragments, dicentric chromosomes) increases initially but disappear completely after 20 months.

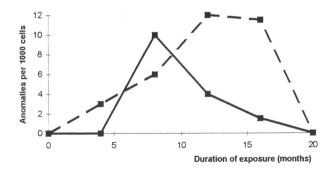

Figure 5. Relationship between duration of exposure and dicentrics (—) and cromosome fragments(-----) per 1,000 cells in rabbit peripheral blood lymphocytes

The male mice maintained in the area of high natural radioactivity received from 13,800 mrad up to 63,250 mrad and the concurrent controls from 34 to 127 mrad according to the duration of exposure (Table 2).

Table 2
Doses of gamma radiation(mrad) recieved by the male mice

Experiment	Duration of exposure	Control	Irradiated
1 st year	3 months	34	13,800
2nd year	4 months	68	15,000
3rd year	6 months	127	45,080
			63,250

The number of litters and of offspring sired after exposure during 6 month period was clearly related to the dose of γ-irradiation received during the exposure period (Fig. 6). Both values increase in a dose-related way up to 45,080 and fall abruptly for the animals receiving 63,250 mrad.

The histological and cytological studies performed on the testes at the end of the mating period did not reveal any difference between the control and the exposed males (Table 3).

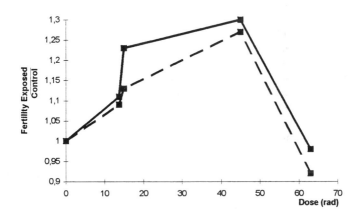

Figure 6. Relation between dose recieved and litters(---) and offspring(—) sired by male mice

Table 3
Results of the cytological and histlogical observations performed on male mice

Experiment	Dose received (cGy)	Weight of the testes (mg)	% of abnormal spermatocytes	Histological results			
				Cross sections with			
				Spermatogonia (%)	Spermatocytes (%)	Spermatids (%)	Spermatozoa (%)
I	0.034	262	0	100	100	100	90
	13.8	270	0	100	100	100	90
II	0.068	264	0	100	100	100	87
	15.000	269	0	100	100	100	88
	0.127	254	0	100	100	100	88
III	45.08	268	0	100	100	100	90
	63.25	265	0	100	100	100	91

2.2. Discussion and conclusions

Our observations confirm that high natural radioactivity can result in the production of structural aberrations in mammals living in those areas. The drastic decrease in the incidence of anomalies occuring after a few months results probably from a selective elimination of the most sensitive stem-cells combined to the fact that the life-span of the rabbit peripheral blood lymphocytes does not exceed 2 months [11]. A short life-span of peripheral blood lymphocytes carrying unstable structural chromosome aberrations appears a common characteristics of the majority of mammals [12] and contrasts greatly with values exceeding three years reported in man it can be inferred, therefore, that one cannot consider that cytogenetic observations performed on wild mammals living in area of high radioactivity (natural or resulting from nuclear assays) would provide reliable informations on the doses received by the animals and on the level of damage produced to their genetic material.

The increase in male fertility observed for animals receiving up to 43,000 mrad could rpresent an example radiation hormesis (13) : stimulating effects of small doses of ionizing radiations is a controversial topic but it should be stressed that our results on the male mice are in agreement with the observations of Newcombe and [14] and [15] who reported an increase of embryo production after low doses of ionizing radiations administered to trout spermatozoa.

REFERENCES

1. Dulieu, H.L, Etude de la stabilité d'une déficience chlorophyllienne induite par traitement au methane sulfonate d'éthyle. Ann.Amélior.des Plantes, 17 (1967) 339-355.
2. Dulieu, H.L., Somatic variations on a yellow mutant in *Nicotiana tabacum L.* ($a_1^+/a_1a_2^+/a_2$), I. Non-reciprocal genetic events occuring in leaf cells, Mutation Research, 30 (1974) 63-70.
3. Dulieu, H.L., (1 975) - Somatic variations on a yellow mutant in *Nicotiane tabacum L.* ($a_1^+/a_1a_2^+/a_2$), Il. Reciprocal genetic events occuring in leaf cells, Mutation Research, 25 (1975) 289-304.
4. Dulieu, H.L.and M.A.Dalebroux, Spontaneous and induced reversion rates in a double heterozygous mutant of Nicotiana tobacum var.xanthi n.c. dose-reponse relationship, Mutation Research, 28 (1975) 66-77.
5. Dulieu, H.L., Antimorphic alleles with various strenghts at the S_u and Y_g loci of Tobacco, The Journal of Heredity, 177 (1986) 301-306.
6. Delpoux, M., Etude expérimentale des effets de la radioactivité naturelle tellurique sur les végétaux. Hypothèses sur l'influence de l'environnement fortement énergétique sur les êtres vivants, Thèse Doct. d'Etat, Université P.Sabatier, Toulouse, France, 1974.
7. Delpoux, M., and M.A.Dalebroux , Genetic effects on the $a_1^+/a_1a_2^+/a_2$ system of Tobacco over a uranous outcrop in Basin of Lodève (Hérault, France), Mutation Research, 82 (1981a) 101-1 10.
8. Delpoux, M., and M.A.Dalebroux, Genetic effects on the $a_1^+/a_1a_2^+/a_2$ system of Tobacco over a uranous outcrop in Lauragais (Aude, France), Mutation Research, 83 (1981b) 375-382.
9. Delpoux, M., La radioactivité et les végétaux, Cptes-rend. Congr. Nat.Soc. franç., Radioprotection, Monte-Carlo, 221 -237, 1984.
10. Delpoux, M., M.A.Dalebroux, H.Dulieu and A.Léonard , L'irradiation naturelle tellurique et les êtres vivants, Bull. Soc. Ecophysiol., France, 11 (1986) 157-184.
11. Decat, G., Léonard, A. Int. J. Radiat. Biol., 38 (1980) 179-185.
12. Léonard, A., Decat, G., Fabry, L., Mutation Res., 95 (1982) 31-44.
13. Gerber, G.B., Ann. Ass. Belge Radioprotection, 19 (1994) 533-556.
14. Newcombe, H. B., McGregor, J. F., Radiat.Res., 51 (19720402-409.
15. McGregor, J.F., Newcombe, H.B., Radiat.Res., 52 (1972) 536-544.

VIII Summary and general discussion

Summary of symposium

Summary of Session A on "Dose estimation"
presented by K.B. Li

This session took place on October 21, 1996. The Chairman was Dr. P. C. Kesavan (India) and the co-chairman was Dr. Kaibao Li (China).

In this session, a keynote lecture on high background radiation area - an important source of exploring the health effects of low dose ionizing radiation, was presented by Prof. Luxin Wei, the President of the 4th International Conference on High Levels of Natural Radiation. An invited lecture on exposure to national radiation sources worldwide was presented by Dr. B. G. Bennett (UNSCEAR). Three other proffered papers were given, in turn, by Drs. K. Nagaoka (Japan), Kedao Wei (China) and Kaibao Li (China).

Prof. Luxin Wei in his presentation remarked that one of the great advantages of radiation epidemiology in the high background area is the possibility of obtaining the results from direct observation without extrapolation of dose-effect relationship from high to low dose levels. This especially true when the families of the subjects have lived in the investigation areas for many generations. He raised two important problems about how to identify the definite confounding factors from a large amount of the doubtful and suggestive agents, and about how to explain the results obtained from radiation epidemiology in the high background radiation areas and control areas, the combination between epidemiology and radiobiology, particularly at the cellular and molecular levels.

Dr. B. G. Bennett, in his presentation, reviewed the exposures from natural radiation sources and some of reported variations based on the latest assessment of radiation exposures in the UNSCEAR 1993 Report. In the area of normal background, the main components and their global-averaged annual effective doses are inhalation of radon, 1.2 mSv, internal dose from ^{40}K, 0.17 mSv, external dose from terrestrial gamma radiation, 0.46 mSv, and 0.38 mSv from cosmic rays. Minor contribution from cosmogenic radionuclides and from ^{238}U and ^{232}Th radionuclides in food and drinking water.

Variation about these average values by factor 2 or 3, are common, and in the case of radon by a few orders of magnitude. In the areas of high background radiation, one or more of these components may exceed the average values by a factor of 10.

D. Nagaoka (Japan) presented his paper on cosmic ray contribution in the measurement of environmental gamma rays dose. Several types of dosimeters were used in the measurements, including scintillation, spectrometer, air-equivalent ionizing chamber, pressured ionizing chamber, thermoluminescence dosimeters, radiophotoluminescence glass dosimeters, and scintillation survey meters. The experimental results show that the responses of these dosimeters to cosmic rays are different. It is considered that the counting rate of the NaI scintillation spectrometer is the most useful to estimate the cosmic ray contribution at ground level.

Dr. Kedao Wei (China) presented the results of national survey on natural radiation background using TLD methods. He concluded:
- the population-weighted average absorbed dose in air from terrestrial gamma radiation indoor and outdoor are 67 nGy/h and 89 nGy/h, respectively;
- a marked characteristic is that the level of natural gamma radiation in south China is higher than that in north China;
- the level of gamma radiation indoor is higher than that outdoor by 33%;
- the time-weighted average absorbed dose rates obtained by environmental monitoring is in agreement with those by individual-wearing measurements.

Dr. Kaibao Li (China) described the methodology of external radiation dosimetry in dose assessment in the high background radiation area in Yangjiang, China, following the lines of UNSCEAR reports. He emphasized the importance of quality assurance and quality control programs in maintaining the quality of dose measurement with adequate accuracy and reliability.

Summary of Session B on "Radioactive aerosols "
presented by C.-S. Zhu

Eight papers concerning the study of radon concentration in the special areas in some countries had provided in this conference and five papers had reported on the Session B: "Radioactive Aerosols " in the afternoon of 22nd October.

Professor M. Sohrabi indicated that the study on the effects of public natural exposure to ionizing radiation are very important for determination of risk factors in his lecture and he suggested and wished to propose criteria for defining different natural radiation areas and to review conventionally known high level natural radiation areas, natural springs used as spas and some radon- prone areas with special regard to dwellings.

Other reports which provided by scientists of Iran introduced the concentration of ^{226}Ra and ^{222}Rn in hot spring and bedrock samples. The Indian scientists gave an interesting paper about the discriminating dosimeter for monitoring radon, thorn and their progeny in monazite areas of Kerala of India.

Chinese scientists provided some papers of the survey results of radon concentration in some areas. A survey of ^{222}Rn concentration indoor and outdoor of cave dwellings in Gansu Province was conducted. The results showed that ^{222}Rn concentration in cave dwellings is far higher than in common dwellings. The study of relationship of the ^{222}Rn concentration and risk of lung cancer are under conducting now in some selected areas in Gansu Province.

Another paper introduced the results of determination of the levels of ^{222}Rn in hot spring water and in air of bathrooms in Liaoning Province. The hot spring water is used only for hydrotherapy. In 86% bathrooms the Rn levels are below 370 Bq m^{-3} in 8% bathrooms. In some places the Rn levels have been detected exceeding 1110 Bq m^{-3}.

The Survey of the levels of radon and its daughters are investigated not only in the air of indoors of different type of dwellings, but also in other environmental media, especially in drinking waters and spring waters in China.

Summary of Session C on "Natural radionuclides"
presented by Dr. M. Sohrbi

The session on "Natural Radionuclides" was chaired by M. Sohrabi (National Radiation Protection Department, AEOI, Iran) and D.C. Lauria (Institute of Radiation Protection and Dosimetry, BNEC, Brazil) covering two invited lectures and five contributing papers.

In the first invited lectures on "India Research on High Levels of Natural Radiation", P.C. Kesavan (Bhabha Atomic Research Center, India) comprehensively reviewed highlights of the studies carried out in Kerala with over 140,000 inhabitants receiving a dose from 15 to 25 mGy/year and in Chinavilai village in the Manavalakuchu of Tamil Nadu with 2,000 inhabitants receiving a dose of 20 to 40 mGy/year. The two areas, according to the criteria proposed by M. Sohrabi (AEOI, Iran) at this conference, with a dose range 20 to 50 mSv/year are classified as HLNRAs. Kesavan, based on studies on dental and skeletal mutations in rats and cytogenetic studies of the newborns and adult population, concluded that no adverse biological effects have been observed in these areas. Also cytogenetic studies with plants indicated that internal radiation is largely responsible for meiotic abnormalities, while external radiation up to a dose of 8.7 cGy has shown no somatic mutations in the sensitive stamen hair system of Tradescantia.The studies of Down's syndrome on 90,000 births carried out during the last three years in HBRAs of Kerala confirmed the earlier studies. The lack of discernible support from such areas for the linear, non-threshold relationship invites more studies between dose and biological effects and the role of apoptosis as well as on the basic question of "paradigm shift" and linear non-threshold relationship between dose and biological effects.

The invited lecture on "Brazilian research in the areas of high natural radioactivity" was presented by A.R. Oliveira (Radiation Medicine Department, Brazilian Nuclear Industries, Brazil). Oliveira with an honest and sincere approach regretted one decade gap existing in the information on Brazilian studies since those carried out previously by Cullen and Penna-Franca in the monazite areas located along the Atlantic Coast (Guarapari, Meaipe) of zones of volcanic intrusive in the interior of the state of Minas Gerais (Pocos de Caldas, Tapira and Araxa), while he stimulated the need for further extensive research in

Brazil. He also introduced a new area in the Amazon Rainforest called Pitinga in the Amazon state which is a village close to a tin mine where many houses are built in that area. Pitinga with a 2000 medium-term inhabitants receiving 3 mSv/a has been under medical supervision which is not expected to show anysignificant increase in cancer rate. The area, according to the criteria proposed by M. Sohrabi (AEOI, Iran) at this conference, is classified as a LLNRA.

Also five contributed papers were presented at the above session: (1) on natural radioactivity in soil samples in some HLNRAs of Iran by M. Sohrabi, M.M. Beitollahi and co-workers (NRPD, AEOI, Iran) showing the highest levels in the HLNRAs of Ramsar and Mahallat; (2) on investigation about the radioactive disequilibium among the main radionuclides of thorium decay chain in foodstuff by D.C. Lauria and co-workers (IRPD, BNEC, Brazil) which has been concluded to be due to storage time of the foodstuff; (3) on transfer of ^{226}Ra, ^{228}Ra, ^{210}Pb and ^{210}Po in aquatic life and food chain by X.T. Yang (Fujian Provincial Institute of Radiation Health, China) showing a higher concentration of ^{226}Ra, ^{228}Ra and ^{210}Pb in bones and scales than those in flesh as well as showing no significant influence of cooking process on the transfer of these elements on the food chain; (4) on intake and initial doses from investigation in two Chinese high radiation areas by H.D. Zhu (Institute of Radiation Medicine, CAMS, China) who concluded that the change of the dietary habits of the people since 1982 requires renewing the previous data taken; and (5) on biosphere effects of radionuclides released from eight coal-fired power plants by Y.H. Mao and Y.G. Liu (Sichuan Institute of Radiation Protection, China) showing an increase in the effective dose of the public living near the plants with no increase in the radioactivity of the river water. The session as a whole had a great contribution on the successful outcome of the conference.

Summary of Session D on "Fallout from nuclear tests"
presented by Dr. L. J. Appleby

The following summary was prepared with the help of Dr. Andr Bouville (RADTEST Scientific Secretary). The session was chaired by Prof. Mao and Prof. Sugahara. Four papers were presented:

"Overview of SCOPE-RADTEST Project", by Ren Kirchmann.

The history and the purpose of the SCOPE-RADTEST program were presented in detail. It was emphasized that RADTEST is enabling the nuclear scientists that were directly or indirectly involved with the nuclear weapons tests to share for the first time previously restricted work involving data, models, and knowledge about the fate of radionuclides released into the environment and their possible human health effects.

"Assessment of Health Effects from the Interactions of Radiation with other Agents", by Werner Burkart.

The difficulties associated with the assessment of health effects resulting from combined exposures to radiation and other agents were presented very clearly. These difficulties were stressed again during the discussion of the paper, which dealt mainly with the combined effects of radiation and benzo-a- pyrene. A better knowledge of the mechanisms involved in the induction of health effects will be necessary in order to predict with reasonable confidence the combined effects of radiation with other agents.

"Exposures Dose from Natural Radiation and other Sources to the Population in China", by Zhu Changshou.

The results of several years of measurements have been summarized in this paper. The average effective dose received by the Chinese population from all sources is estimated to be 2.6 mSv a^{-1}, including 2.4 mSv a^{-1} from natural sources. Three components were discussed in detail: (1) the dose from cosmic rays, which is particularly high in Tibet, (2) the dose from indoor radon, which seems to have been overestimated, and (3) the dose from medical irradiation, which is relatively small because of the low frequency of use of X-ray examinations and of nuclear medicine.

"The Shifting and Control Measures of Radionuclides of Nuclear Fission Products in Crops and Soils", by Zhao Wenhu

Radioecological experiments that have been conducted in China to study the behavior of fission products in crops and soils are described in this paper. Unfortunately, in most of those experiments, which appear to have been conducted in the 1960s, only measurements of total beta or gamma measurements were carried out; therefore, the valuable information that can be derived from these experiments is limited. Only a few experiments, presumably conducted recently, dealt with specific radionuclides such as Sr-90, Fe-59, and Co-60.

A general discussion followed those four presentations. Yuri Izrael stressed the importance of the SCOPE-RADTEST Project and compared the doses resulting from fallout from nuclear weapons tests with those due to natural radiation background.

Summary of Session E on "Epidemiological study"
presented by S. Akiba

Session E was chaired by Drs. Kato and Raie and started with an excellent lecture, titled "Methodology of Epidemiological study in Low Dose Radiation Exposure", given by Dr Kato. He emphasized the importance of establishing a study with sufficient population size and reliable information on radiation dose, neither of which can easily be attained in the epidemiological study examining the health effects of high background natural radiation (HBR). He illustrated us the mechanisms of biases due to incomplete and differential case ascertainment, inaccurate diagnosis, and various confounding factors. The session consisted of two parts: the first one was for the health effects of radiation, and the second, for chromosome studies. In the first part of the session, there were nine proffered papers. Seven of them were reported by the researchers participating in the Yangjiang study led by Dr. Wei and his Japanese collaborators.

Drs. Yuan and Morishima reported various results obtained from their dosimetry work in Yangjiang study. What was most important from epidemiological point of view was their conclusion that reliable individual lifetime radiation dose could be estimated using information on sex-and-age-dependent occupancy factors, which were obtained from their field surveys, and the indoor and outdoor radiation doses. Since it is practically impossible to directly measure the individual radiation doses of all the members of study population, even if they are alive, their conclusion gave the Yangjiang study a good prospect of success in evaluating the dose-response relationship between various health effects and individual natural radiation doses and reduce biases involved in the dichotomous comparison between the HBR and control areas.

Dr. Sun described the study subjects and the method of mortality follow-up, which was started in Yangjiang in the 1970s and study subjects. He also explained the way the study group used computerized database and established a cohort, which consisted of the study subjects who were alive at the beginning of the year 1979.

There were four papers on the studies examining possible health effects of natural radiation in Yangjiang. Dr. Tao presented the results obtained from mortality study during the period between 1979-1990. He concluded that the cancer mortality in the HBR area was slightly lower than that in the control area. The difference wasn't, however, statistically significant. Site specific analysis revealed that cancers of the lungs, liver and stomach were less common in the HBR area than in the control area. On the other hand, the mortality of leukemia and cancer of the nasopharynx was slightly elevated in the HBR area than in the control area. None of those differences between the HBR and control areas was statistically significant. These results suggested a possibility that various factors other than natural radiation might have different distributions in the HBR and control areas. Dr. Zha and his colleague addressed this problem in their study. They compared dietary habits, and smoking and drinking habits in the two areas. The data obtained from a survey interviewing 302 and 83 subjects in the HBR and control areas, respectively, showed that there were no differences in the distribution of those factors in the HBR and control areas. However, the number of subjects may not be large enough to detect differences. The representativeness of the sampled subjects was not clear, either. Dr. Akiba, the rapporteur of the session, reported the elevated risk of childlrood cancer excluding leukemia in the HBR when compared to that in the control area. The appropriateness of the expected number used in his analysis was discussed in the discussion after his presentation. He also presented the data on leukemia risk in infants, who are younger than 1 year old. At this moment it is difficult to draw any conclusion from the data presented by him. Dr. Zou presented the paper from

Dr. Zha's group. Some of the results obtained by their study were already mentioned in this summary. He also reported life expectancy, which was similar to each other in the HBR and control areas. It may be worthwhile mentioning that the life expectancy in those areas at age 70 or older were not much different from that in Japan. Dr. Li compared the prevalence of dementia in Yangjiang study. Here again, there was no difference in the the two areas.

Dr. Gangadharan from India introduced the study in Kerala. In the study area, tumor registry was established in 1990 and has become able to produce reliable cancer incidence data in the HBR and control areas, recently. Over the past several years, his group conducted interview survey of 100,000 and 400,000 subjects in the HBR and control areas, respectively, to obtain information on various factors including occupation, education, life-style, and medical history. I am sure all the participants in this conference are anxiously waiting for the results to be obtained from his study. During the discussion after hispresentation, he mentioned that one of the drawbacks in the questionnaire used in his survey was the lack of question on the history of medical exposure, which was obtained in sampling surveys in Yangjiang study. The fact indicated that the exchange of information in various aspects of HBR studies is necessary to improve our study designs. Dr. Raie presented the results on lung cancer in northern areas in Iran, including Mazandaran. The important feature of her study was the comparison of histological distribution of lung cancer in the HBR and control areas. Apparently, the prototype of her study was the study conducted by CE Land and his coworkers, including me, in the atomic bomb survivors in Hiroshima and Nagasaki. In their study prominent pathologists were invited from Japan and the US to review pathological slides of lung cancer and determine diagnosis. Using the data thus obtained they compared histological distributions between different radiation dose groups, separately for smokers and non-smokers. Although her study has several drawbacks in terms of pathological review and the information on smoking, further studies will make contribution in enhancing our knowledge on this topics.

The second part of the session was devoted for chromosome study. Dr. Hayata introduced us state of the art techniques that are used in recent studies of chromosome aberrations. He emphasized the importance of using standardized methodology and monitoring the data quality, particularly in low dose studies, including HBR studies. Drs. Nakai and Jiang presented the data obtained by the study supervised by Dr. Hayata. The important finding in their studies is the linear relationship between cumulative individual radiation doses and the frequency of particular types of chrornosome aberrations, i.e., dicentrics and rings. Dr. Chen suggested the need to take into account the effect of smoking in the evaluation of the relationship between natural radiation doses and chromosome aberrations in Yangjiang study. In the study conducted in Iran, Dr. Raie compared the frequencies of chromosomal aberrations and sister chromatid exchanges in Magandaran and Teheran. In the discussion, the use of standardized methodology in chronrosome study was emphasized again.

In conclusion, the results presented in the session of epidemiology gave us new insight into the various effect of natural radiation on the health of residents and the chromosome of their lymphocytes. In order to improve our knowledge in this field, it is necessary to promote the exchange of information on the various aspects of methodology in epidemiological studies and the use of standardized methodology in chromosome studies.

Summary of Session F on "Radiation hormisis"
presented by Professor P.C. Kesavan.

The session was chaired by Prof. Mao and Prof. Sugahara. Six papers were presented. The Session was started with a thought-provoking lecture by Professor T. Sugahara on the radiation paradigm regarding health risk from exposure to low dose radiation. His presentation was supported by relevant data derived from laboratory and epidemiological studies. The available epidemiological data largely indicate a decrease ("hormetic effect") rather than an increase in cancer incidence at low doses. Therefore, the view that "stochastic effects cannot be completely avoided" at even very low doses lacks support from epidemiological and experimental studies from High Level Natural Radiation Areas.

The epidemiological observations are supported by the fact that carcinogenesis is not a single-step somatic mutation, but a multistage process. Further, several cellular and molecular events are known to intervene the sequential events (i.e., initiation, propagation, progression) of carcinogenesis and modify the end result (occurrence of cancer). The elegant experiments of Professor M. Watanabe suggest that *in vitro* mutagenesis and cell transformation possibly involve basically different molecular and cellular events. There appears to be an increasing acceptance of the view that transformation is epigenetic.

Professor Sugahara also focused on the point that HBRA studies in China reveal that a slight increase in the chromosomal aberrations in the adult populations is not accompanied by an increase in the cancer incidence. On the other hand, there is a discernible reduction in the mortality of cancer in the HBRA populations.

It came up during the discussions that cell to cell (intercellular) communication is essential for adaptive response and possibly hormisis. Its implication is that biological defense mechanisms are possibly initiated in the neighboring cells and not in the cells in which energy absorption could indeed be very substantial as deduced from microdosimetric considerations.

Professor Shu-Zheng Liu made an extensive review on the mechanistic aspects of the stimulatory effects of low dose radiation on the immune status. His review provided a possible basis for the reports of hormesis with High Background Natural Radiations, presented in the Session F of the 4th ICHLNR. In simple terms, his work has shown that activation and expression of genes responsible for immune function are stimulated at low doses, but somewhat depressed at high doses. The point is that the molecular and biochemical studies show that low and high doses produce not the same, but opposite types of effects. If so, the "linear, no-threshold" model which assumes that low and high doses produce exactly the same type of effects needs to be modified suitably or even replaced.

Professor Shu Zheng Liu described molecular signal transduction pathway in T-cells. The role of T-cell receptor (TCR), signal transduction, activation of phospholipase C (PLC), inositol triphosphatase (IP-3) and secretion of IL-2 were elaborated. The effects of low and high doses of radiation on p53 and the resulting G_1-block were presented. At low doses, p53 level is not increased, and there is not G_1-block. At high doses, p53 level is enhanced, and G_1-block occurs. Also, it was shown that Bcl-2 is unregulated (and apoptosis is inhibited) at low doses. It was found that Bcl-2/BAX ratio, ras and for protein expressions were increased and p53 was lowered after dose radiation, just the opposite of the effects induced by high doses of radiation.

The effectiveness of enhanced immunologic reactivity was supported by challenge experiments. For instance, pre-exposure of of C57BL/6J mice to 75mGy X-rays suppressed the induction of thymic lymphoma induced by fractionated high doses of radiation in a regimen of 1.75 Gy per week in four successive weeks.

A basic question based on several studies especially those of Professor Watanabe is whether cancer is a stochastic effect. It may be that "genetic instability" and transformation are induced via epigenetic pathway and low doses (0.02 Gy X-rays 5 hr before challenging dose) can prevent the radiation-induced genetic instability. It may well be that a better resolution of "genetic" and "epigenetic" pathways may elucidate the scenario better, and throw light on several questions such as the possible relationship between chromosome aberrations and cancer incidence etc. .

Summary of Session G on "Adaptive response and genetic effects"
by Dr. Chiyoko Satoh

In the Session G, "Adaptive Response and Genetic Effects", a lecture by Professor Pelevina and three proffered papers were presented. Dr. Pelevina from Russia presented data on adaptive responses and radiosensitivities in HeLa cells, mice and human beings induced by exposure in the contaminated areas of the Chernobyl accident. The cells and the mice were exposed by γ-ray from the contaminated area with low dose-rate (2.4 cGy/day) for 1-4 days and 5-10 days, respectively. Then, the cells were incubated at 37ºC in Moscow. After additional acute gamma-ray irradiation with doses of 1-5 Gy, number of HeLa cells with micronuclei, number of giant cells and cell survival, as markers of the adaptive response or the radiosensitivity, were counted for many cell generations. The radiosensitivity to the acute irradiation which depends on the doses of chronic and acute irradiations was observed. When the mice were exposed to the acute γ-ray irradiation with doses of 3.0-9.0 Gy and their survival was examined, an enhanced radiosensitivity was observed. Lymphocyte of individuals suffered from the Chernobyl disaster were irradiated by gamma-ray with a conditioned dose of 5.0 cGy after 24 hr of PHA stimulation. They were irradiated with a challenged dose of 1.0 Gy after 5 hr of the conditioned dose of irradiation. No adaptive response was observed but an enhanced radiosensitivity was observed. In contrast, lymphocytes of control individuals from Moscow showed an adaptive response but no radiosensitivity. It is concluded that the chronic low level irradiation in the contaminated areas induced the genomic instability in the HeLa cells, mice and human lymphocytes.

Dr. Watanabe from Japan reported the genetic instability in mouse cells induced by low-dose of radiation. He irradiated mouse m5s cells with X-rays of 10 Gy or 15 Gy. After irradiation, the cells were inoculated into plastic dishes and were cloned twice successively. As markers for the genetic instability, cloning efficiency, frequency of colonies with giant cells, frequency of chromosome aberrations (dicentric and breaks) and mutation rate at the *hprt* locus were determined. When these values were determined, the cells had divided at least 40 times after the irradiation. An reduced cloning efficiency and an increased frequency of colonies with giant cells were observed. The frequency of the chromosome aberrations increased over 30 times. The mutation rate at the hprt locus increased by 5-28 times of the spontaneous mutation rate. Pre-irradiation of X-rays of 0.02 Gy reduced the frequency of the radiation induced genetic instability described above. He suggested that target for the genetic instability should not be a single gene or a single gene family, but various genetic and epigenetic mechanisms for the genetic instability should exist.

Dr. Satoh from Japan introduced two techniques used in her laboratory for the detection of fresh germinal mutations. In the examination of minisatellite loci in children of atomic bomb survivors and control children, she detected many mutations in both groups of the children but she detected no difference between the mutation rate of the exposed group and that of the control group. However, there is a possibility that the effect of an acute external radiation is different from that of a chronic internal radiation on the mutation rates at the minisatellite loci. Therefore, she believes that it is worthwhile to examine the germline mutation rates at the minisatellite loci of inhabitants of high natural radiation area. She introduced the second technique, i.e., two-dimensional gel electrophoresis of DNA. This technique can detect over 2,000 fragments from genes from a DNA sample without using any probes. Results of a preliminary study showed that this is a very promising technique for the detection of germinal mutations.

Dr. Delpoux reported experimental studies of the genetic effects of high levels of natural radiation in South-France. He showed that a double heterozygote of Tobacco for two loci (a1 and a2) determining leaf color is a good tool to detect mutations. He detected a linear dose-effect relationship over the dose rate of 35 x

10^{-2} µGy/h, but found a threshold in lower dose rate. In experiments using mice and rabbits, he reached a conclusion that one cannot consider that cytogenetic observations obtained from wild animals living in the high natural radiation area would provide reliable informations on the levels of damage produced in their genetic materials.

IX Concluding remarks

Concluding remarks

by Dr. B.G. Bennett

All interesting and important feature of our environment is the occurrence of high levels of natural radiation. The four international conferences that have been held on this topic demonstrate that this has been an evolving issue. At the first conference in Brazil in 1975 the emphasis was on the occurrence of these anomalies, the geological and physical features and the measurements of external exposures. At the time of second conference in India in 1981, emphasis was placed as well as on internal exposures, and the study in Kerala was described in detail. At the third conference in Iran in 1990, exposures from natural radionuclides in ground water and building materials were described with special attention to the situation in Ramsar. At the fourth conference just concluded here in China, the results of health studies, particularly in Yangjiang County, were presented and many biologists have spoken of mechanisms of response and cellular adaptive behaviors.

Many changes occurred that have led to improvements in the scientific quality of the results of studies in high background areas. Improved exposure measurements with personal dosimeters or TLDs have provided more detailed and accurate assessment of actual exposures. The inhomogeneities of external exposures in some cases, such as in India, have become apparent. Improved epidemiology has also been achieved. In the large studies in India and China better registration and follow-up of cancer incidence and deaths are occurring. Attention is given to selection and measurements in the control areas and to the possible influences of confounding factors. The difference of only a factor of exposure levels in the study and control areas in China means that a significant difference will be difficult to see in the epidemiological study, but the large and stable populations give some promise of a worthwhile result. The collaboration of China and Japan greatly strengthens the project and sets an important example for the conduct of such studies elsewhere.

There have been changes, as well, in the background areas themselves, thus requiring need to update the assessments. For example, more substantial housing and some back-filling of land have been occurring in India. The village of Guarapari in Brazil has been transformed into a resort area with consequent changes in the radiation background levels resulting from the construction activities.

There have also been some features of stability with regard to the high background areas. The importance of the study of natural radiation exposures persists. This is the most significant exposure of individuals and is the baseline on which all man-made exposures occur. The implications for low-level risks are well recognized. So far, the main outcome has been less cancer mortality in the high background areas. It should be noted, however, that the difference is not statistically significant. Thus, we must be careful not to refute the linear-nonthreshold model or to explain with adaptive responses a difference that does not statistically exist. We must also not be selective with the results we quote. Remember, the infant and children cancer moralities were higher in the high background area in China, although again the differences were not statistically significant.

The participation at this conference of scientists from many countries attests to the wide interest in the subject of this conference. On behalf of all sponsoring international organizations: the United Nations Environment Program (UNEP), the United Nations Scientific Committee on Effects of Atomic Radiation Nations Environment Program (UNSCEAR), the International Atomic Energy Agency (IAEA) and the World Health Organization (WHO), I thank you for your contributions to this conference. I would like to encourage you all to continue your studies, maintain high scientific standards, involve international collaborators to strengthen the scientific basis of your studies, publish your results in scientific journals and make your results available to UNSCEAR. The outcome of your endeavors will be a better understanding of our environment and improved assessment of the sources and effects of ionizing radiation. For all of us, I express our gratitude to the organizers for providing such good hospitality and excellent scientific conference.

X Start of International Committee on High Levels of Natural Radiation and Radon Areas(ICHLNRRAs)

International Committee on High Levels of Natural Radiation and Radon Areas (ICHLNRRAs)
by Dr. Werner Burkart

During the 4th International Conference on High Levels of Natural Radiation and Radon Areas held in Beijing, China from 21-25 October 1996, it was suggested that an International Committee on High Levels of Natural Radiation and Radon Areas needs to be formed to achieve the following:
1- To harmonize research protocols in different areas to arrive at a definite conclusion.
2- To harmonize different existing definitions and criteria of HLNRAs.
3- To classify such areas worldwide for harmonized studies in particular as regard to epidemiological and biological studies.
4- To make follow up actions for organization of workshops, future conferences, meetings, etc., in order to periodically review the data from different HLNR areas and to make appropriate recommendations.
5- To stimulate think-tanks for promotion of science in this humanitarian field to protect mankind.

To achieve above, a Founder Committee called International Committee on High Levels of Natural Radiation and Radon Areas (ICHLNRRA) was established during the conference. The Committee consists of the following members (in alphabetical order);

- B.G. Bennett, Secretary of UNSCEAR, Vienna
  Phone: 43- 1 -21345/4330
  Fax: 43-1-21345/5902
- Werner Burkhart, Prof. and Director of Leiter des Instituts fur Strahlenhygiene, Germany
  Phone: 08913 16 03-200
  Fax: 08913 16 03-202
- P.C. Kesavan, Prof. and Director of Biosciences Group, Bhabha Atomic Research Center, India
  Phone: 91- 22-556-8084
  Fax: 91-22-556-0750
- A. R. Oliveira, Prof. and Head of Department of Safety and Environment of Brazilian Nuclear Industries, Brazil
  Phone: 5521 536-1664

Fax: 5521 537-9428
- M. Sohrabi, Prof. and Director (Convenor), National Radiation Protections Department, Iran
  Phone: 0098-21-634023
  Fax: 0098-21-8009502
- T. Sugahara, Prof. and Director (Emeritus), Health Research Foundation, Kyoto, Japan
  Phone: 81-75-702-1 14 1
  Fax: 81-75-702-2141
- L. Wei, Prof. and Chief of High Background Radiation Research Group, PRC
  Phone: 86-10-62389930
  Fax: 86-10-62012501

In the first meeting held on October 24, 1996, the followings were decided;
- A questionnaire to harmonize research protocols be kindly prepared by UNSCEAR and be sent to members of the Committee as well as to other research groups.
- The Committee would arrange meetings, workshops and conferences periodically. Prof. Sugahara kindly invited the Committee members to meet in a workshop on HLNRAs in May 1998 during the final meeting of the joint research between China and Japan.
- Prof. Oliveira also offered to have a meeting at the new HLNRA in Brazil to harmonize the research protocol for that area.
- Prof. Sohrabi also announced that a Research Center for Studies of HLNRAs is being established at the HLNRAs of Ramsar. After operations, a meeting can also be held in Ramsar.
- Prof. Sohrabi was selected as the Secretary of the International Committee.
- Prof. A. Rannou from IPSN, CEA in France and Prof. F. Steinhausler from Austria were proposed also to be the member of the Committee.
- Prof. Werner Burknart proposed to host the next conference in Munich, Germany in the year of 2000. The conference should also include radon and its daughter products.
- Membership of the Committee is open for further nominations.
- A proposed title for next conference is : 5th International Conference on High Levels of Natural Radiation and Radon-Prone Areas", Munich, Germany, September-October, 2000.
- The Insignia(Emblem) designed for the 3rd International Conference on High Levels of Natural Radiation in Ramsar, Islamic Republic of lran, 3-7 November, 1990 was also kindly adopted by the 4th ICHLNR. So this Insignia was also proposed to be the formal "Insignia" of the International Committee to be also used for the future conferences.

The above was approved at the General Assembly of the International Committee on October 25, 1996.

XI Appendix

Opening remarks in fourth International Conference on "High Levels of Natural Radiation", Beijing, 21-25 October, 1996

by Minister Chen Minzhang, Ministry of Health of the People's Republic of China

Ladies and gentlemen,

First of all, please allow me to extend, on behalf of the Ministry of Health of the People's Republic of China and the Conference Organizing Committee, the warm welcome to the distinguished guests present at today's conference. I would also like to wish the 4th International Conference on High Levels of Natural Radiation a great success with the support from each and everyone of you! China is a developing country with a huge population. The government attaches great importance to the people's health care and the medical science development, encouraging and supporting various activities for academic exchange and cooperation in the medical field. The nuclear medicine and the radiation protection are also issues attracting our continuous concern. With the development of the nuclear energy, more and more concerns have been given to the effects of low dose-rate irradiation on people's health by various governments and public, while the high level natural radiation areas provide an ideal place for research on this subject, since the information from direct observation may be achieved from these areas for answering the above questions. Therefore, the research on the high level natural radiation areas is of significance for development of nuclear energy and application of radiation resources. The Yangjiang County in Guangdong Province of China is one of well-known high level natural radiation areas throughout the world. With the prime concern and support of the Chinese government, the professional team led by Prof. Wei Luxin have been making a continuous observation and study on the health status of the residents in this area for more than 20 years, and they have made research achievements in successive stages, which have been highly valued by the colleagues and the academic bodies concerned. Since 1991, the study in the Yangjiang high level natural radiation area stepped into a new stage of the Sino-Japanese cooperative research which is still under way. We hope the wide-range exchanges in this international conference will promote the understanding and the friendship among the scientists and further enhance the international cooperation and development on this topic.

Finally, may I wish the Conference a great success!
Thank you all!

by B. G. Bennett, Secretary, UNSCEAR (Representing also UNEP)

The United Nations Environment Program (UNEP) welcomes the opportunity to co-sponsor the 4th International Conference on High Level of Natural Radiation. UNEP is interested in all aspects of the environment, and it works with scientists throughout the world in cooperation with institutes, governments and United Nations agencies to understand the various features of our environment and to take appropriate measures to preserve and protect it. At present, the foremost issues of interest to scientists and governments are problems of air and water pollution, global warming from CO_2 emissions, depletion of ozone layer, deforestation, land degradation and erosion and toxic industrial waste discharges to the environment. Radiation issues are also of concern, for which UNEP relies on the scientific evaluation of UNSCEAR. The United Nations Scientific Committee on the Effects of Atomic Radiation (UNSCEAR), whom I also represent, is particularly interested in this conference. UNSCEAR will be one of the main recipients of the information that will be provided here. The UNSCEAR assessments of natural radiation doses have given greatest attention to estimating the worldwide average levels of exposures. But there is an interest and awareness, as well, of the areas which fall in the upper tails of the wide distribution of natural radiation levels - the high background areas.

In its 41 years of activity, UNSCEAR has issued 12 substantive reports. All have included evaluations of natural radiation exposures. It has been an interesting and evolving issue. In the first reports of 1958 and 1962, the main components of exposure were recognized: cosmic rays, terrestrial gamma rays and potassium-40 in the body. Each of these components contributes about one third of the total dose, estimated at that time to be 1 mSv.

In the reports of 1964, 1966, 1969 and 1972, the absorbed doses to 3 tissues were evaluated: gonads, bone lining cells and bone marrow. In the UNSCEAR 1977 Report, the absorbed doses to the lungs from radon decay products was evaluated and included in the summary tables along with other organ doses. In the UNSCEAR 1982 Report, the effective dose equivalent was evaluated. The significance of the absorbed doses to the lungs from radon exposures then became obvious, since this component contributed about one half of the total of the worldwide effective dose equivalent. The value of the annual effective dose equivalent was estimated to be 2 mSv.

In the UNSCEAR 1988 Report, the estimate was further adjusted to 2.4 mSv, with account taken of the population distribution with altitude for cosmic ray exposures and of continued monitoring of radon indoors indicating higher average concentrations.

In the UNSCEAR 1993 assessment of natural radiation exposures, changes in terminology and in the tissue weighting factors were made, reflecting new recommendations of the International Commission on Radiological Protection. The effective dose was then the quantity evaluated. The dosimetry of radon decay products in the lungs was discussed, and further evaluation was made of this important component of exposure. The doses to children were considered, and evaluations were made of the doses from industrial uses of materials containing natural radionuclides.

Areas of high levels of natural background radiation were recognized many years ago, and the doses in these areas have been reviewed in the reports of UNSCEAR. In the first report in 1958, the Committee reported on the high background areas in India and Brazil and mentioned the elevated background areas of the United States, China and several other countries. In the next report of 1962, the Committee added to this list the Pacific island of Nuie, the Nile delta in Egypt, areas of countries with high levels of radium in drinking water and Bad Gastein in Austria where high levels of radon in air were reported. In the 1977 report of UNSCEAR, areas of Italy, Iran, Madagascar and Nigeria were added to the list of high background areas. Further details and evaluations were included in the 1982 and 1988 reports of UNSCEAR.

Although a great deal of interest has been focused on the areas of high natural background radiation by UNSCEAR and by the previous conferences on this topic and also by many papers in the scientific journals, there has not been a systematic evaluation of all of the populations involved. It has been thought that this has involved a relatively small proportion of world's population, but this assumption should be examined. Perhaps one outcome of this conference could be a detailed listing of the various high background areas and the populations living there. This information would go immediately into the next report of UNSCEAR.

It is a great opportunity for scientists interested in high levels of natural radiation to meet together at this conference to exchange information and to establish and renew friendships. It is a great pleasure for those of us who come from abroad to visit China and see some of the fascinating beauty of this country and experience the hospitality of its people. We look forward to the scientific discussions to be held here this week. Dr. Wei Luxin and his colleagues have already demonstrated their very able competence to organize this conference and to provide every convenience and comfort for us. I would like to thank them very much for that. We are very pleased to be here, and we look forward to a very interesting and successful conference.

Closing remarks

by Dr. Wu Dechang

Ladies and gentlemen, dear friends:

It is my great honor to be here and make closing remarks. First of all, I would like to express our sincere thanks to all the reporters, who have made very good scientific summaries. And we also extend our sincere thanks to all distinguished guests, who gave us very warm speaking. I think all of you would agree with me that we have had a successful meeting, a successful symposium in the past couple of days.

Firstly, we have communicated information broadly and intensively. We have 32 foreign distinguished scientists and around 30 Chinese scientists from 9 countries, that is Belgium, Brazil, France, Germany, India, Iran, Japan, Russian People's Republic of China and United Nation's representative, Dr. Bennett, who is the representative of UNSCEAR and UNEP, which include all the main research groups worldwide, such as those from Brazil, India, Iran and China. All the papers presented dealt with the current status of the research of HBRA.

Secondly, we also discussed the future prospect of the research. We emphasized that we should pay more attention to combining the epidemiological study with radiobiology, particularly at the molecular and cellular levels, which is very important for elucidating the consequences of low dose exposure. We should also pay attention to the natural radiation of radon and its progeny exposure. In this symposium we also heard one of Chinese colleagues report that in the Long-dong Area of Northwestern China there are two million inhabitants living with high level of radon in combination with exposure to smoking. I think this is a critical group for further study. As far as I know, there is a collaborative group carrying out this research just now.

One important question raised is how to minimize the uncertainties of risk assessment of low level exposure. This issue involves not only the epidemiological survey, the dose estimation, but also the statistical method we used, the control group we selected, the confounding factors we faced , the model of dose response and time projection model we adopted. Another critical scientific question we discussed is whether the dose response relationship of low level exposure is linear non-threshold or linear quadratic S shape with threshold. I think all the questions mentioned above are hot points for further research.

Thirdly, we all expressed eagerly the wish to strengthen the collaboration between different groups of HBRA research, and proposed to establish a collaborating Committee in order to make a worldwide plan and share mutual experiences, to harmonize and promote the progress of this research.

Now I should say we have achieved successful results. The proposal has been approved by the General Assembly. I just want to indicate three points: first, we have established a Founding Committee; Second, we decided to hold the next conference in Munich, Germany in 2000; and third, before that in 1998, as a regular meeting in the China-Japan joint project, we shall hold a mini-workshop to discuss and communicate the research work. So, I think we are all satisfied with all these conclusions.

In closing, I would like to emphasize that we seem to have covered a broad complex area with extremely useful discussion in a short period of time and we have attained good results. So, may I have the honor on behalf of the organizing committee to extend our sincere thanks to the co-sponsors: IAEA, WHO, UNEP, UNSCEAR and Health Research Foundation of Japan, the Society of Radiological Medicine and Protection of Chinese Medical Association for their co-sponsorship, to all the distinguished scientists for their contribution to the symposium, to the International Health Exchange Center of Ministry of Health and Laboratory of Industrial Hygiene of Ministry of Health for their organization, to all the Conference officers including Program Committee and Local Committee, esp. Secretary-General, for their hard work and good arrangements, and to the International Convention Center and Grand Continental Hotel for their good Service.

And finally, we are grateful to Dr. Sugahara, who has made strenuous efforts to publish the proceedings of this symposium, which is to be published in March or May of next year.

Ladies and gentlemen, we are being dismissed. Thank you again, and see you again at next conference. Thank you again for your attention.

Dr. Wei Luxin: Now, I declare the 4th International Conference on High Levels of Natural Radiation closed. Thank you very much. See you at next conference.

Speech at closing ceremony

by Dr. T. Sugahara

Mr. Chairmen, ladies and gentlemen;

At the closing of this meeting, I'd like to apologize for my insufficient knowledge about this field of studies in general as a chairman of International Scientific Committee. But supported by the Chinese colleagues who organized excellent program, we had a very successful meeting. I appreciate the contribution of Chinese colleagues for the success of this meeting very much. We organized the International Committee during the Conference, so we are sure now that we can continue this kind of work and this kind of meeting. I think we are now at the starting point. In this respect I am very happy to be here with you and celebrate the start of new International Committee with you. Finally because of my age , I'm not sure whether I will be able to attend the next meeting. Dr. Wei is in the same condition as I am. So, I eagerly want to see you again in the next meeting with Dr.Wei with good health and high activity.

Thank you very much.

by Dr. A. R. Oliveira

Dear colleagues, Chairmen Professors Wei Lu-xin and Wu De-chang:

First of all, I would like to say that I am deeply honored to participate in the 4th ICHLNR and I thank you all for the warm reception, generosity and friendship. The Conference was very important for all of us under different standpoints, especially for the impressive contributions made by the dynamic Chinese scientists in areas of High Natural Radioactivity, and for the fine studies performed by the Laboratory of Industrial Hygiene since 1972. I would like to stress the research work performed by our colleagues from Japan, India and Iran, and for the personal insight of Dr. Bennett (UNSCEAR) to the Conference. Finally, I would like to suggest that we all could keep in touch, if possible establishing reciprocal exchange relations in the field of High Background Radiation Areas. For my part, I wish to reiterate that the personal relations -- of friendship, respect and admiration - will not fade with time, since I shall keep them alive regularly with relevant news and information regarding our work.

by Dr. M. Sohrabi

My great profit over 1400 years ago said: "Learn science even if you have to go to China". At that time of course coming to China took a very long time through the Silk Road and it was rather impossible for any scholar to come from Persia to China at easy. Now our countries are very close.

I do not want to talk here again the same as Prof. Kesavan and Dr. Bennett spoke on high levels of natural radiation. During the week we have heard enough of it. But all I want to say is that "Yes"; I still want to come back to China again to learn more. This is in fact my fourth time being invited to China. The first time in 1991 to present an invited lecture at the Symposium on Nuclear Energy and Radiation Protection on the occasion of the "30th Anniversary of the Establishment of the Radiation Protection Research Institute at Taiyuan". That was the first wide interactions I had with the Chinese colleagues. I also traveled to China two times in 1993 to present papers at the 16th International Conference on Solid State Nuclear Track Detectors, 7-11 September and IRPA Regional Congress: Asia Congress on Radiation Protection, 18-22 October. After that in 1996 for the International Symposium on Nuclear Energy and Environmental held in Beijing last week after that now at this conference. Based on such experiences, I could easily say that besides learning scientific matters in China, you can learn hospitality, friendship and thoughtfulness from them as well sincerity in cooperation.

I would like to really express my thanks to all those who assisted in organization of this conference, especially Prof. Luxin Wei whom I had the honor to meet at the International Conference on HLNRAs in Ramsa, Iran in 1990 and Prof. Tao who was also very effective and helpful at this conference. I would also like to thank Prof. Sugahara and his colleagues who supported the conference, to Secretaries who type these lines of words and in principle all those who made us comfortable and made this conference a success.

Unfortunately I can not speak Chinese, but I learned the meaning of a very nice Chinese statement from Dr. Bennett*s speech at the Closing Ceremony of the last week*s conference; I do not know the Chinese words but it state that: "Now we have come to the end of the meeting with our minds richer from learning and our stomachs also full from Chinese hospitality". Thank you very much for you kind attention.

by Dr. Liu Shuzheng

Mr. Chairman. Ladies and gentlemen, dear friends:
First of all, I would like to congratulate on the successful conference convened in the past few days. I am very happy to be here for several reasons. Firstly. I am very much pleased for having joined the project of studies on the health conditions of the inhabitants in the area of high natural radiation background in China. This

has been a team work under the direction of Dr. Wei Luxin. We participated in the research in the late seventies and early eighties. We were in charge of the Immunology-Biochemistry Program. The early unexpected findings have led to a series of experimental studies in animals in our laboratory which have actually pushed the research work on low dose radiation effects in our group. One of the findings is the stimulation of the T lymphocytes in the peripheral blood of the inhabitants exposed to low level radiation which has led to the studies on the manifestations and mechanisms of this stimulatory effect as presented in my lecture yesterday. The other is the increased capability of DNA repair in the peripheral blood lymphocytes in the exposed individuals which has led to a series of studies in our lab on the cytogenetic adaptive response induced by low dose radiation. Secondly. I am very happy to be invited to give a presentation of our work and to have the opportunity to discuss the data of my laboratory with all the international colleagues attending this conference. Thirdly, I' m very happy to be here to have a chance to get acquainted with so many distinguished scientists from abroad in addition to meeting with many of my old friends, both international and domestic. I hope we'll have further interchange of data and of ideas in the future to promote, to push forward the research in radiation biology as a whole, and specifically the biological effects of low dose radiation in order to make contributions in this very important field. Thank you very much.

by Dr. P.C. Kesavan

Fist of all, I would like to thank Prof. Wei Luxin, Chief of High Background Radiation Research Group, Laboratory of Industrial Hygiene, Ministry of Health, China, and International Health Exchange Center of Ministry of Health, PRC for inviting me to the 4th International Conference on High Levels of Natural Radiation (4th ICHLNR) in Beijing. In this regard, I must also thank Prof. Tsutomu Sugahara, Chairman, Health Research Foundation, Kyoto, Japan for bring us together for collaboration and exchange of information in this important area of research. My visit here has given me an opportunity not only to see a bit of this beautiful country, but also to participate in discussions on the biological and health effects of high background natural radiation. Data from epidemiological and experimental studies presented by eminent scientists at this conference seem to suggest a need for change of basic concepts in radiobiology. A time has come to dispassionately re-evaluate the "liner, no-threshold" relationship between dose and effect. We are at the close of the 20th century, and let us usher in the 21st century with brave new ideas about low dose chronic exposure and biological defense mechanisms. Let us ask ourselves how far the "liner, no-threshold" paradigm has really been purposeful to radiation risk estimations. The meeting in Beijing has taken one more positive step towards the much-required "Paradigm Shift".

With a gathering of eminent scientists and thinkers not only from different countries but also from various academic and Governmental organizations, a rich diversity of view is bound to exist. Without such healthy but friendly divergence of views good science for the benefit of mankind cannot be promoted. It is in this spirit that I tend to think differently from Dr. B. G. Bennett who mentioned that presently the major emphasis for health effect considerations has been concerned with man-made radiation. To my mind, the biological response to natural and man-made radiation could not be different. There is much to learn about health effects from the inhabitants of High Level Natural Background Radiation Areas. As all of us know the LET, the dose, and dose-rate are however more important. It must also be pointed out that biological systems are possibly more used to (adapted to) ionizing and non-ionizing radiation than oxygen even. We all are aware that biological systems on this planet originated and flourished under anaerobic conditions, for millenniums and with the advent of photosynthetic algae oxygen appeared in our Earth's atmosphere and proved toxic to several anaerobic organisms. But then oxygen did not cause the extinction of all organisms; on the contrary, aerobic organisms well equipped to utilize oxygen for oxidative phosphorylation (an efficient energy metabolic pathway) appeared on this planet.

My purpose of stating this is to reassure ourselves that time has come to acknowledge that radiation hormesis and radioadaptive responses are no more a figment of imagination of a few radiobiologists, but a realization of the course of organic evolution. The sooner we realize that low and high doses of ionizing radiation produce very opposite molecular and cellular pathways of action, the better we would be equipped to erase "nuclear phobia" from the minds of general public and accelerate the applications of nuclear energy for peace and prosperity of mankind.

In conclusion, the scientific outcome of this conference has been very fruitful, and I think it will be a catalyst for the subsequent meetings. I thank the host Chinese for their warm hospitality and friendship. And I take home with me new ideas and warm friendship of every participant from different parts of the world.

by Dr. I. Pelevina

This is the first time I came to China. I feel it's very wonderful. It's wonderful people, it's wonderful sights, it's wonderful conference, it's wonderful castle and temples. So all the things I have seen was wonderful. So I thank you for your invitation to this conference and it's very useful for me to have heard many your papers, and to know about Chinese science. Science in China is very high level, I thank you for that so. Thank you very much.

Author index

Abe, S. 103
Afanasjev,G. 373
Akiba,S.241,249, 255, 411
Aledavoud, M. 325
Aleschenko, A. 373
Alves, R. N. 119
Amidi, J. 129
Appleby,L.J. 161, 410
Apsalikov, K. N. 191
Asakawa, J. 38

Beitollahi, M. M. 69, 129
Bennet, B.G. 15,419, 424
Burkart W. 167, 421
Bolourchi, M. 129
Boroujeny, Z. T. 277
Burkartt, W. 167

Cao, Z.-Y. 43, 89
Castanho, A. M. 119
Chen, D.- Q. 301, 307, 317
Chen , J.-X. 141
Chen, N.-G. 355
Chen, W.-Y. 141
Chen, X.-Y. 141
Cheng, X. -S. 75
Cheng, Y.-Y. 355
Coelho, M. J. 119
Conti, L. F. 133
Cui, G.-Z. 31

Dai, L.-L. 317
Dalebroux, M. 397
de Almeid,C.E.V. 119
Deng, Z.-H. 75
da Silva, H. E. 119
Dulieu, H. 397
Dulpoux, M. 397

Endo, S. 191

Fang, G.-Q. 97
Fujinami, N. 207
Fujita, S. 307

Gangadharan, P. 271
Gao, J.-H. 31
Gao, P.-Y. 85
Godoy, M. L. P. 133
Gotlib, V. 373
Guo, S.-X. 85
Gusev, B. I. 191

Hacon, S. 133
Hanash, S.M. 385
Hayata,I.293,301, 307
He, S.-Y. 197
He, W.-H. 249, 255, 263
Honda , K. 25
Hoshi, M. 191
Hou, J.-L. 197
Hou, L.-X. 185
Huang, Z.-H. 75
Huo, J.-L. 85

Iida, T. 103

Javaroni, J. H. 119
Jayadevan, S. 271
Jayalekshmi, P. 271
Jiang, D. -Z. 39
Jiang,T.301, 307, 317, 379
Jin, Y. -H. 97, 103
Julião, L. 133
Jung, T. 167

Kato,H.215, 241, 249, 255, 263

Kayata, S. 391
Kesavan,P.C.111,414, 431
Kirchmann, R. 161
Kishikawa, M. 283
Kodaira, M. 385
Kodama, S. 391
Koga, T. 223, 235
Krishnan, N. M. 271
Kuick, R. 385
Kudrjachova, O. 373

Lasemi, Y. 69
Lauria, D. C. 133
Leonard, A. 397
Li, J. 283
Li, B.-C. 43
Li, K. -B. 49, 407
Li, S. 317
Li, S.-R. 197
Li, Y. 75
Li, Z. 355
Liang, Z.-Q. 43, 89
Lin, J. -M. 263
Lin, X. -J. 255
Lin, Z.-X. 249, 263
Liu, A.-H. 75
Liu, F.-W. 197
Liu, G.-M. 197
Liu, S.-Z. 341, 430
Liu, X.-F. 197
Liu, Y. 179
Liu, Y.-G. 81, 155
Liu, Y.-S. 241, 249
Lu, W.-T. 355

Ma, Q. -L. 89
Mani, K. S. 271
Mao, Y. -H. 81, 155
Mine, M. 283
Minzhang, C. 423
Miyano, K. 25

Morishima, H. 223, 235, 301, 307

Nagaoka, K. 25
Nagatomo, T. 191
Nakai,S.223,235, 301, 307, 317
Nakane, Y. 283
Neel, J.V. 385
Ni, J.-Q. 43, 89
Nouilhetas, Y. 133

Oliveira,A.R. 119, 429

Padmanabhan,V. 271
Pelevina, I. 373, 432
Pi, Z.-Z. 197

Raie, R. M. 277, 325
Rozenson, R. I. 191

Satoh, C. 385, 415
Semenova, L. 373
Serebryanyi, A. 373
Sreedevi, A. N. 271
Shang, Z.-R. 185
Shen, H. 223
Shimo, M. 97
Sobhani, E. A. 69
Sohrabi, M. 57, 69, 129, 409, 430
Sugahara, T.223, 235, 249, 301, 307, 317, 331, 391, 429
Sugasaki, H. 283
Sun, Q.-F. 223, 235, 241, 249, 255, 283
Sun, W. 85
Suzuki, K. 391

Taherzadeh
Takada, J. 191

Takahashi, N. 385
Takizawa, Y. 379
Tao, Y.-F. 85, 241, 249, 255, 283
Tatsumi, K. 223, 235
Tchaijunusova, N. J. 191
Tian, D.-Y. 85
Tomonaga, M. 283
Tu, Y. 39

Wang, B.-R, 85
Wang, C.-Y. 301, 317
Wang, X.-Z. 197
Wang, Y. 85
Wang, Z.-Y. 103
Warner, Sir F. 161
Watanabe, M. 391
Wei, K.-D. 31
Wei, L.-X. 1, 223, 239, 249, 283, 301, 307, 317
Weng, D.-T. 141
Wu, D. 426

Xu, S.-M. 185
Xu, T.-G. 197
Xu, B.-R. 85
Xu, L. 97

Yamamoto, M. 191
Yao, J. 355
Yao, S.-Y. 379
Yamashita, J. 379
Yang, S.-Q. 359

Yang, X.-T. 141
Yang, Y.-H. 263, 283
Yang, Z.-S. 365
Yao, S.-Y. 317
Yuan, Y.-L. 223, 235, 239, 307, 317

Yuan, Y.-R. 301

Zeng, G. 355
Zeng, Q.-X. 75
Zeng, Z.-Y. 355
Zha, Y. 255
Zha, Y.-R. 223, 249, 255, 263, 283
Zhang, S. 355
Zhang, S.-Z. 85, 249
Zhang, L.-A. 179
Zhao, S.-A. 49
Zhao, W.-H. 185
Zou, J.-M. 359

Subject Index

(Each number show the first page of paper including subject indicated.)

Adaptive response appeared in
    chromosome aberration 379, 391
    death 391
    micronuclei 373
    mutation 391

Biological effects of low doses of radiation
    cancer incidence 1, 241, 249, 255, 271, 277
    chromosome aberrations 1, 111, 301, 307, 317, 325, 379
    Down's syndrome 1, 119
    genetical effects 391, 397
    immune response 341, 355, 359
    infant leukemia 255
    leukocyte 197
    lung cancer 277
    micronuclei 111
    motility 241
    senile dementia 283
    sister chromatid exchange 325
    unscheduled DNA synthesis 341, 365

Chernobyl accident 373
Criteria of
    radon-prone area 15, 57
    high level radiation area 1, 15, 57
Cumulative dose 307

Effects of combined exposure by
    alcohol 263
    chemicals 167
    diet 1, 15, 133, 141, 149, 185, 263, 273
    life style 263, 273
    cigarette 1, 167, 263, 273, 317
    mechanism of interaction 167
External radiations from
    altitude 15, 89
    building materials 31, 43, 75, 223
    cosmic rays 15, 25
    hot spring 57, 69, 379
    life style 89
    medical care 119, 179
    radioactive aerosols 97
    terrestrial radiation 15, 31, 43, 69, 81, 85

Fallout 179, 185, 191, 197, 207

Internal radiations from
    indoor radon 15, 31, 43, 69, 81, 85
    food 1, 15, 133, 141, 149, 185, 263

Genetic instability 391

Methodology of epidemiology
    uncertainty 49, 215
    database 241

Natural high radiation area 15, 57
    Azarbayjian, Iran 277
    Gansu, China 85
    Gilan, Iran 277
    Hainan, China 75
    Kerala, India 111, 271
    Laurageais, France 397
    Mazanadaran, Iran 277, 325
    Minas Gerais, Brazil 119
    Qinghai, China 43, 89
    Ramsar, Iran 129
    Sichuan, China 81
    Suzhou, China 39
    Wuhang, China 103
    Xinglong, China 75
    Yangjian, China 1, 49, 223, 241, 249, 317

New radiation paradigm 331

Nuclear tests
    by USSR 179, 185, 191, 197
    by China 207

Population dose 49, 223, 235, 307

Power plants
    coal plant 155
    nuclear power plant 179, 185

Radiation from
    soil 129, 185, 223
    water 155
    food 1, 15, 133, 141, 149, 185, 197, 263

Techniques to estimate biological effects of low dose radiation
    chromosome aberration 293, 301, 307, 317, 391, 397
    gene mutation 385, 391

RADTEST 161

UNSCEAR 15